AF806085

H E R C U L E S

Higher European Research Course for Users of Large Experimental Systems

NEUTRON AND SYNCHROTRON RADIATION FOR CONDENSED MATTER STUDIES

Volume III

APPLICATIONS TO SOFT CONDENSED MATTER AND BIOLOGY

Editors

José Baruchel
Jean-Louis Hodeau
Mogens S. Lehmann
Jean-René Regnard
Claire Schlenker

Springer-Verlag

Berlin, Heidelberg,
New-York, London,
Paris, Tokyo, Hong-Kong,
Barcelona, Budapest

Les Editions de Physique

Z.A. de Courtaboeuf
B.P. 112
91944 Les Ulis Cedex A
France

Cover photograph by Studio de la Revirée

ISBN 3-540-57693-2 Springer-Verlag Berlin Heidelberg New-York
ISBN 2-86883-212-1 Les Editions de Physique Les Ulis

Printed in France

The HERCULES Course is organized by :

Institut National Polytechnique de Grenoble (INPG)

Université Joseph Fourier de Grenoble (UJF)

Université Paris-Sud

Centre National de la Recherche Scientifique
Formation Permanente

Institut National des Sciences et Techniques Nucléaires (INSTN)

It is supported by :

- Ministère de l'Enseignement Supérieur et de la Recherche
- Centre National de la Recherche Scientifique (CNRS)
- Commission of the European Communities
- Commissariat à l'Energie Atomique (CEA)
- European Synchrotron Radiation Facility (ESRF)
- Institut Laue-Langevin (ILL)
- Laboratoire Léon Brillouin (LLB)
- Laboratoire pour l'Utilisation du Rayonnement Electromagnétique (LURE)
- Institut des Etudes Scientifiques Avancées de Grenoble

Secrétariat HERCULES
Marie-Claude Simpson
Maison des Magistères Jean Perrin - CNRS
BP 166 - F-38042 GRENOBLE CEDEX 9

FOREWORD

The present volume is the third and last in the series based on the HERCULES Course on Neutron and Synchrotron Radiation for Condensed Matter Studies. This course, co-organized by Universities in Grenoble and Paris, CNRS and INSTN has taken place in Grenoble once a year since 1991, close to and with the support of the European Synchrotron Radiation Facility (ESRF) and the Institut Laue-Langevin (ILL).

Volume I of the HERCULES series provides the lecture notes of the common part of the Course (four weeks) and deals with the theory, instruments and methods relevant to the fields. It was published in March 1993 and gives the basic information required by a research scientist who wishes to undertake studies using neutron or synchrotron radiation at a Large Facility.

Volumes II and III which correspond to the optional parts of the Course (last two weeks) are devoted to applications to the physics and chemistry of solids, and to soft condensed matter and biology, respectively. The present volume covers selected applications of neutron and synchrotron radiation to soft condensed matter (introduction plus 7 chapters) and biology (introduction plus 10 chapters). The fields are extremely wide, so only some selected topics representative of recent interest and developments have been reported. The authors have been selected on the basis of their high scientific level and teaching skills. An index covering the content of the three volumes has been included.

The HERCULES volumes represent the fruition of three HERCULES sessions. Neither the course nor the publication of these volumes would have been possible without the contributions of many scientists, especially the members of the Scientific Advisory Committee and of the Organizing Committee. We would like to acknowledge our gratitude to all of them.

Special thanks are due to Erik Geissler for his greatly appreciated scientific and linguistic advice. We are also extremely grateful to Marie-Rose Guillermet for the technical preparation of the typescripts and to Marie-Claude Simpson, the Administrative Assistant of the HERCULES Course.

José Baruchel, European Synchrotron Radiation Facility
Jean-Louis Hodeau, Laboratoire de Cristallographie, CNRS
Mogens S. Lehmann, Institut Laue-Langevin
Jean-René Regnard, Université Joseph Fourier and CENG - DRFMC
Claire Schlenker, Institut National Polytechnique de Grenoble and CNRS - LEPES

Contributors

Prof. **Jens ALS-NIELSEN**
Risø National Laboratory, DK 4000 Roskilde, Denmark
and ESRF, BP 220, F-38043 Grenoble Cedex, France

Prof. **Marc BEE**
Lab. de Spectrométrie Physique, Département de Chimie
BP 87, F-38042 St. Martin d'Hères, France

Prof. **David M. BLOW**
Blackett Lab., Imperial College
Prince Consort Road, London SW7 2BZ, United Kingdom

Prof. **Georg BÜLDT**
Forschungszentrum Jülich, IBI-2, Department of Structural Biology
D-52425 Jülich, Germany

Dr. **Jean CHARVOLIN**
Institut Laue-Langevin, BP 156, F-38042 Grenoble Cedex 9, France

Dr. **Stephen CUSACK**
EMBL, BP 156, F-38042 Grenoble Cedex 9, France

Prof. **Jean DOUCET**
Lab. de Physique des Solides, Université Paris-Sud, Bât. 510,
F-91405 Orsay Cedex, France

Dr. **Bela FARAGO**
Institut Laue-Langevin, BP 156, F-38042 Grenoble Cedex 9, France

Prof. **Roger FOURME**
LURE, Université Paris-Sud, Bâtiment 209 d
F-91405 Orsay Cedex, France

Prof. **C. David GARNER**
Dept. of Physics, Manchester University
Manchester M 139 PL, United Kingdom

Dr. **Erik GEISSLER**
Lab. de Spectrométrie Physique, BP 87
F-38042 St. Martin d'Hères, France

Prof. **John HELLIWELL**
Dept. of Chemistry, Manchester University
Manchester M 139 PL , United Kingdom

Dr. **Kristian KJAER**
Risø National Laboratory, DK 4000 Roskilde, Denmark

Prof. **Meir LAHAV**
Structural Chemistry Dept., Weizmann Institute
Rehovot, 76100, Israel

Prof. **Jean-François LEGRAND**
Institut Laue-Langevin, BP 156, F-38042 Grenoble Cedex 9, France

Dr. **Mogens Steen LEHMANN**
Institut Laue-Langevin, BP 156, F-38042 Grenoble Cedex 9, France

Prof. **Lester LEISEROWITZ**
Structural Chemistry Dept., Weizmann Institute
Rehovot, 76100, Israel

Prof. **Anne-Marie LEVELUT**
Université Paris-Sud XI, Lab. de Physique des Solides
Bâtiment 510, F-91405 Orsay Cedex

Prof. **Dietmar SCHWAHN**
IFF de KFA Jülich, Postfach 1913
D-5170 Jülich, Germany

Prof. **Heinrich B. STUHRMANN**
GKSS - Geesthacht BMBH, Max-Planck Str.
D-2054 Geesthacht, Germany

Dr. **Annette TARDIEU**
Centre de Génétique Moléculaire, CNRS
F-91198 Gif-sur-Yvette

Dr. **Peter A. TIMMINS**
Institut Laue-Langevin, BP 156, F-38042 Grenoble Cedex 9, France

Dr. **Clive WILKINSON**
EMBL, BP 156, F-38042 Grenoble Cedex 9, France

VOLUME III

CONTENTS

SOFT CONDENSED MATTER

BIOLOGY

SOFT CONDENSED MATTER

SOFT CONDENSED MATTER

INTRODUCTION

J. CHARVOLIN

It has become clear in recent years that several types of condensed matter previously considered as distinct, such as polymers, gels, liquid crystals, amphiphile assemblies and colloids, share the same basic properties: weak interactions between the structural units, large thermal fluctuations, a large number of internal degrees of freedom and a macroscopic softness indicative of a low elastic modulus. A suitable general name for this category of substance is "soft condensed matter". The units forming the structures of such materials are either large macromolecules held together by covalent bonds, or assemblies of molecules built with weaker intermolecular forces.

Because of these common basic properties, all these systems exhibit very rich polymorphism and a wide range of behaviour. They are a source of puzzling fundamental questions the answers to which require novel approaches mainly based on topology and statistical mechanics. Applications are extremely widespread, ranging from the well known uses of polymers, which by now affect virtually all facets of daily life and vast areas of technology, to aspects of the food and drug industries. It should also be emphasized that recent investigations in structural biology, which is now considering not only molecules but also their association within the cell, suggest that various biological media, e.g. chromosomes, cytoplasm, membranes and viruses, are structurally very close to organised structures found in soft condensed matter such as liquid crystals, gels, films and colloids, and that interactions within and between these associations are controlled by colloidal forces.

Their common features make all these systems relevant to the same experimental approaches, among which synchrotron radiation and neutron scattering play an important role, for reasons summarized below:

a) The structural units have sizes and interaction distances ranging from a few tens to a few thousands of Ångström units. The structures are therefore to be determined by small angle scattering methods, which are particularly easy with cold neutrons and long wavelength X-rays.

b) The units of structure are assemblies of different atoms or molecules, mostly in solution. The systems therefore have a rather complex distribution of scattering centres. This can be determined by taking advantage of the different neutron scattering length densities of protons and deuterons, through selective deuteration in contrast variation methods. At a smaller scale, the immediate neighbourhood of particular atoms may be described using EXAFS methods.

c) Owing to the large size of their structural units, the dynamics of the systems are controlled by characteristic times that are larger than those of atomic systems by a factor of at least 10^3. This is a range that is now accessible to neutron inelastic scattering through the development of back scattering and spin echo methods.

d) Because of the weak interactions between the structural units, their organization may be quite sensitive to weak forces: structural changes under constraints can easily be studied on thick samples in complicated configurations with applied external fields by virtue of the penetrating power of neutrons, and kinetics of organizational changes may be followed owing to the very intense flux of synchrotron radiation.

Recent investigations

In the case of polymers, the complete range of possible scattering experiments has been used with remarkable success. Like diffraction for the investigation of crystalline materials, small angle scattering has become a choice method for large scale studies of structure and morphology of polymers. Present topics encompass e.g. the characterisation of new polymers such as liquid crystalline materials, investigations of phase transitions like spinodal decomposition in blends, the study of elastomers with an aim to understand the molecular basis of rubber elasticity, and research into fibre formation during gel spinning. Pioneering investigations have also been undertaken into the behaviour of polymers exposed to non equilibrium conditions, such as flow fields or polyelectrolytes in electric fields.

Dynamic experiments on polymers in general aim at a molecular understanding of mechanical properties, or, more specifically, of their viscoelastic behaviour. So far these experiments have been mainly restricted to linear chains. Experiments have resolved the role of hydrodynamic interactions in dilute and semi-dilute solutions. One of the recent large successes in dense polymer systems was the microscopic observation of topological constraints in polymer melts and concentrated solutions. Pioneering experiments were also performed on star-like branched polymers, where a relation between polymer architecture and dynamic response was discovered, as well as on networks, where e.g. a first direct observation of the cross-link dynamics gave some insight into fundamental aspects of rubber elasticity. Finally, another important class of studies is concerned with the microscopic identification of the various relaxation mechanisms that are of great importance for the mechanical properties of polymer solids.

For other soft materials, only specific aspects have been investigated. For instance, dynamics in thermotropic liquid crystals, structures of aggregates and their organizations in systems of amphiphiles, structures of heterogeneous particles and behaviour under shear of colloidal particles in solution. An obvious orientation for the near future is therefore to extend investigations to a wider range of cases but, also, new opportunities are appearing which we emphasize in the next section.

Future Developments

In the field of soft condensed matter, as in others, it appears that judicious combinations of components may extend the range of existing properties, or may create new ones, in a way that is quite promising for industrial applications. Examples are provided by the control of forces between colloidal particles by polymer absorption, the control of interfacial curvatures and flexibility in films of amphiphiles by addition of alcohols, reversible gelation of polymers in solution by colloidal particles, new materials with large sized ordered structures formed by copolymers, composites built of polymers and fibres, the controlled growth of small inorganic solid particles in micelles of amphiphiles, and the control of the hydrophilic-lipophilic balance of macromolecules by amphiphiles. Correct structural descriptions of all these combinations, with precise location of all their components, would be well nigh impossible were it not for the contrast variation method of neutron scattering experiments.

It is also being discovered that the weak interaction potentials in these multi-component systems have rather complex shapes. Dynamic studies can provide a test of these potentials. The increase in resolution attained with the development of spin echo spectrometers will help in obtaining information about dynamic behaviour in a manner that complements light scattering at shorter time scales.

These complex potentials may exhibit several minima corresponding to energies of about kT and permitting interesting structures to exist according to the mechanical and thermal history of the sample, possibly with limited lifetimes. Complete exploration of the whole potential surface and of the structures will be facilitated by short recording times that stem from high radiation flux, detector sensitivity and also the development of on-line non-equilibrium equipment.

One of the basic and also specific problems in the dynamics of soft matter is the importance of geometrical or topological interactions directly related to the large scale molecular and supramolecular structures. In polymer dynamics e.g. they manifest themselves as 'entanglements', while in the case of amphiphilic structures they show up as large scale steric interactions induced by fluctuations.

Now that the first observations of such topological interactions on a molecular level in polymer melts have been made, the route is open to microscopic clarification of the molecular nature of an 'entanglement' as well as to an exploration of its impact on rubber elasticity. The functioning of polymer networks, rubbers, and gels may be further scrutinized by an investigation of the motion of specifically labelled parts such as meshes, paths or cross-links. The role of inherent inhomogeneities also awaits clarification.

Another important challenge is the glass transition and the glassy state. Investigation and molecular identification of the various relaxation phenomena associated with the glass transition may lead to a better understanding of its very nature, as well as help to unravel the molecular processes on which mechanical properties such as flow and ductility of solid polymers are based. In this context also the macroscopic study of structure-property relations in general should be addressed. They aim at the so far much neglected non universal features of polymer systems and try to relate structural details, such as short range order and packing of polymers, as well as different relaxation mechanisms, to mechanical properties. This calls for combining scattering experiments with mechanical studies and computer simulations, and could lead to a kind of molecular engineering for polymers.

A particular study related to the above points is that of structural changes induced when the samples are subjected to various velocity fields. This is an extremely important field, in view of the influence of rheological behaviour on many natural phenomena (e.g. blood circulation) and industrial applications (e.g. flows, polymers fibre and foil extrusion). A full understanding of the complexity of the behaviour observed with colloidal and polymer materials (e.g. viscoelasticity, thixotropy, rheopexy, drag reduction) cannot be reached by developing the phenomenological approach of classical rheology alone.

To go further, rheological measurements and structural studies must be performed simultaneously, i.e. in the case of scattering experiments, samples should be irradiated in the presence of different velocity fields in viscosimeters of various geometries. One can envisage relaxation, cyclic or steady state experiments. Furthermore, the investigation of polymers or amphiphiles in random media such as porous systems should be emphasized. As a consequence of the large internal surfaces and the related interaction mechanism together with finite size effects, the phase diagrams of the solutions will be strongly affected. As far as transport mechanisms are concerned, these materials may be imagined as random bulk systems giving rise to qualitatively new phenomena. Besides its scientific appeal, a study of such systems could conceivably lead to an understanding of processes involved in fields ranging from electrophoresis to tertiary oil recovery.

The above considerations involve bulk samples only, but behaviour at interfaces should be examined as well. Most often, interactions at the interface are strong enough to compete with weak interactions in the bulk, and this may induce local perturbations of the structure (e.g. formation of monolayers of amphiphilic molecules or absorption/depletion layers of polymers at the air/solution interfaces, transition to a smectic state at the air/nematic interface, contact layers in wetting phenomena). In certain cases these perturbations can propagate deep into the bulk (e.g. anchoring phenomena of liquid crystals on a solid substrate, which are at the basis of display systems). In other more extreme cases the interaction with the interface may be much stronger (e.g. polymers attached to solid surfaces or diffusing into materials, phenomena which are at the basis of adhesion properties in flues or soldering). One may also add the description of fracture surfaces. All these examples are of the utmost importance in our everyday life or in industry, and are far from being fully understood yet. In all these cases we need to know the concentration profiles normal to the interfaces and the lateral organizations parallel to them. Access to such information is provided by specular reflectivity and surface diffraction studies.

One of the major difficulties of such experiments is the very small amount of matter at an interface, which implies very intense fluxes. This is certainly a domain where X-rays are most efficient. However, although the flux of a neutron source will never compete with that of a synchrotron radiation facility, neutrons still have an important complementary role to play in the use of contrast variation for locating different species. Another domain where neutrons might play a role, when their feasibility has been demonstrated, is that of inelastic studies in these interfacial layers.

CHAPTER I

STRUCTURE OF POLYMER SOLUTIONS AND GELS

E. GEISSLER

I.1. Introduction

Strictly speaking, polymer solutions can either be in the liquid or solid (gel) state. Here, however, we adopt the currently accepted convention of reserving the term polymer solution for liquids. Both solutions and gels scatter radiation because their components are unevenly distributed throughout space. Owing to the connected nature of polymers, their segments do not possess all the degrees of freedom that are available to smaller solvent molecules, and their spatial arrangement is the resultant of random thermodynamic forces as well as of permanent elastic constraints. These notes introduce in turn the basic thermodynamic concepts underlying scattering processes, then solutions of flexible polymers, and finally gels.

I.2. Thermodynamics

In what follows, 'light' is the incident and detected radiation, irrespective of whether it is visible or not, or even neutrons. The total intensity of light scattered by a uniform fluid is determined by its free energy. The Gibbs free energy at constant volume of simple one-component fluids contains two intensive variables, pressure (P) and temperature (T), thus

$$dG = -SdT + VdP$$

where S is the entropy and V the volume of a fixed mass of the fluid.

In binary mixtures, the number N_1 and N_2 of molecules of each component can vary. We define component 1 to be the polymer and component 2 the solvent. On passing from the pure component to the mixture, each molecule carries with it a difference in chemical potential, μ_1 or μ_2, , such that the free energy increment becomes (Landau and Lifshitz, 1959)

$$dG = -SdT + VdP + \mu_1 dN_1 + \mu_2 dN_2 \tag{I.1}$$

N_1 and N_2 are not independent. They are constrained either by the condition of constant sample volume or of constant osmotic pressure Π. In the latter case

$$dG = -SdT + VdP - \Pi du \tag{I.2}$$

where u is the volume occupied by the solvent and a constant mass of polymer at constant density (V being still the volume of the sample at fixed total mass with variable density). The variables T, P and u (or rather, certain linear combinations that are practically the same as these variables) are mutually independent, and each fluctuates with a mean square amplitude that is determined by the absolute temperature and by the curvature of the free energy density at thermodynamic equilibrium.

$$<\Delta u^2> = kTu/(\partial^2 G/\partial u^2) = -kTu/(\partial\Pi/\partial u) \quad (I.3)$$

Another, equivalent, expression of (I.3) is found by taking the constant osmotic pressure condition. The relation between u and the polymer volume fraction φ, namely

$$\varphi = \frac{m}{u\,d_p}$$

where d_p is the density and m is the mass of polymer enclosed within the volume u, yields

$$<\Delta\varphi^2> = kT\varphi/(\partial\Pi/\partial\varphi) \quad (I.4)$$

The total intensity of the coherently scattered light from these fluctuations is

$$I_{coh} = a(AA^*/r'^2)(b_p - b_s)^2 kT\varphi/(\partial\Pi/\partial\varphi) \quad (I.5)$$

in which $b_p=d_p(\Sigma b_{ip}/m_p)$ is the scattering length density of the polymer and b_s that of the solvent. Σb_{ip} and m_p are respectively the sum of the coherent scattering lengths and the molecular weight of a monomer sub-unit; a is an apparatus constant. The b_i are calculated from the composition of the constituent molecules and from tables of scattering lengths of the nuclei (Lovesey, 1984).

I. 3. Polymer Solutions

We consider linear polymer chains consisting of N flexible links of length l. l is not necessarily the length of a single backbone bond or a single monomer, but may be a temporary association of monomers that together confer the required flexibility to the unit. It is sometimes referred to as the Kuhn step length, or the statistical step length.

Following scaling theory (de Gennes, 1979; des Cloizeaux and Jannink, 1987), we distinguish schematically between three conditions of polymer solution: a) dilute, b) semi-dilute, and c) concentrated. The dilute regime is of less interest here, except that its osmotic behaviour helps derive that of regime b). Also, as there is no well defined boundary between b) and c), this interval calls for investigation.

In regime a, the characteristic dimension of a polymer coils is its radius of gyration, R_G (the root mean square distance of the segments from the centre of mass), or, equivalently, the end-to-end distance of the coil, R. For a random coil (Flory, 1953), $R_G^2=R^2/6$. Since there is only a constant factor between these two radii, they are taken as equivalent for present purposes. The osmotic pressure of a dilute solution is just proportional to the number of such coils per unit volume

$$\Pi_{dilute} = N_A kTc/Nm_p \quad (I.6)$$

where N_A is Avagadro's number, and N is the number of monomers per polymer chain (sometimes called the degree of polymerisation). The polymer concentration in mass per unit volume is $c=d_p\varphi$, where d_p is the density of the pure polymer and φ is the polymer volume fraction. In systems in which the attraction between polymer and solvent is the same as that between polymer and polymer, the coils adopt a Gaussian configuration and their radius is

$$R_0 = \sqrt{N}\,l,$$

where the subscript $_0$ is for the reference state. These *theta* systems are described in more detail below.

A more common situation is that of a *good solvent,* in which attraction between solvent and polymer is greater than between polymer and polymer. In this case the coil expands because the polymer segments repel each other: instead of a *random walk* as in a theta system, the polymer segments now follow a *self avoiding walk*. The effect can be understood by examining the free energy of a single coil (Flory 1953, de Gennes 1979). The elastic energy is proportional to the square of the expansion factor R_α/R_0 of the coil along each orthogonal direction $\alpha = x, y, z$. When the coil is swollen isotropically to radius R, the Gibbs free energy G is given by

$$\frac{G}{kT} = \frac{1}{2}\left(\frac{R_x^2}{R_0^2} + \frac{R_y^2}{R_0^2} + \frac{R_z^2}{R_0^2}\right) + Nv\varphi^*$$

$$= 3R^2/2R_0^2 + Nv\varphi^*, \quad (I.7)$$

The second term in I.7 is a mean-field sum of the individual repulsive energies kTv (v is the *excluded volume*) between one segment and another in a coil. There are N such monomers, and the probability of their being in contact is the product of N with the mean polymer concentration inside the coil, φ^* ($\approx Nl^3/R^3$). Since $R_0^2 = Nl^2$ is a constant, minimisation of expression I.7 with respect to R gives

$$3R/Nl^2 - 3vN^2l^3/R^4 = 0,$$

hence

$$R \propto N^{3/5}.$$

The exponent of N is more generally written ν, the *excluded volume exponent,* with $\nu = {}^1/_2$ in the theta condition.

An analogous calculation may be made for the theta temperature, at which the two-body excluded volume interaction v vanishes. In this condition, the only residual repulsions are *three-body* interactions, the characteristic coefficient of which is w. In the same mean-field approximation as above, the expression for the free energy now becomes

$$G_\theta/kT = 3R^2/2R_0^2 - \ln(R^3/R_0^3) + Nw(\varphi^*)^2, \quad (I.8)$$

where the second term, negligible in the good solvent case, is the entropic contribution from the volume occupied by the coil. The last term, describing the three-body repulsions, is proportional to the number of segments and, now, to the *square* of the concentration in the coil. Substituting for φ^* and differentiating with respect to R as above, one finds

$$R_\theta/Nl^2 - 1/R_\theta - 2wN^3l^6/R_\theta^7 = 0. \quad (I.9)$$

Eq I.9 admits of two solutions:

1 Expanded coil (R_θ large) $R_\theta \approx N^{1/2}\, l$,

and

2 Collapsed coil (R_θ small) $R_\theta \approx (2w)^{1/8} N^{1/2}\, l$,

both of which exhibit the excluded volume exponent $^1/_2$.

In the semi-dilute regime, the individual coils overlap and become indistinguishable. This implies that the osmotic pressure no longer depends upon N. It must, however, depend on the concentration φ, and is assumed to obey a power law of the form

$$\Pi_{sd} = A\varphi^x,$$

where A is a constant, and the exponent x is to be determined. At the boundary between the dilute and semi-dilute regimes, the separate coils start to overlap and the osmotic pressure $\Pi_{sd} = \Pi_{dilute}$. The overlap condition occurs at a polymer volume fraction equal to that inside a coil, i.e.

$$\varphi^* = Nv_p/R^3, \quad \text{(I.10)}$$

where $v_p \approx l^3$ is the volume of the constituent monomers. Hence, at $\varphi=\varphi^*$,

$$A\varphi^{*x} = N_A kT\varphi^*/N$$

and therefore

$$N^{x(1-3\nu)} = N^{-3\nu},$$

thus yielding $x = 3\nu/(3\nu-1)$

i.e. $\Pi_{sd} = A\,\varphi^{9/4}$ (I.11)

in good solvent (excluded volume) conditions.

In the dilute regime, the individual polymer coils are on average far apart, behaving like the molecules of an ideal gas: the osmotic pressure is essentially proportional to the number of coils per unit volume of solution, c/M, where M is the molecular weight of the polymer (eq I.6). In addition, repulsive interactions arise between coils, analogous to those in a real gas. The total osmotic pressure in the dilute regime can then be described by a virial expansion

$$\Pi = RTc/M + A_2 c^2 + \ldots \quad \text{(I.12)}$$

The scattered intensity from eq I.5 becomes

$$Kc/I(Q, c) = 1/[M\, P_{11}(Q)] + 2A_2 c\, P_{12}(Q), \quad \text{(I.13)}$$

where K is a constant comprising the contrast factor and experimental parameters, and $P_{11}(Q)$ and $P_{12}(Q)$ are *intra*- and *inter*-molecular scattering functions, respectively (Flory and Bueche 1958).

$P_{11}(Q)$ is defined so that $P_{11}(0)=1$. Its behaviour at small values of Q can be deduced from the general expression for the scattering intensity (c.f. C. Williams, Volume I, Chapter X),

$$P_{11}(Q) = \int_0^\infty g_{11}(r)\, e^{-i\mathbf{Q}\mathbf{r}}\, d\mathbf{r} \quad \text{(I.14)}$$

where $g_{11}(r)$ is the *intra*coil correlation function, i.e. the probability that if a polymer segment is present at r=0, there will be another at r > 0. For small Q, expression I.14 can be expanded to give (the odd terms vanishing because of symmetry)

$$P_{11}(Q) = \int_0^\infty g_{11}(r)\left[1 - \tfrac{1}{2}Q^2r^2\cos^2\theta + ...\right]dr$$
$$= 1 - \tfrac{1}{2}Q^2 <r^2><\cos^2\theta> + ...$$

where θ is the angle between **Q** and the vector **r** connecting two segments of the polymer chain, and the brackets denote spatial averages. The spatial average of $<\cos\theta^2>$ is $^1/_3$, yielding

$$P_{11}(Q) = 1 - Q^2<r^2>/6 + ... \tag{I.15}$$

or, as an approximation (Guinier, 1939)

$$P_{11}(Q) = \exp(-Q^2R_G^2/3) \tag{I.16}$$

where $R_G = \sqrt{<r^2>}$ is the radius of gyration described above. Equation I.15 is quite general.

By measuring the scattering intensity as a function of concentration and Q (in the range $QR_G \leq 1$) and extrapolating to the limit $c \to 0$, it can be seen from eqs I.13 and I.16, that the molecular weight M and the radius of gyration R_G of the polymer in solution can be deduced. At values of Q larger than $1/R_G$, smaller scale features of the polymer chains start to contribute to the scattering pattern, such as the persistence length, and the cross-sectional radius of the chain. These questions will not be pursued here, but details can be found elsewhere, eg Glatter, Kratky (1982).

In the semi-dilute concentration range, the total scattering intensity from a system of polymer chains can also be estimated from eq I.5, which yields

$$I_{coh}(0) \propto a(AA^*/r'^2)(b_p - b_s)^2kT\varphi^{-0.25} \tag{I.17}$$

in good solvent conditions.

For scattering vectors Q greater than zero, the distribution (pair correlation) function g(r) of the polymer must be found. Assume that the origin is a point somewhere along the polymer chain. The distance between the origin and another point on the chain is defined by the number of intervening sub-units N ,

$$r \propto N^\nu .$$

Thus the mass of polymer enclosed in the sphere of radius r is

$$\propto r^{1/\nu}$$

$$p(r) \propto r^{1/\nu} . r^{-3}$$

i.e.

$$p(r) \propto r^{-1}, \tag{I.18}$$

$$\Pi = -(N_AkT/v_1)[\ln(1-\varphi) + (1-N^{-1})\varphi + \chi\varphi^2 + w\varphi^3],$$

assuming a Gaussian coil ($\nu = ^1/_2$). Now $p(0) = \varphi$, the polymer volume fraction, so that g(r) is essentially given by (I.18). However, owing to overlap between different coils in the semi-dilute condition, screening between the different elements arises, such that g(r) falls to its average value more quickly than indicated by I.18:

$$g(r) \propto (\xi/r)\exp(-r/\xi) \tag{I.19}$$

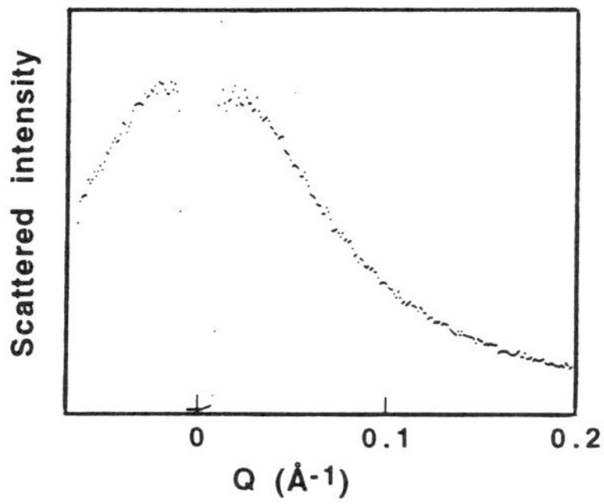

Fig I.1. — Scattering spectrum from a polyvinyl acetate/toluene solution with small angle X-ray scattering (linear detector). The gap in the data near Q=0 is caused by the beam stop which absorbs the direct beam.

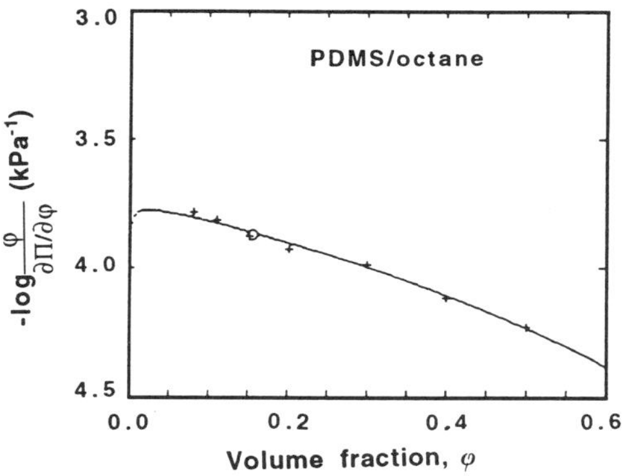

Fig I.2. — Measurements of the osmotic compressibility of polydimethyl siloxane in octane as a function of polymer volume fraction φ, using small angle X-ray scattering (+) and small angle neutron scattering (o)(Horkay et al., 1993). The continuous curve is the best fit to a Flory-Huggins type expression (Flory 1953) for the osmotic pressure :

$$\Pi = -(N_A kT/v_1)[\ln(1-\varphi) + (1-N^{-1})\varphi + \chi\varphi^2 + w\varphi^3],$$

where v_1 is the molar volume of the solvent. This fit yields χ=0.318±0.009; w=0.157±0.020.

As usual, the scattering intensity is proportional to the Fourier transform of the correlation function g(r). Thus

$$I_{coh}(Q) \propto Nb_{coh}{}^2\int(\xi/r)\exp(-r/\xi)\exp(i\,\mathbf{Q.r})d\mathbf{r}$$
$$= a\,(b_p-b_s)^2[kT\varphi^2/K_{os}]/(1+Q^2\xi^2) \quad \text{(I.20)}$$

where $K_{os} = \varphi(\partial\Pi/\partial\varphi)$ is the osmotic modulus. In the good solvent case (ν=$^3/_5$) at $Q\xi\leq 1$, the result is hardly changed, but at higher Q values the intensity decreases more slowly than Q^{-2} ($I \propto Q^{-5/3}$). Fig I.1 shows a typical spectrum from a polymer solution obtained by small angle X-ray scattering.

In addition to yielding information on the correlation length, ξ, such spectra, from their absolute intensity, also give the osmotic pressure Π (or, rather, the modulus K_{os}) of the solution. Figure I.2 shows an example of this for a set of polydimethyl siloxane/octane solutions.

I. 4. Gels

Gels are formed either by inserting cross-links into an assembly of already existing polymer chains or by co-polymerising bifunctional with tri- or higher functional monomers. To the naked eye, lightly cross-linked swollen gels hardly differ from the equivalent liquid solution, but when a laser beam is passed through the sample, the gels usually display noticeably stronger scattering. This is caused by regions where the cross-linking density is higher or lower than average: hence the concentration to which the gel swells locally is permanently greater or less than the mean, and the refractive index is not uniform.

Early theories of polymer gels assumed implicitly that their structure was topologically regular (James and Guth, 1943; Flory, 1953). Neutron scattering measurements (Benoît et al. 1976) in swollen polymer gels found the following characteristics:

1) When the cross-links consist of marked (e.g. deuterated) molecules, their mean distance of separation R_s varies as the gels swells, approximately as

$$R_s \propto \varphi^{-1/3}, \qquad (I.21)$$

where the subscript s stands for *spatial*. This result corresponds to an *affine* deformation, i.e. the relative increase in separation between cross-links is identical to that of the whole sample.

2) When individual chains are marked, however, the radius of gyration of one chain R_t (t indicating *topological*) varies much more weakly.

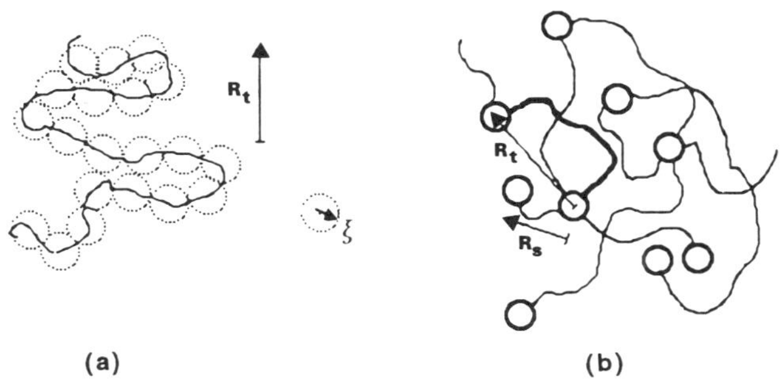

Fig I.3. — Representation of the characteristic distances prevailing in a swollen gel: R_t, the radius of gyration of a given chain, ξ the thermodynamic correlation length, and R_s the nearest neighbour separation between cross-link points. In b) the marked chain is shown as a thick line (Geissler and Hecht, 1981).

To understand this behaviour consider Fig I.3, where a chain of mass

$$M=Nm \qquad (I.22)$$

is distinguished, m being the mass of a monomer of length b.

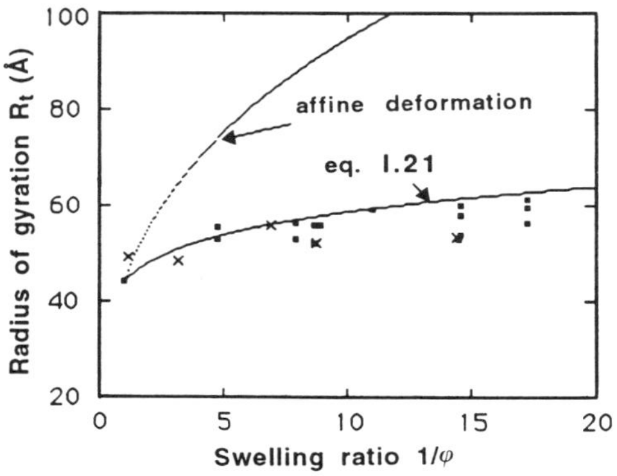

Fig I.4. — Radius of gyration of deuterated polystyrene chains in protonated polystyrene network, swollen in CS_2, as a function of swelling ratio, between the dry state (φ=1) and the fully swollen state ($1/\varphi$=17) (squares:Bastide et al. 1984; $\times$: Geissler et al.1983). The two continuous curves are respectively those for an affine deformation (upper curve: eq I.16, $R_t \propto \varphi^{-1/3}$) and eq I.21 (lower curve, $R_t \propto \varphi^{-1/8}$).

A given marked chain in the presence of a solvent swells to form a succession of correlation spheres of radius ξ ("blobs") as in Fig I.3a, which behave like links in a Gaussian chain. If there are n such spheres, each containing g monomers. Then

$$R_t^2 = n\xi^2, \qquad (I.23)$$

where

$$ng = N.$$

Also

$$g = c\xi^3/m, \qquad (I.24)$$

where c, the polymer concentration in g/cm^3, is

$$c = Mf/2R_s^3,$$

f being the number of polymer chains connected to each cross-link point.

Elimination of M and c from the above equations yields

$$R_t^2\xi \approx R_s^3. \qquad (I.25)$$

Assuming, in agreement with I.21, that R_s^3 varies with concentration as φ^{-1}, and that for good solvent conditions ξ varies as $\varphi^{-3/4}$, then we find from (I.25) that

$$R_t = R_0\varphi^{-1/8}, \qquad (I.26)$$

where R_0 is the radius of gyration of the coil in the dry state. Eq I.26 is in moderate agreement with the experimental observations (Fig I.4).

A corollary of the above argument is that if a swollen network is stretched uniaxially, then the individual chains in the network will also be expected to deform, but their deformation will not be affine. A swollen gel that is subsequently stretched is in the same state as if it had been swollen anisotropically. Thus small deformations of the network, under either swelling or shear strain, are accomplished by unfolding and reorienting the constituent polymer chains rather than by their individual deformation.

The scattering pattern of swollen gels is not in general as simple as that of solutions. Fig I.5 shows an example of a spectrum obtained from a liquid solution of polyvinyl acetate in toluene (dots), together with those of a gel, obtained by random cross-linking, measured at three states of swelling. The fully swollen gel (circles), at approximately the same concentration as the liquid, scatters more strongly at small values of Q. The partially deswollen gels (× and +) have a behaviour intermediate between liquid and gel.

The excess intensity observed at the lower Q region of the gel spectrum over that of a liquid (eq I.20) is caused by static variations in the polymer concentration, imposed by the permanent cross-links.

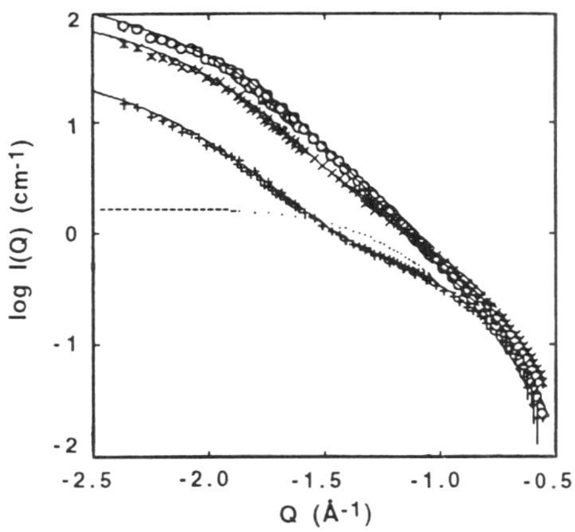

Fig I.5. — SANS scattering curves for a PVAc gel at different swelling degrees in toluene; o: φ = 0.146 (fully swollen); ×: φ = 0.208; +: φ = 0.399. Continuous curves are fits of $I_d(Q)+I_s(Q)$ to the data using eqs I.31 and I.32. Dots: spectrum of a PVAc/toluene solution at φ = 0.114 (dashed line is the low Q extrapolation of these data from eq I.20). The apparent slopes in the quasi-linear region of the more swollen gels are roughly -2.1 and -1.7 respectively.

One approach to the understanding of static heterogeneities is based on a theory of Daoud and Leibler (1988) for polydisperse solutions of branched polymers. These authors point out that in *dilute solution* branched polymers have a radius of gyration R that varies as $N^{1/4}$, rather than as $N^{1/2}$ for linear polymers. Therefore the swelling equation equivalent to I.7 is

$$G/kT = R^2/N^{1/2} + vN^2/R^3. \tag{I.27}$$

This gives

$$R \propto N^{1/2}$$

When the same solution becomes more concentrated, however, the smaller polymers penetrate the larger ones and the excluded volume interaction v between the branches of a given cluster is screened by the intervening smaller clusters. This causes the swelling equation to be modified to

$$G/kT = R^2/N^{1/2} + (1/N_w)vN^2/R^3, \tag{I.28}$$

where N_w is the weight-average degree of polymerisation. Randomly polymerised branched polymers are highly polydisperse and it is expected that in a mixture containing molecules where N is large, then $N_w \approx N^{1/2}$. Eq I.28 then yields

$$R \propto N^{2/5}$$

for the radius of the larger molecules in the distribution; i.e. as solvent is removed from the solution, they are penetrated by smaller molecules and therefore *shrink*. The polydisperse solution thus contains two types of molecule, large clusters that are penetrated by much smaller ones, and smaller masses that have not yet been penetrated. As the solution becomes more concentrated, the threshold between these populations shifts to lower masses, until, when all the solvent is removed, all the clusters mutually interpenetrate. Bastide et al. (1990) applied this model to describe the swelling of gels, obtaining the following fractal expressions

1) for small wave vectors, an intensity that varies as

$$I(Q) \propto \varphi^{-5/3}/(1+Q^2\xi^2), \tag{I.29}$$

where the effective cluster cut-off size is $\xi \approx \varphi^{-5/3}$

2) for large Q vectors

$$I(Q) \propto \varphi Q^{-8/5}. \tag{I.30}$$

Slopes compatible with (I.30) do arise (eg Figure I.5 for the less swollen sample), but measured values often differ significantly from $-^8/_5$. Furthermore, the Q range to which this feature belongs is severely restricted, and the more the gels deswell, the narrower does the Q-range become.

The above model is illuminating for the static structure of the gel, but yields no information about the thermodynamics. A promising avenue to investigate this question is to explore the total scattering properties of gels.

An important prerequisite to further analysis is to determine whether the cross-links themselves, on account of their different chemical composition, contribute to the observed excess scattering. In lightly cross-linked networks (ca 1% cross-link per network monomer unit) in which the foreign molecules are evenly distributed, such a contribution must be small. But clustering can make them visible. A convenient technique to detect such foreign matter is contrast variation in conjunction with neutron scattering, using solvents of differing degrees of deuteration. According to eq I.5, the difference in scattering length density between polymer and solvent, $(b_p\text{-}b_s)$, is proportional to the square root of the scattered intensity. Figure I.6 shows a series of spectra obtained on the D11 instrument at the ILL, Grenoble, for a set of identical polydimethyl siloxane (PDMS) gels swollen in octane. The vertical arrays of circles are separate spectra with different values of b_s, the circles being the individual measured values of

$\sqrt{I(Q)}$. The straight lines are the least square fits to $\sqrt{I(Q)}$ *vs* b_s at each value of Q. The fact that they all pass through zero at a relative octane-D content 0.09 is a sign that the gel is chemically homogeneous in the Q-range explored (0.018Å≤Q≤ 0.14Å).

Even though the cross-links are too dilute to modify the chemical composition of the gel, they can certainly modify the local polymer concentration. Clusters of cross-links occur from which emanate many polymer

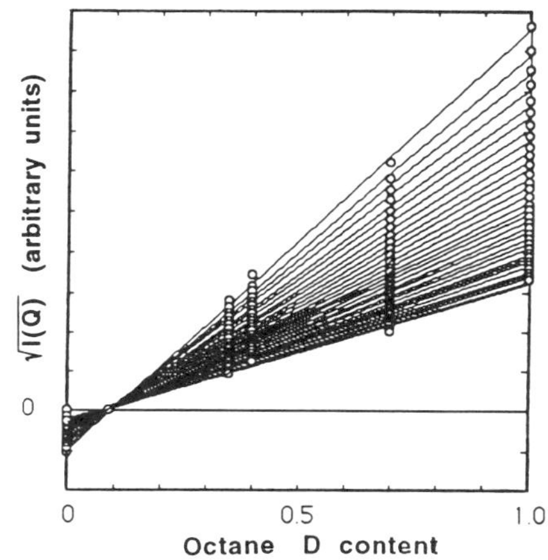

Fig I.6. — Contrast variation of signal from PDMS gels swollen in octane. Straight lines are least squares fits for each value of Q to the six intensities I(Q) with differing octane-D content. The I(Q) are normalised by the transmission factor of the sample. The fact that all the lines pass through zero at octane-D content 0.09 indicates that the signal comes only from the PDMS (an octane mixture of 9%D+91%H has the same scattering length density as PDMS).

chains leading to other less densely cross-linked points. Such clusters will have a concentration distribution with a mean square amplitude $<\delta\varphi^2>$ and some mean radius R. The polymer in the loosely cross-linked regions, moving almost freely as in an uncross-linked solution, can then be expected to scatter according to eq I.20, so that the resulting overall scattering intensity of the gel will take the form

$$I(Q)=a(b_p-b_s)^2\left(\frac{kT\varphi^2}{K_{os}}\frac{1}{1+Q^2\xi^2}+<\delta\varphi^2>f(Q.R)\right) \qquad (I.31)$$

Various trial functions have been used to represent f(QR). One possible form is the stretched exponential

$$f(QR) \propto \exp[-(QR)^s] \qquad (I.32)$$

where the exponent $s < 1$. Whatever function is fitted to the experimental spectra, it is, however, generally found that the values of the correlation length ξ are longer in the gel than in the solution of the same overall concentration (Mallam et al. 1989, Horkay et al. 1991).

An important property of the probability function describing the static concentration fluctuations is that its second moment is finite. [This condition is not satisfied by the Lorentzian $1/(1+Q^2\xi^2)$ of the liquid-like component, since at large Q it decreases more slowly than Q^{-3}, but this fact is of no consequence in the low resolution region, $Q\xi \leq 1$, of interest here.] The second moment is of importance in the following discussion. The correlation function is (Porod, 1982)

$$g(r) = [\varphi(0)\varphi(r) - <\varphi>^2]$$

$$= \frac{1}{2\pi^2(b_p - b_s)^2}\int_0^\infty I(Q)\frac{\sin(Qr)}{Qr}Q^2 dQ$$

in a sample of unit volume. When r=0, we obtain

$$\left[<\varphi(0)\varphi(0)> - <\varphi>^2\right] = <\Delta\varphi^2> = \frac{1}{2\pi^2(b_p - b_s)^2}\int_0^\infty I(Q)Q^2 dQ \qquad (I.33)$$

i.e. the mean square fluctuation in concentration is proportional to the second moment of the total intensity, the so-called *scattering invariant*. Rearrangement of the components of the sample within the same volume does not change its value. For systems with two components having well defined boundaries (Porod, 1982)

$$<\Delta\varphi^2> \equiv \varphi(1-\varphi).$$

The concept of the invariant fails when scattering intensities from the region $Ql > 1$ are included, l being the size of the monomer. The reason is that at high resolution, owing to their internal structure, the monomers no longer scatter like uniform objects. The present discussion is valid, however, for partial invariants, in which *differences* in intensity are measured in the low Q region.

Use of an exponential form for the static concentration fluctuations yields a finite value for their associated amplitude $<\delta\varphi^2>$. For most of the gels so far examined $<\delta\varphi^2>/<\varphi>^2$ is less than 0.2. Similar values of $<\delta\varphi^2>$ were found for swollen polyacrylamide gels using dynamic light scattering (Mallam et al. 1989).

Within our restricted sense of the invariant, we can separate the total mean square concentration fluctuation into a static and a dynamic part

$$<\Delta\varphi^2> = <\delta\varphi^2> + <\Delta\varphi^2>_{dyn} \qquad (I.34)$$

where the left hand side is the total invariant for a given sample. The question is now, what are the structure factors associated with each of these two components?

The thermodynamics of a gel are determined by the swelling pressure ω, i.e. the pressure exerted by the gel in absorbing free solvent molecules from the outside (Horkay, Zrínyi, 1982). This is the difference between the *mixing pressure* Π_m (equivalent to the osmotic pressure of a polymer solution of infinite molecular weight) and the elastic modulus G_e. Thus

$$\omega = \Pi_m - G_e \qquad (I.35)$$

The swelling pressure is measured by inserting the gel in a semi-permeable membrane and placing the resulting packet in a polymer solution of known osmotic pressure. The gel expels (or absorbs) solvent until the swelling pressure is equal to the external osmotic pressure. The concentration of the gel at equilibrium is then measured by weighing. The same procedure is also used for investigating biological samples (c.f. A.Tardieu, this Volume, Chapter IX)

The effective bulk osmotic modulus of a gel (i.e. including both osmotic and elastic effects) is $\tilde{K}_{os} \equiv \varphi \partial \omega / \partial \varphi$, while the *longitudinal osmotic modulus*, which describes the restoring force for osmotic plane waves (corresponding to the scattering of plane-wave radiation), is

$$M_{os} = \tilde{K}_{os} + (4/3)G_e \tag{I.36}$$

G_e is a function of cross-linking density, and is zero in liquid polymer solutions.

Thus, for the dynamic part of the scattering spectrum (Tanaka et al. 1973)

$$<\Delta\varphi^2>_{dyn} = kT\varphi^2/M_{os} \tag{I.37}$$

Hence, for gels, the intensity of the dynamic scattering contribution in eqs I.20 and I.31 should be replaced by

$$I_{dyn}(Q) = a(b_p-b_s)^2[kT\varphi^2/M_{os}]/(1+Q^2\xi^2) \tag{I.38}$$

where in principle M_{os} can be compared with the macroscopic measurements of the swelling pressure and shear modulus of the gel.

A plausible form for the static structure factor is required. The static intensity variation in eq I.30 for random gels, $I(Q) \propto \varphi Q^{-8/5}$, is the consequence of a distribution of fractal objects in the gel. In fact, as noted previously, scattering functions from random gels display the same fractal behaviour only over a rather short range of transfer wave vectors. Fig I.5 showed an example of a randomly cross-linked polyvinyl acetate (PVAc) gel fully swollen in toluene, along with two less swollen gels of the same composition: the apparent slope depends on the degree of swelling. In addition, the fractal form (I.25) is inappropriate for our purpose since its second moment $<\delta\varphi^2> \propto \int Q^{-8/5} Q^2 dQ$ diverges. Indeed, the second moment of any fractal form having an exponent $\geq$-3 must diverge. This difficulty can be overcome by using a stretched exponential form I.32, the second moment of which is finite and is equal to

$$\Gamma(3/s)/sR^3,$$

where $\Gamma(x)$ is the gamma function. (Other fitting functions are of course possible).

With this procedure, the total theoretical intensity

$$I(Q) = I_{dyn}(Q) + I_{stat}(Q)$$

i.e.

$$I(Q) = a(b_p-b_s)^2 \left\{ \frac{kT\varphi^2}{M_{os}} \frac{1}{1+Q^2\xi^2} + \frac{2\pi^2 s R^3}{\Gamma\left(\frac{3}{s}\right)} <\delta\varphi^2> \exp\left[-(QR)^s\right] \right\} \tag{I.39}$$

can be fitted to the observed spectrum by non-linear fitting procedures, and the results compared with the values of M_{os} obtained from macroscopic measurements. This function gives good fits over a wide Q range to the scattering spectra of a set of randomly cross-linked polyvinyl acetate (PVAc) gels

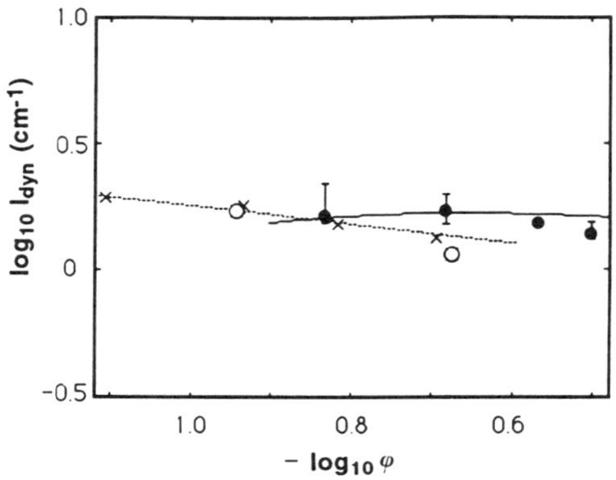

Fig I.7. — Observed dynamic component of the scattering intensity (absolute units) from semi-dilute PVAc solutions (x: SAXS; O: SANS) and gels (•) at different degrees of swelling in toluene (Horkay et al. 1991). Broken line: intensity of solution calculated theoretically from known variation of osmotic pressure; continuous line: intensity calculated theoretically from macroscopic swelling pressure ω and shear modulus G_e measured for these gels.

swollen in toluene, provided a value close to 0.7 is taken for s. For the dynamic component, comparison of the observed scattering intensity with that calculated from the swelling pressure at different degrees of swelling in toluene is shown in Fig I.7. In the most swollen gels the ratio of the static to the dynamic intensities at Q=0 is large (>100), so that the uncertainty in this estimate of the dynamic intensity becomes appreciable.

The excess static intensity $<\delta\varphi^2>/<\varphi>^2$ has a macroscopic counterpart in the swollen gel. For a gel at equilibrium, the swelling pressure ω is uniform throughout the sample; i.e., from eq I.35

$$\omega = \Pi_m - G_e = \text{constant},$$

where

$$G_e = (kTN/V)\varphi^{1/3} = G_0\varphi^{1/3},$$

N being the number of cross-links in a volume V of the dry sample (Treloar 1975).

The cross-links are not uniformly distributed, but tend to form clusters. We assume an average cluster size v containing n cross-links. Then

$$n/v = N/V.$$

Such non uniformities in cross-linking entail non uniformities in the concentration φ, which varies between one region and another in the sample in such a way that

$$\delta\Pi_m - \delta G_e = 0.$$

But

$$\delta\Pi_m = \varphi(\partial\Pi_m/\partial\varphi)(\delta\varphi/\varphi)$$
$$= K_{os}(\delta\varphi/\varphi),$$

and

$$\delta G_e = \delta(kTn\varphi^{1/3}/v),$$

Hence

$$(K_{os}-G_e/3)(\delta\varphi/\varphi) = (kT/v)\varphi^{1/3}\delta n.$$

In a randomly cross-linked sample, the average value of $<\delta n^2> = n$. Assuming small excursions $\delta\varphi$, and taking averages, one then gets

$$<\delta\varphi^2>/<\varphi>^2 = (G_e^2/n)/(K_{os}-G_e/3)^2. \quad \text{(I.40)}$$

For these gels (Horkay et al. 1991) scaling expressions of the form eq I.11 give acceptable fits to eq I.35 for the swelling pressure, i.e.

$$\Pi_m = A\varphi^p$$

with

$$p = 9/4$$

At swelling equilibrium with the pure solvent, $\omega = 0$, and, from I.35 the polymer volume fraction is

$$\varphi_e = (G_0/A)^{3/(3p-1)}.$$

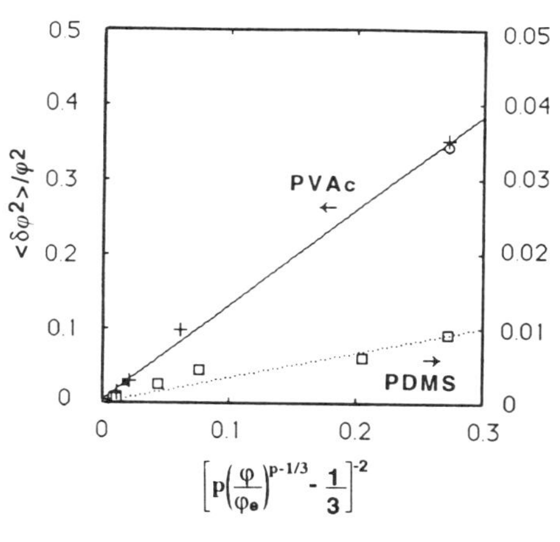

Figure I.8. — Upper curve: plot of $<\delta\varphi^2>/<\varphi>^2$, the total excess intensity due to the static scattering component in PVAc gels swollen in toluene. Crosses: measurements for gel 6/50 at various degrees of swelling. Circle: fully swollen gel 9/400. These gels contain 1 cross-link molecule for each 50 or 400 vinyl acetate monomers respectively (for sample description see Horkay et al. 1991). Squares (right hand scale): $<\delta\varphi^2>/<\varphi>^2$ for end-linked PDMS gels swollen in toluene (for sample description see Mallam et al. 1991).

Hence, from I.40

$$<\delta\varphi^2>/<\varphi>^2 =(1/n)\,[p(\varphi/\varphi_e)^{p-1/3} -1/3]^{-2}. \qquad (I.41)$$

Figure I.8 shows the values of $<\delta\varphi^2>/<\varphi>^2$ measured for two different gels swollen in toluene (randomly cross-linked PVAc and end-linked PDMS), plotted as a function of the right hand side of eq I.41. The neutron scattering data, although sparse, are consistent with the macroscopic thermodynamic parameters. For the randomly cross-linked PVAc gel the proportionality factor n in eq I.41 is approximately unity; for the end-linked PDMS gel the number n is close to 30. By giving an estimate for n, these measurements thus provide information on the statistical distribution of the cross-links in a gel that is not otherwise available.

I. 5. Conclusions

SAXS, and more particularly SANS, have been able to solve questions on the structure of polymer solutions that were previously open only to speculation.

The thermodynamic properties of swollen flexible networks seem to be reasonably described by a fluid-like component that coexists with denser clusters of branched polymers. Owing to the disordered arrangement of the chains in such networks, the individual chains or groups of chains do not move in the same way as the overall macroscopic sample. Instead, deformations of the sample on a macroscopic scale are the result of local unfolding of interpenetrating chains or clusters. The static structure of the networks depends upon the details of the cross-linking as well as on the degree of swelling.

REFERENCES

BASTIDE,J., DUPLESSIX,R., PICOT,C., CANDAU,S. 1984-Macromolecules, 17, 83

BASTIDE,J., LEIBLER,L., PROST,J., 1990-Macromolecules 23, 1821

BENOIT,H., DECKER,D., DUPLESSIX,R., PICOT,C., REMPP,R., COTTON,J.P., FARNOUX,B., JANNINK,G., OBER,R. 1976-J.Polym.Sci. Polym. Phys. Ed. 14, 2119

DAOUD,M., LEIBLER,L., 1988-Macromolecules 21, 1497

des CLOIZEAUX, J. and JANNINK, G. "Les Polymères en Solution" (Editions de Physique) 1987.

FLORY, P.J. "Principles of Polymer Chemistry" (Cornell) 1953

FLORY, P.J.; BUECHE,A.M. 1958 J. Polymer Sci. 27, 219.

GEISSLER,E., HECHT,A.M. 1981-C.R.Acad.Sci.(Paris) 293 Série II, 865

GEISSLER,E., DUPLESSIX,R., HECHT,A.M. 1983-Macromolecules 16, 712

de GENNES,P.G. "Scaling Concepts in Polymer Physics" (Cornell) 1979

GLATTER,O., KRATKY,O. "Small Angle X-ray Scattering" (Academic Press) 1982

GUINIER, A. 1939 Ann. Phys. 12, 161

CHAPTER II

POLYMER BLENDS AND PHASE SEPARATION

D. SCHWAHN

II.1. Introduction

Besides of their technical importance polymer blends are ideal model systems to study equilibrium and non-equilibrium behaviour as critical phenomena, relaxations between two equilibrium states, and phase decomposition. With respect to critical phenomena polymer blends especially allow to study the crossover between two universality classes of meanfield and 3d-Ising approximation. In most other systems as metallic alloys, ferroelastica, super conductors the critical behaviour of only one universality class is observed. In addition, theoretical concepts for phase transition and relaxation behaviour can be studied in detail because systems are available with a large range of relaxation times from seconds to several days. In this lecture first it is tried to show up the thermodynamic basis for binary mixtures as found in standard textbooks. Secondly the respective relations for polymer mixtures are presented, and finally the present state of the subject as it appears to us is roughly described. We only discuss linear polymers building the most simple class. In the chapter of phase separation we limit ourselves to the case of early state spinodal decomposition.

II.2. Thermodynamics of polymer blends

The thermodynamic properties of a blend can be evaluated from the Gibbs free energy of mixing

$$\Delta G_m = G_m - \left[(1 - \phi) G_a + \phi G_b \right] \tag{II.1}$$

which is the difference of the Gibbs free energy of the mixture G_m and of the pure components G_a and G_b weighted by their volume fractions $(1-\phi)$ and ϕ, respectively. ΔG_m is a function of temperature T, volume fraction ϕ of the polymer b, and the pressure P. ΔG_m is given by

$$\Delta G_m (T,\phi,P) = \Delta H_m (T,\phi,P) - T \Delta S_m (T,\phi,P) \tag{II.2}$$

ΔH_m is the enthalpy of mixing and ΔS_m the entropy of mixing weighted by the temperature T. ΔG_m must be negative for a miscible system as shown in Figure II.1.

II.2.1. Flory-Huggins model

Flory and Huggins independently have given a formulation of the Gibbs free energy of mixing (Flory, 1953), namely

$$\frac{\Delta G_m}{RT} = \frac{(1-\phi)\ln(1-\phi)}{\overline{V}_{a,w}} + \frac{\phi \ln\phi}{\overline{V}_{b,w}} + \phi(1-\phi)\,\chi \qquad (II.3)$$

ΔG_m is normalized by RT with R the gas constant and is given in units of mol/cm^3.

The first two terms are the entropy of mixing $\Delta S_m/R$ and the last one is $\Delta H_m/RT$ with the Flory-Huggins parameter χ. The entropic part is determined by the molecular weight averages of the molecular volume $\overline{V}_w$ of both polymers and their volume fractions in the mixture. The form of $\Delta S_m \alpha - \Sigma\, \phi_i \ln \phi_i$ is known f.i. for binary ideal gas mixtures and is found in thermodynamic textbooks. The characteristic difference of polymer blends with respect to mixtures of gases, low molecular liquids, or metallic alloys is the large size of the polymers, given by $\overline{V}_{a,w}$, and $\overline{V}_{b,w}$, which are typically two to three orders of magnitude larger than the corresponding atomic sizes. This gives us a first characteristic feature of polymer blends, namely their small value of ΔS_m is a small driving force for mixing two different polymers.

The Flory-Huggins parameter was originally introduced as an energy parameter, namely (Flory, 1953)

$$\chi = \frac{z-2}{R} \cdot \frac{\Delta \varepsilon}{T}, \quad \Delta\varepsilon = \varepsilon_{ab} - \frac{1}{2}(\varepsilon_{aa} + \varepsilon_{bb}) \qquad (II.4)$$

χ is a segmental quantity which does not depend on molecular weight for not too small polymers. χ and ΔS_m in eq. II.3 and II.4 are derived under the assumption that fluctuations of ϕ can be neglected. The Flory-Huggins theory is therefore a meanfield theory. ε_{aa}, ε_{bb}, and ε_{ab} are interaction energies between the corresponding bonds of the monomers and $\Delta\varepsilon$ is half the change of energy if two a-b bonds have been created onto the costs of one a-a and one b-b bond. $\Delta\varepsilon < 0$ or $\Delta\varepsilon > 0$ mean the preference of ordering (a-b bonds are preferred) and phase separation (a-a and b-b bonds are preferred), respectively. However, experiments have shown that χ is a function of temperature, which means that χ is a segmental free energy

$$\chi = \frac{\chi_h}{T} - \chi_\sigma \qquad (II.5)$$

with an additional entropic term χ_σ whose origin is related to the free volume. χ as a function of T, ϕ, and P is used as a fit parameter to interpret the experimental results. It is a phenomenological number whose physical interpretation is still subject of present research by experiment and theory.

II.2.2. Phase stability and phase diagram

For the discussion of phase stability we need the Euler equation

$$\Delta G_m = (1-\phi)\,\Delta \mu_a + \phi\, \Delta \mu_b \qquad (II.6)$$

with $\Delta \mu_i = \mu_i - \mu_i^o$, the chemical potentials of mixing per unit volume of polymer a and b. μ_i and μ_i^o are the chemical potentials of the component i in the mixed and pure state, respectively. The Gibbs condition for stability reads

$$\delta(\Delta G_m) = 0 \text{ and } \partial^2(\Delta G_m) / \partial \phi^2 > 0 \qquad (II.7)$$

In Figure II.1 ΔG_m is plotted versus ϕ for a partly miscible system. There are the regions I, II, and III which are the stable, metastable ($\delta(\Delta G_m) < 0$, $\partial^2(\Delta G_m)/\partial\phi^2 > 0$) and unstable ($\delta(\Delta G_m) < 0$, $\partial^2(\Delta G_m)/\partial\phi^2 < 0$) ones, respectively. In region I all fluctuations in ϕ increase ΔG_m. In region II fluctuations with small amplitudes are stable because of $\partial^2(\Delta G_m)/\partial\phi^2 > 0$ while fluctuations with sufficient large amplitudes become unstable. Finally, in region III all long wavelength fluctuations lead to a decrease of ΔG_m because of $\partial^2(\Delta G_m)/\partial\phi^2 < 0$.

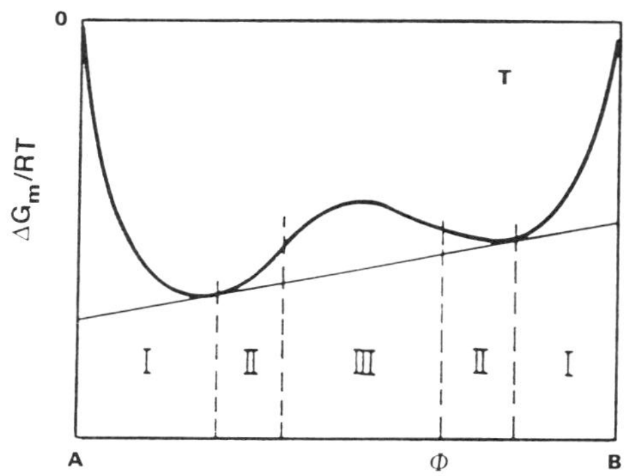

Figure II.1. — Gibbs free energy of mixing at a temperature T as a function of composition. The mixture is miscible in region I and decomposes into two phases in regions II and III. Regions II and III are separated where $\partial^2(\Delta G_m)/\partial\phi^2 = 0$.

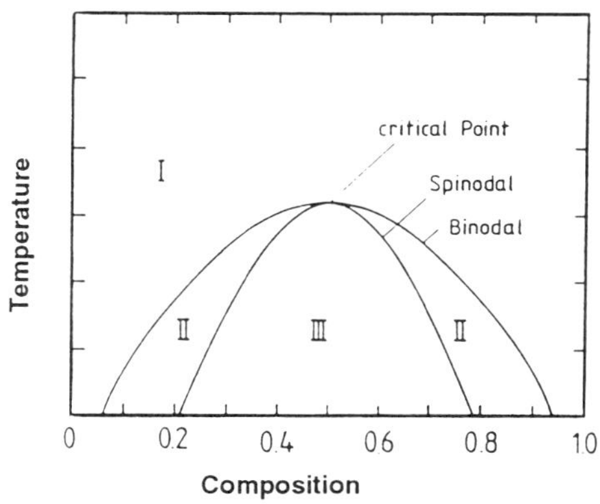

Figure II.2. — Schematic phase diagram of a binary mixture (see also Figure II.1). At high temperatures above the binodal the system is miscible (region I). The spinodal separates the unstable (II) and metastable (II) regions.

In Figure II.1 ΔG_m of eq. II.3 is plotted together with the Euler relation of eq. II.6 as a straight line. According to the tangent rule the phase boundary of the two phase-region or binodal is determined by the intercepts of both curves. According to the equilibrium condition the chemical potentials of both components must be the same in both phases. In the decomposed state the straight line of the Euler equation gives ΔG_m of the system, and its difference to ΔG_m of the mixed state is the driving force of the decomposition process.

The intercept between metastable and unstable region is defined by

$$\partial^2 \Delta G_m / \partial \phi^2 = 0 \tag{II.8}$$

and is called spinodal. In Figure II.2 a schematic phase diagram with T and ϕ as axes is plotted. The stable, metastable, and unstable regions are indicated as in Figure II.1. Spinodal and binodal touch each other at the critical point, the only point where stable and unstable region are linked together.

II.2.3. The spinodal in polymer mixtures

According to the definition of the spinodal in eq. II.8 we have to discuss

$$\partial^2 \Delta G_m / \partial \phi^2 = 2(\Gamma_s - \Gamma) \tag{II.9}$$

with

$$\Gamma_s = \frac{1}{2}\left[\frac{1}{(1-\phi)\, V_{a,w}} + \frac{1}{\phi\, V_{b,w}}\right] \tag{II.10}$$

and

$$\Gamma(T,\phi) = \partial^2/\partial\phi^2\left[\phi(1-\phi)\,\chi(T,\phi)\right] \tag{II.11}$$

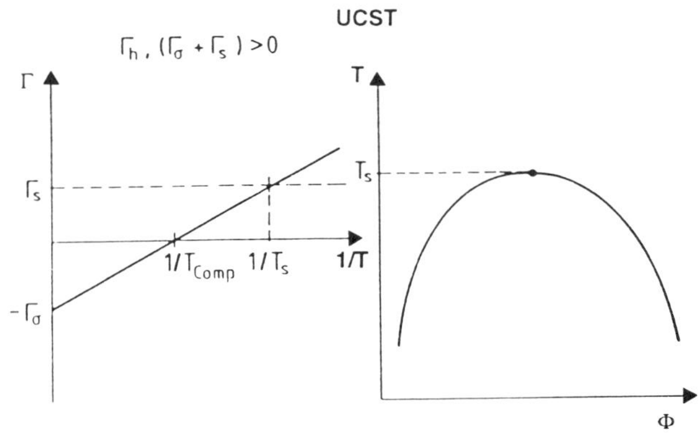

Figure II.3a. — Flory-Huggins parameter and corresponding spinodal for an UCST (upper critical solution temperature) system. According to the Gibbs condition of stability $\Gamma < \Gamma_s$ in the one phase regime.

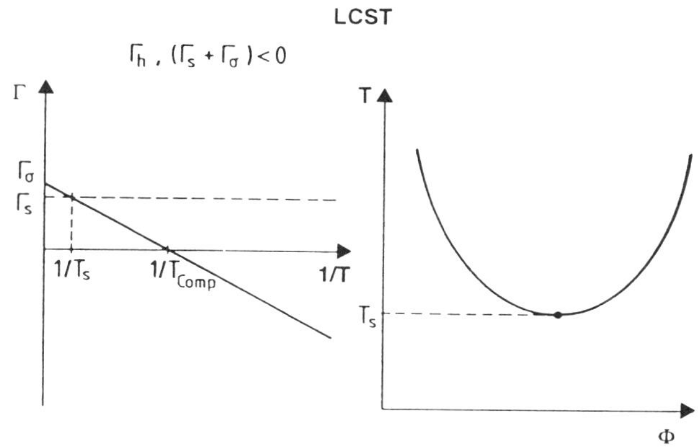

Figure II.3b. — Same as Figure II.3a for a LCST (lower critical solution temperature) system.

It is useful to discuss Γ instead of χ, because χ is usually a function of ϕ and additionally in scattering experiments we measure $\partial^2 (\Delta G_m)/\partial \phi^2$ in eq. II.9. It reads

$$\Gamma(T,\phi,p) = \frac{\Gamma_h(\phi,P)}{T} - \Gamma_\sigma(\phi,P) \qquad \text{(II.12)}$$

with Γ_h and Γ_σ the enthalpic and entropic contributions, respectively. The stability conditions in eq. II.7 together with eq. II.9 then reads: $\Gamma < \Gamma_s$ for the stable and metastable region, $\Gamma = \Gamma_s$ definition of spinodal, and $\Gamma > \Gamma_s$ for the unstable regime. This is demonstrated in Figure II.3a and b. In Figure II.3a the so-called UCST (upper critical solution temperature) case is discussed. Γ_h and Γ_σ are positive, e.g. at high temperature above the compensation temperature T_{comp} Γ is negative and therefore less than Γ_s above T_s. Below T_s $\Gamma > \Gamma_s$ and the system decomposes by spinodal decomposition as will be discussed later. Phase transition at low temperature seems quite natural and is f.i. known from metallic alloys. The driving force for decomposition is the Γ_h term while the entropic terms Γ_σ and Γ_s are dominant above T_s.

In Figure II.3b we discuss the so-called LCST (lower critical solution temperature) case. Here we have a homogeneous mixed state at low temperatures and phase-decomposition at high temperatures. This is a quite unusual case. The conditions for the LCST case are $\Gamma_h < 0$ and $\Gamma_s + \Gamma_\sigma < 0$ with $\Gamma_\sigma < 0$ and its absolute value larger than Γ_s. A negative Γ_σ does mean that the entropy related to the free volume is larger for the pure phases of the a and b polymers in comparison to the mixed state. A negative Γ_h as the always positive Γ_s prefers the homogeneous phase. However, at high temperatures the entropic terms become dominant and because of $|\Gamma_\sigma| > \Gamma_s$ Γ_σ is the entropic driving force of the phase decomposition. In this way the free volume and the entropy of the system is increased by phase decomposition (Patterson, 1969).

A last remark of this chapter should be the molecular weight dependence of the phase boundary. For not too small polymers Γ is a segmental quantity i.e.

independent of the molecular weight, and only $\Gamma_s \alpha 1/V_w$ in $\partial^2 \Delta G_m/\partial \phi^2$ of eq. II.9 changes with V_w. If V_w becomes larger the spinodal line shifts to higher temperatures for UCST and to lower temperatures for LCST systems. Because an increased V_w diminishes the entropy of mixing, the two phase region becomes larger on costs of the one phase regime. We have found two more characteristic features of polymer blends, namely: (1) phase decomposition at high temperatures and (2) the possibility to manipulate the phase boundary by molecular weight.

II.3. Structure factor and experimental examples

The structure factor S(Q) as a function of the scattering vector Q

$$Q = 4\pi/\lambda \sin \vartheta/2 \qquad \text{(II.13)}$$

(λ wavelength of neutron, ϑ scattering angle) is the result of a scattering experiment. It is related to the macroscopic cross-section

$$\frac{d\Sigma}{d\Omega}(Q) = k \,.\, S(Q) \qquad \text{(II.14)}$$

with the contrast factor k for neutrons

$$k = \Delta \rho^2 / N_a \qquad \text{(II.15)}$$

$\Delta \rho$ is the difference of the coherent scattering-length densities of both polymers a and b, and N_a is Avogadros number. $d\Sigma/d\Omega$ measures the heterogeneity of ϕ and therefore the statistical composition fluctuations. S(Q) only contains information about the sample and is given in units of cm^3/mol.

The structure factor at Q = 0 is obtained quite generally from the fluctuation-dissipation theorem (Doi et al, 1989) and reads

$$S^{-1}(0) = \partial^2 (\Delta G_m/RT) / \partial \phi^2 \qquad \text{(II.16)}$$

and according to eq. II.9 and II.12

$$S^{-1}(0) = 2(\Gamma_s - \Gamma) = 2\Gamma_h (1/T_s - 1/T) \qquad \text{(II.17)}$$

it is the link between scattering experiment and thermodynamic potential ΔG_m. The Q dependent structure factor has been formulated in the random phase approximation by de Gennes (1979)

$$S^{-1}(Q) = S_a^{-1}(Q) + S_b^{-1}(Q) - 2\chi \qquad \text{(II.18)}$$

with the structure factors of the polymers a and b

$$S_i = \phi_i V_i g_d (R_i^2 \,.\, Q^2) \quad i = a,b \qquad \text{(II.19)}$$

and the Debye function $g_d(x) = (2/x)\ [1 - (1 - \exp(-x))/x]$ for a Gaussian chain and the radius of gyration R_i. An expansion of eq. II.18 to lowest order in Q gives

$$S^{-1}(Q) = S^{-1}(0) + A\,Q^2 \qquad \text{(II.20)}$$

with S(0) in eq. II.17 and

$$A = \frac{1}{3}\left[R_a^2 / (1 - \phi) V_a + R_b^2 / \phi V_b \right] \qquad \text{(II.21)}$$

which measures the flexibility of the polymer, because the statistical segment length of a chain is given by $\sigma \propto R_g \sqrt{V}$. S(Q) in the approximation of eq. II.20 is measured at Q less than about $1/R_g$ and $S^{-1}(0)$ is obtained by a linear extrapolation of Q to zero in a so-called Zimm-representation of the experimental data, plotted in $S^{-1}(Q)$ versus Q^2. The flexibility of polymers is not further discussed here.

As an experimental demonstration of the presented Flory-Huggins theory we will show SANS (small angle neutron scattering) results of a d-PS/PVME system (d-PS: deutero-polystyrene, PVME: polyvinylmethylether) of critical composition investigated in the one phase regime (Schwahn et al., 1992). Figure II.4a shows a Zimm plot i.e. $S^{-1}(Q)$ versus Q^2 of S(Q) at four different temperatures. All experimental points are represented by straight lines according to eq. II.20 from which the two parameters A and $S^{-1}(0)$ can be evaluated as a function of temperature. In Figure II.4b $S^{-1}(0)$ is plotted versus 1/T for 3 d-PS/PVME systems. All three systems are mixed with the same PVME of V_w = 62800 cm^3/mol molecular volume but with different d-PS polymers, namely $\overline{V}_w(I)$ = 693000, $\overline{V}_w(II)$ = 335000, and $\overline{V}_w(III)$ = 205000 in cm^3/mol. The composition of the d-PS component is ϕ(I,II) = 0.13 and ϕ(III) = 0.2. The slope of $S^{-1}(0)$ in Fig. II.4b gives $2\Gamma_h$ (eq. II.17) and is the same for all systems while the spinodal temperature T_s, where $S^{-1}(0)$ = 0 is found to be different. T_s is highest for system III with the smallest V_w(d-PS) and lowest for system I with the largest V_w(d-PS). This behaviour is expected from the Flory-Huggins theory as discussed above in context with the entropy of mixing ΔS_m.

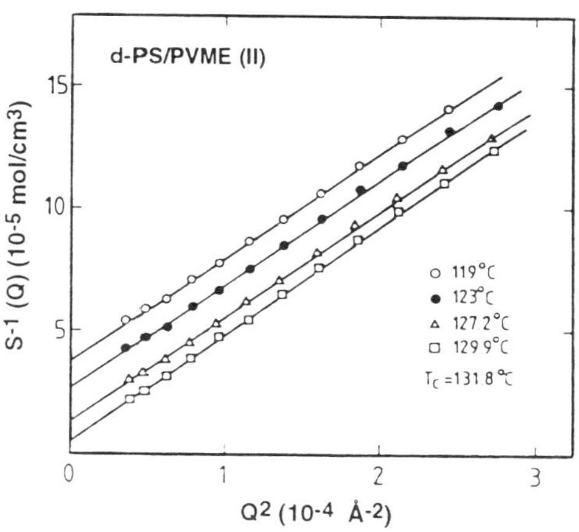

Figure II.4a. — Zimm plot of static structure factor (susceptibility) for the critical mixture d-PS/PVME in equilibrium at four different temperatures. $S^{-1}(0)$ at T = T_c demonstrates that d-PS/PVME is a LCST system discussed in Figure II.3b.

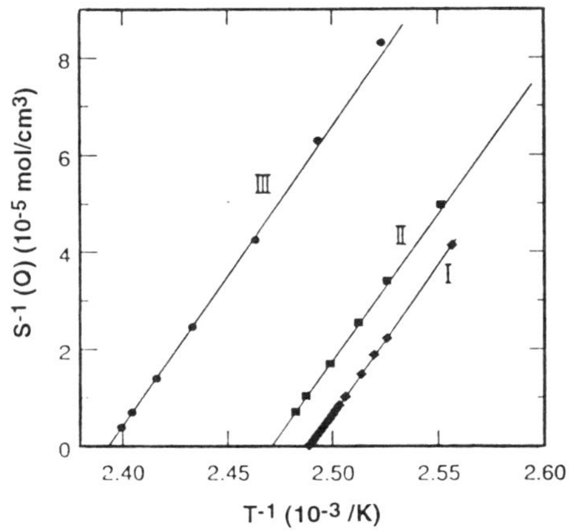

Figure II.4b. — Extrapolated $S^{-1}(0)$ versus $1/T$ for three d-PS/PVME systems with different molecular weights of the d-PS component. $S^{-1}(0)$ always follows a straight line as expected from meanfield approximation and the Flory Huggins model. The spinodal temperature is found at $S^{-1}(0) = 0$. From Schwahn et al. (1992), with permission.

II.4. Composition fluctuations near the critical temperature T_c

According to eq. II.17 we have found

$$S^{-1}(0) \propto \tau^{\gamma} \qquad \text{(II.22)}$$

with the meanfield critical exponent of the susceptibility $\gamma = 1$ and the reduced temperature defined for polymer blends by : $\tau := (1/T_c - 1/T)/(1/T_c)$. The experimental data in Figure II.4b are very well described by the meanfield approximation of eq. II.17 and II.21, and therefore the Flory-Huggins theory is applicable. However, if one approaches the critical temperature closely the fluctuations become so strong and long ranged with a large correlation length ξ, that their contribution to the Gibbs free energy cannot further be neglected. There exists a transition temperature T^* at which a crossover from the universality class of meanfield approximation to another universality class, namely the 3d-Ising model occurs. Asymptotically near T_c, where the 3d-Ising model is valid, one observes $S^{-1}(0) \propto \tau^{\gamma}$ with $\gamma = 1.24$ (Huang, 1987). Such a crossover is shown in Figure II.5 for the PS/d-PB (d-PB: deutero polybutadiene) mixture with a $\overline{V}_w \simeq 2000$ cm^3/mol (Janβen et al., 1992). The observed $\gamma = 1.26$ is only slightly larger than the expected one. For this system a ($T^* - T_c = 5.4$ K) temperature regime is observed where the Flory-Huggins theory is not valid. T_c is slightly lower than the extrapolated T_c^{MF} , e.g. fluctuations stabilize the mixture. The temperature width ($T^* - T_c$) or the reduced temperature τ^* is estimated by the Ginzburg criterion (Bates et al., 1990).

$$\tau^* = \frac{c}{54} \, \frac{\left[\dfrac{1}{V_a\,(1-\phi)^3} + \dfrac{1}{V_b\,\phi^3}\right]^2}{(1 + \Gamma_\sigma / \Gamma_c)\,\Gamma_c\, A^3 N_A^2} \qquad \text{(II.23)}$$

with c $\simeq 0.3$ as a universal constant. All other symbols already have been defined. τ^* scales with $1/V$ which means that the Flory-Huggins theory is better fulfilled for higher molecular weight systems! The proposed value for c is still under discussion in literature as is seen in (Janßen et al., 1992).

The scattering data in Figure II.5 are described by two asymptotic laws, namely the meanfield approximation far away from T_c and the 3d-Ising approximation valid very near T_c. Quite recently we have found in the literature (Belyakov et al., 1992) a crossover function for $S^{-1}(0) = f(\tau)$ describing our experimental S(0) as f.i. shown in Figure II.4b and II.5 over the whole measured temperature range. We have learned that the analysis only with the asymptotic laws gives to some extent erroneous results for Γ_h f.i. and that the range of meanfield approximation is strongly overestimated. These are quite recent developments (Meier et al., 1993).

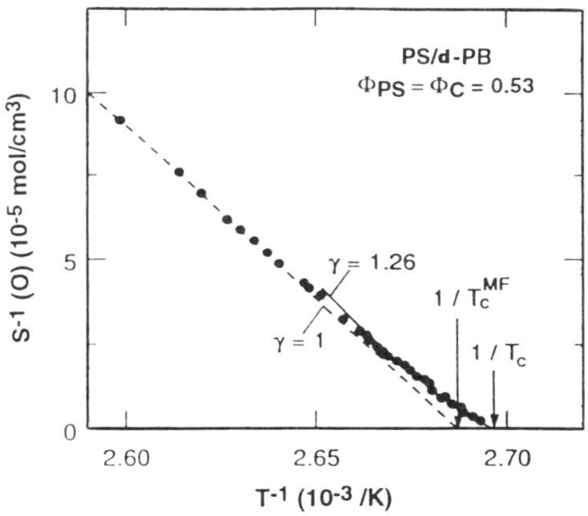

Figure II.5. — $S^{-1}(0)$ as in Figure II.4b for the system PS/d-PB. In this system near T_c we observe a transition from meanfield to Ising critical behaviour. In the Ising regime the Flory Huggins theory is not valid any more. From Janßen et al. (1992), with permission.

II.5. Early state of spinodal decomposition

If one has measured the equilibrium structure factor $S_{T0}(Q)$ at a temperature T_0 in the one phase region and then changes the temperature very fast to a value T in the one or two phase regime one observes a time dependent $S_T(Q,t)$, e.g. one observes the changes of the degree of composition fluctuations. If T is also in the one phase regime $S_T(Q,t)$ becomes $S_T(Q)$, the equilibrium structure factor at T after some time. This so-called relaxation process is described by the Cahn-Hilliard-Cook equation (CHC) (Cahn, 1961) (Cook, 1970)

$$S_T(Q,t) = S_T(Q) + [\, S_{T_0}(Q) - S_T(Q)\,]\, L^2(Q,t) \qquad \text{(II.24)}$$

with $L(Q,t) = \exp(-Rt)$, $R = Q^2 D_{coll}$, and $D_{coll} = \Lambda / S_T(Q)$ (Pincus, 1981).

According to the CHC-theory the time dependence of the various Q modes is exponentially and determined by the relaxation rate R, which is given by the collective diffusion constant D_{coll} and the square Q. D_{coll} itself is given by the Onsager coefficient Λ and the structure factor at T. If T is in the unstable part of the phase diagram $\partial^2 (\Delta G_m)/\partial \phi^2$ becomes negative according to eq. II.7 and formally one extrapolates a negative $S_T(Q)$ for Q smaller than a critical value Q_c (Schwahn et al., 1992). This does mean a negative D_{coll} and R for $Q < Q_c$ and an exponential increase of fluctuation modes of $Q < Q_c$. This is the process of spinodal decomposition. In Figure II.6a an èxperimental example of spinodal decomposition in the system d-PS/PVME (II) is given (Schwahn et al., 1992). S(Q,t) is measured between 43 and 143 seconds, the measurement time for each curve was 10 seconds, and T is 1.5K above T_c. If one analyses S(Q,t) at various Q values with eq. II.24 one indeed finds an exponential growth rate as demonstrated in Figure II.6b. A constant maximum growth rate found at $Q_m \simeq 1.5 \cdot 10^{-3}$ Å^{-1} is typical for spinodal decomposition. The CHC-theory describes the spinodal phase separation only in a limited time window. At later times, which is for our example in Figure II.6a/b at t > 150 sec, the decomposition process becomes nonlinear, e.g. couplings between different fluctuation modes occur, and the domains start to coarsen (Hashimoto, 1988). This is observed experimentally by a shift of Q_m to smaller values. On the other hand at small times we observe a non-exponential growth rate which seems to be typical for polymer blends (Schwahn et al., 1990). The non-exponential time rate depends on Q. For larger Q values the non-exponential time behaviour is stronger while in the limit of Q = 0 the time rate becomes exponential. At Q = 0 the polymer is "seen" by the neutrons as a point because no destructive interference occurs and only the motion of the center of mass is visible. At $Q > 1/R_g$ only parts of the molecule are "seen" by the neutron. Therefore, in the beginning of phase

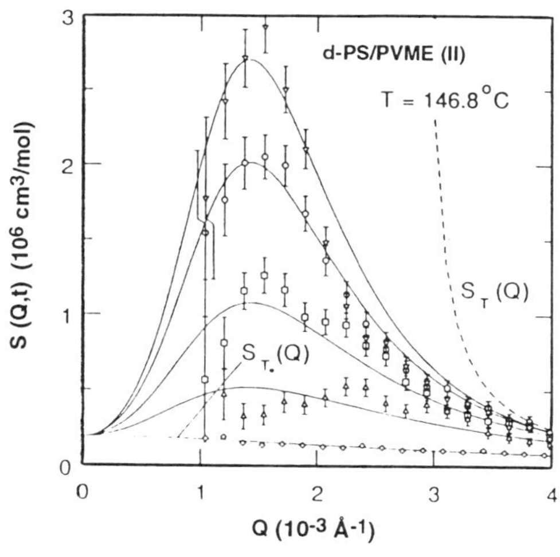

Figure II.6a. — Time dependent structure factor S(Q,t) during the early state of spinodal decomposition in d-PS/PVME. The interference peak at $Q \simeq 1.5 \cdot 10^{-3}$ Å^{-1} shows that the decomposition process occurs with a dominant wavelength of ϕ which is $\Lambda = 2\pi/Q \simeq 4200$ Å.

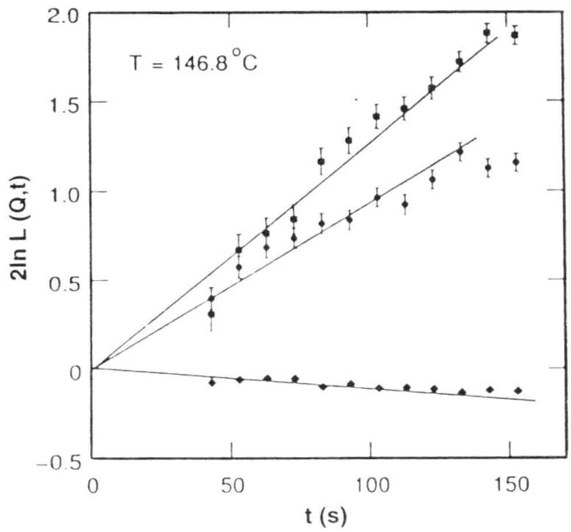

Figure II.6b. — Growth rate at three Q values of the spinodal decomposition measured in Figure II.6a. Within statistical errors the data follow a straight line indicating an exponential growth rate of the fluctuation modes. The slope of the straight lines gives the relaxation rate R(Q). From Schwahn et al. (1992), with permission.

separation internal "Rouse" modes ($<x^2> \alpha\ t^{1/2}$) of the polymer seem to determine the process and after some characteristic time the center of mass motion becomes visible ($<x^2> \alpha\ t$) (Doi et al., 1989).

General Readings

BINDER K. (1990) "Materials Science and Technology", Vol. 5, VCH, Weinheim, Germany

DOI M. and EDWARDS S.F. (1989) "The Theory of Polymer Dynamics", Clarendon Press, Oxford

FLORY P.J. (1953) "Principles of Polymer Chemistry", Cornell University, Ithaca

de GENNES P.G. (1979) "Scaling Concepts in Polymer Physics", Cornell University, Ithaca

HASHIMOTO T. (1988) Phase Transitions 12, 47

HUANG K. (1987) "Statistical Mechanics", 2. edition, John Wiley & Sons

References

BATES F.S., ROSEDALE J., STEPANEK P., LODGE T.P., WILTZIUS P., FREDRICKSON G.H., HJELM R.P., 1990 - Phys. Rev. Lett. 65, 1893

BELYAKOV M.Y., KISELEV S.B., 1992 - Physica A 190, 75

CAHN J.W., 1961 - Acta Metallogr. 9, 795

COOK H.E., 1970 - Acta Metallogr. 18, 297

JANSSEN S., SCHWAHN D., SPRINGER T., 1992 - Phys. Rev. Lett. 68, 3180

MEIER G., SCHWAHN D., MORTENSEN K., JANSSEN S., 1993 - Europhys. Lett. 22, 577

PATTERSON D., 1969 - Macromolecules 11, 672

PINCUS P., 1981 - J. Chem. Phys. 75, 1996

SCHWAHN D., HAHN K., STREIB J., SPRINGER T., 1990 - J. Chem. Phys. 93, 8383

SCHWAHN D., JANSSEN S., SPRINGER T., 1992 - J. Chem. Phys. 97, 8775

CHAPTER III

PARTIALLY CRYSTALLIZED POLYMER SYSTEMS

J.F. LEGRAND

III.1. Introduction

This chapter is intended to show what kind of information can be extracted from neutron and X-ray scattering experiments with examples of crystallization of polymer systems. The process of formation of crystals in polymeric materials has been a long standing problem since many polymers of industrial interest are partially crystallized (e.g. polyethylene, polyamides, polyesters, Teflon, etc.). These materials contain long and entangled polymer chains whose parts may be included in different crystalline regions. This forms a three-dimensional network made of tiny crystallites linked by non-crystallized portions of chains and such a microstruture is of prime importance in regard to the mechanical properties of the polymer.

The contrast in density between the crystalline regions and the amorphous matrix gives rise to small angle scattering of both X-rays and neutrons, and the structure of the crystallites can be analyzed by wide angle diffraction techniques. In addition, deuterium labelling has permitted information to be obtained by neutron scattering on the conformation and the spatial extension of individual polymer chains inside the material. Using synchrotron radiation, further structural details have been obtained by high resolution diffraction, and changes in the microstucture have been investigated by time resolved studies.

III.2. Scattering features in the medium Q-range

The major technique to decide whether a polymer is partially crystallized or not is powder diffraction. Indeed the observation of Bragg peaks in the powder diffractogram is a clear indication of the presence of crystalline regions in the material. Figure III.1a shows the powder diffraction pattern of a semi-crystalline polymer at room temperature (Bellet-Amalric et al.,1993). Obviously, the number of observable Bragg peaks is much smaller than from an ordinary crystalline powder. This is mainly due to the fact that the Bragg reflections are rather broad on account to the small size of the coherent crystalline regions in polymers (of the order of 10 nm). Upon heating one eventually observes melting of the crystallites. Then the scattering pattern (Fig. III.1b) only shows a broad maximum of the structure factor as in other amorphous polymers (Frick et al., 1989, Mitchell, 1989). Coming back to Figure III.1a it may be observed that at low temperature this amorphous "halo" is also present below the first Bragg peak although less intense and a little shifted towards higher Q. This arises from the remaining amorphous matrix in which the crystallites are embedded. We remark however that the total scattered intensity cannot be analyzed by a simple superposition of the Bragg peak over an amorphous halo comparable to that observed above the melting point. Indeed, the structure of the amorphous phase is noticeably modified upon crystallization especially for a high degree of crystallinity.

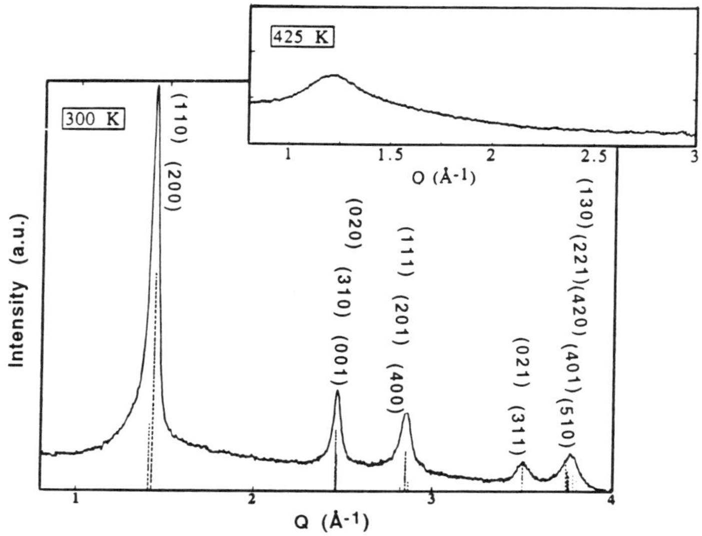

Figure III.1. — Neutron powder diffractograms of P(VDF-TrFE) copolymer of composition 80/20 mol%

a) in the semi crystalline state obtained at room temperature after cooling from the molten phase. The bars indicate the positions and relative intensities of the Bragg peaks calculated from crystal structure data

b) in the amorphous state above the melting temperature of the crystallites.

Another example is given for thin polyimide films, which polymerize and crystallize by curing at elevated temperature. Figure III.2. shows the growth of Bragg reflections and the modification of the amorphous structure factor upon crystallization.

It is worth noting concerning "powder" diffractograms that samples of semi-crystalline polymers frequently exhibit a texture, which means that the crystal axes are not oriented at random in the specimen. Alternatively, diffraction studies of well oriented specimens are often required for the determination the crystal structure (Hall, I.H., 1984, Forsyth V.T. 1993). Different techniques such as stretching, rolling or epitaxial growth produce an uniaxial texture (fibre symmetry) and, in some cases, doubly textured specimens can also be prepared (Stein et al., 1975, Wittmann et al., 1991). Figure III.3. shows an example of fibre diffraction pattern with the corresponding crystalline structure. One may note in the centre of the photo the remaining amorphous halo whose almost circular shape indicates poor orientation of the polymer chains in the amorphous phase.

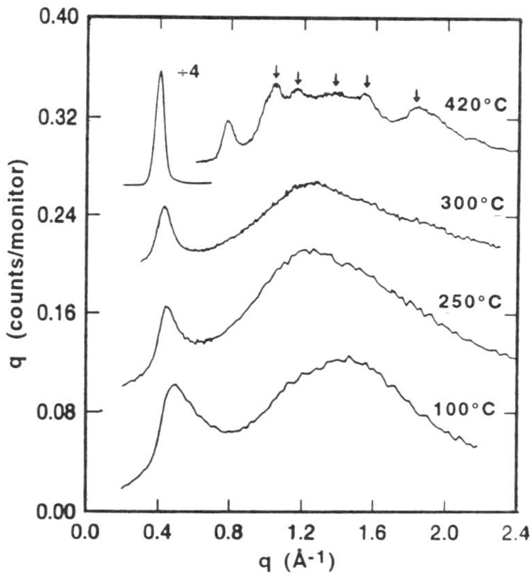

Figure III.2. — X-ray powder diffraction of a 280 nm PMDA-ODA polyimide film ("Kapton") cast on a silcon wafer and prepared by curing para-PAE precursor at different temperatures. These mesurements were performed using Synchrotron Radiation under grazing incidence conditions (Factor et al. 1991).

As a general rule in polymers, the crystalline structure consists of parallel chains aligned along the **c**-axis. Depending on the chemical structure of the polymer, different sequences of conformations may be observed and hence different packing

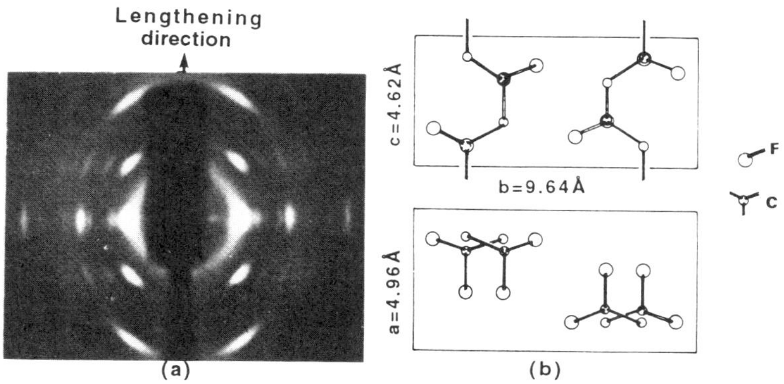

Figure III.3. — X-ray fibre pattern of the "α" modification in a PVDF sample oriented by rolling and schematic diagram of the corresponding centered orthorhombic unit cell (Delzenne, 1986).

structures in the (**a,b**) plane. Therefore stereo-regularity of the chains is required for crystallization to occur, and defects such as crosslinks, branchings, cis- double bonds and chain-ends are rejected to the amorphous regions. Moreover, if the density of such defects is too large or if the chains include large side-groups with random sequences of L and R carbons ("atactic" structure) the polymer may remain completely amorphous (e.g. Polystyrene, PMMA, PDMS,...). Indeed crystals of size smaller than a few nanometers are unstable due to an unfavourable balance between surface energy and volume energy. Let us note finally that a sufficient mobility of the chains is required for crystal growth and that polymers which are able to crystallize at high temperature may remain amorphous if quenched below their glass transition temperature. (e.g. PVC, polyesters,...).

III.3. Small angle scattering features

Owing to the small size of the crystallites and to their relatively high concentration, semi-crystalline polymers exhibit typical correlation peaks in the small Q-range of X-ray or neutron scattering. Figure III.4. displays two typical SAXS profiles recorded with the same polymer sample after different thermal treatments. It demonstrates that the average interparticle distance is stongly dependent on the

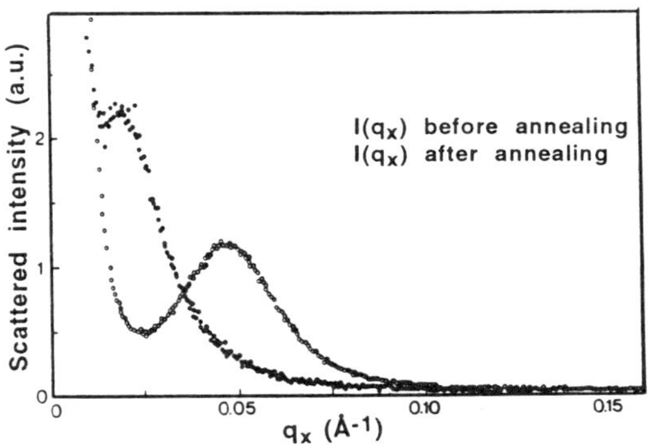

Figure III.4. — Small angle X-ray scattering profiles from a P(VDF-TrFE) copolymer film of composition 70/30 mol% and of thickness 11 μm. In the initial state the crystallization is obtained by solvent evaporation at room temperature. In the final state the specimen was annealed at 130°C for 30 minutes (Legrand, 1989).

growth conditions of the crystallites. Such scattering profiles are observed both with X-rays and neutrons because they arise from the contrast in density between crystalline and amorphous regions.

With oriented polymer films or fibres a four dot diffraction pattern is often observed in the small angle scattering region. As shown in Figure III.5. this is attributed to the stacking of lamellar crystals separated by amorphous regions and inclined with respect to the draw direction. The results obtained in polyethylene (Keller et al.,1971, Grubb et al., 1975) and in PVDF show that the **c**-axes are well

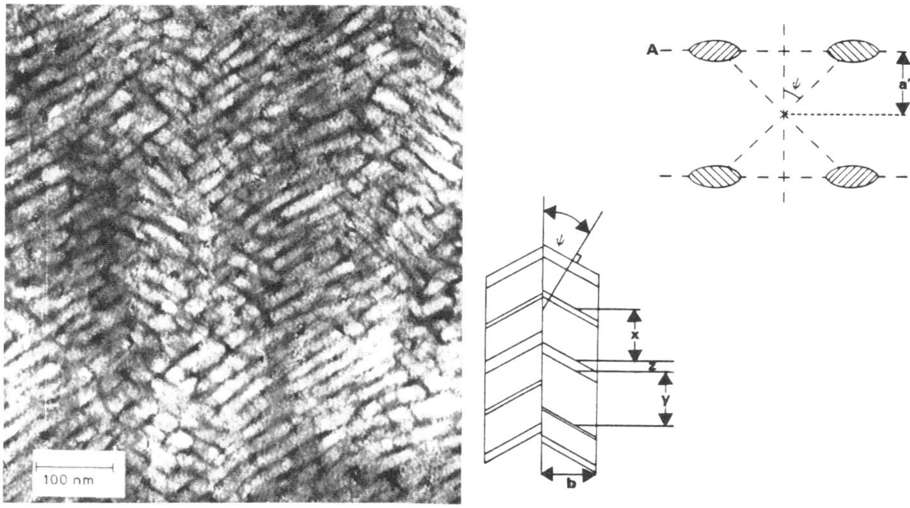

Figure III.5. -- Electron microscope photo of the crystalline lamellae in ultra-drawn polyethylene and schematic diagram showing the interpretation of the four dot small angle scattering pattern. The "long period" of the stacking $\overline{x}=\overline{y+z}$ is given by $\overline{x}=2\pi/a''$. Note also that the smaller is the width b of the lamellae the more extended are the dots along the line AB (Keller, 1979 and Hall 1984).

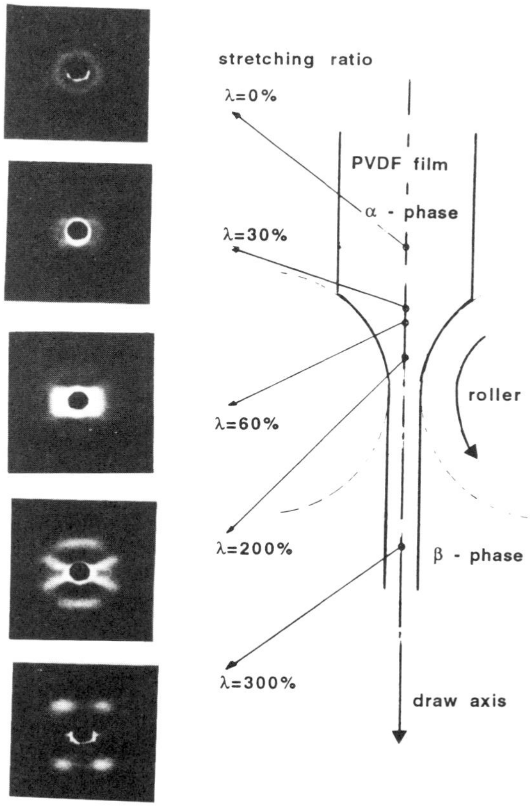

Figure III.6. — Gradual change of SAXS patterns produced by the rolling process of a PVDF plate and recorded with an incident X-ray beam parallel to the generator of the cylindrical rollers. The long period of the lamellar stacking is 140Å in the initial state and 90 Å (along the draw direction) in the final state (Legrand,1993).

aligned along the draw direction but that the crystal-amorphous interfaces are more often parallel to (111) or (201) planes than to (001) planes. Finally, Figure III.6. displays a SAXS investigation of the microstructural changes produced by the rolling of a PVDF plate. Starting with lamellar crystals of the α modification arranged in a spherulitic texture, these are destroyed by the large deformation and an oriented growth of new crystals of the β modification is observed. One may note equatorial diffuse stripes at intermediate draw ratio (≈ 200%), which are attributed to the formation of micro-fibrils in the polymer (Delzenne, 1986, Legrand,1993).

III.4. Neutron scattering studies of chain folding crystallization

The key process which has permitted an understanding of the morphology of semi-crystalline polymers and the lamellar shape of the crystals is chain folding crystallization demonstrated by electron microscope studies of solution-grown single crystals of polyethylene (Wittmann et al. 1985). For bulk polymers crystallized from the melt or from solutions the chain conformation in the solid state has been investigated by neutron scattering using deuterium/hydrogen contrast (Figure III.7).

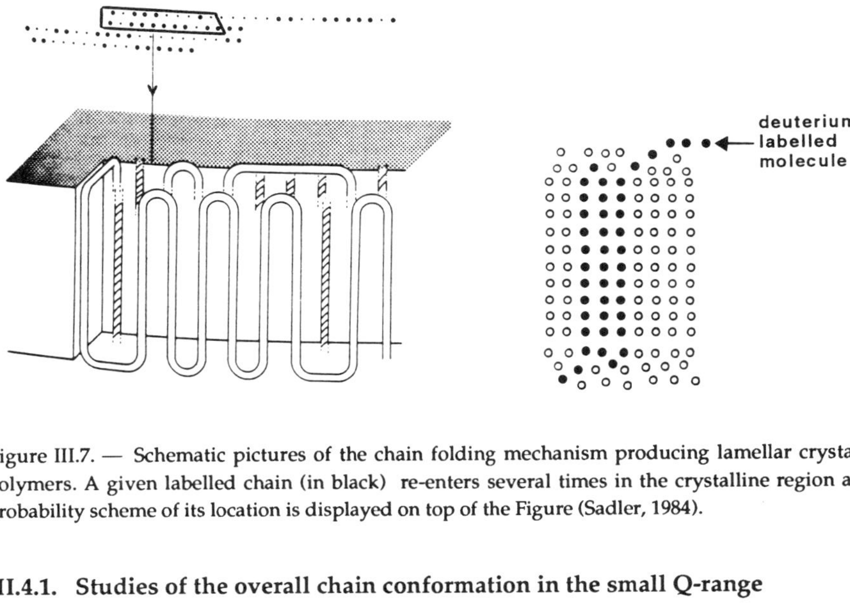

Figure III.7. — Schematic pictures of the chain folding mechanism producing lamellar crystals in polymers. A given labelled chain (in black) re-enters several times in the crystalline region and a probability scheme of its location is displayed on top of the Figure (Sadler, 1984).

III.4.1. Studies of the overall chain conformation in the small Q-range

As long as segregation effects can be avoided by appropriate thermal treatments, the conformation of the labelled chain can be determined by SANS . In the molten state, if a few polymer coils are deuterated, the SANS intensity follows the "Zimm behaviour" :

$I(0)/I(Q) = 1 + Q^2R_g^2/3$ with $R_g^2 = \langle x^2\rangle + \langle y^2\rangle + \langle z^2\rangle$ the radius of gyration of the coil.

For polyethylene samples crystallized from the melt it has been shown that Zimm behaviour is also observed in the solid state with almost no change in the radius of gyration of the labelled chain. In particular the scaling relation : $R_g \sim (M_w)^{1/2}$ holds at

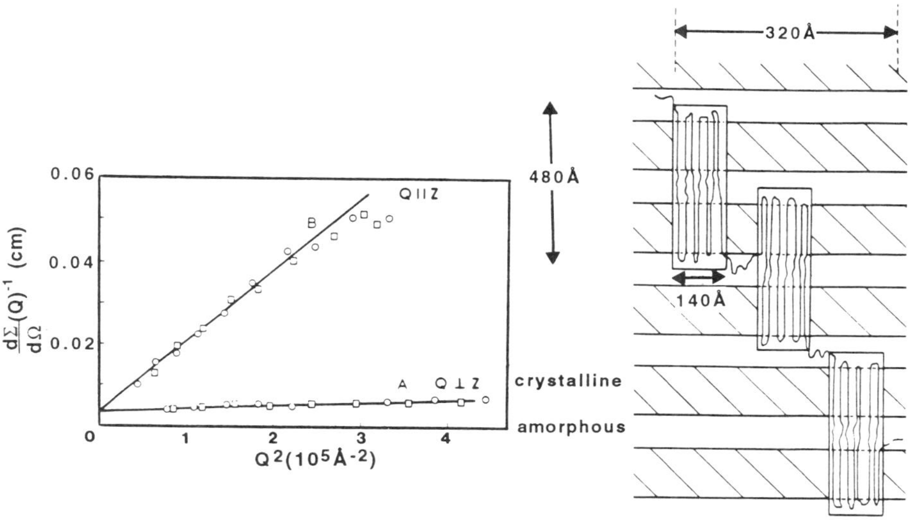

Figure III.8. — Zimm plot of the inverse scattered intensity along the meridian(R_g=480Å) and equatorial directions (for a polypropylene film with deuterium labelled chains (the radii of gyration are R_g=480Å and R_g=140Å respectively), and schematic drawing of the structure of an individual chain (Ballard et al., 1982).

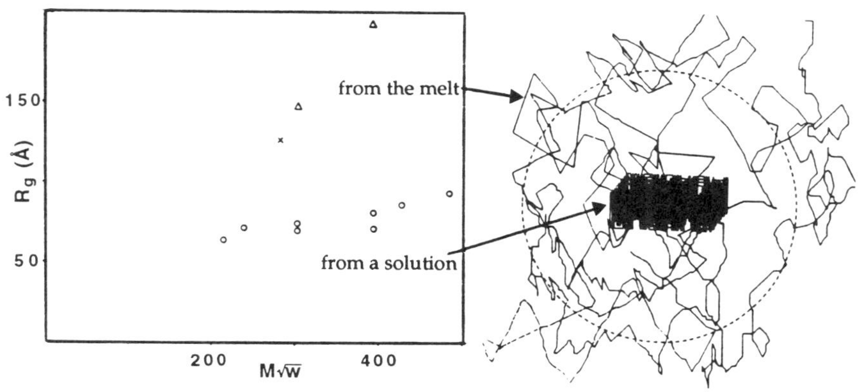

Figure III.9. — Comparison between the radii of gyration of deuterated polyethylene chains crystallized from the melt (x,Δ) and from solution (o), and schematic drawing of an overall chain conformation in these two situations(Sadler 1984 and Barham et al.,1984).

least for high molecular weights M_w (Sadler et al., 1979, Ballard et al.,1979). This means that, while the local structure of the chain is changed by the folding and the crystallization, its overall extension across the material remains essentially unchanged; a result which points out the role of the entanglements in the structure of dense polymer systems. For polymer samples oriented by stretching, the radius of gyration is increased along the draw axis but much less than expected by affine deformation, as shown by the results of Figure III.8 concerning an oriented sample of polypropylene with draw ratio 8/1 (Ballard et al., 1982). In contrast, for polymer crystallized from solutions, the radius of gyration of the labelled chains is much smaller than for melt crystallization as shown in Figure III.9. (Sadler et al., 1979, Barham et al., 1991). This means that owing to fewer entanglements in the semi-dilute solution a given chain can diffuse and participate in the crystallization of essentially one lamella.

III.4.2. Study of the local chain conformation in the intermediate Q-range

Fifteen years ago one of the great debates about polymer crystallization concerned the chain folding topology at the surface of the lamellar crystals : A. Keller (1979) proposed a model of regular adjacent re-entry, while P.J. Flory et al. (1978) proposed a "switchboard model" with almost random re-entry. From neutron scattering results on quench crystallized polyethylene, D. M. Sadler et al. (1982) demonstrated that at least 70 % of the foldings produce adjacent re-entry. The results of Figure III.10. show how this conclusion has been drawn.

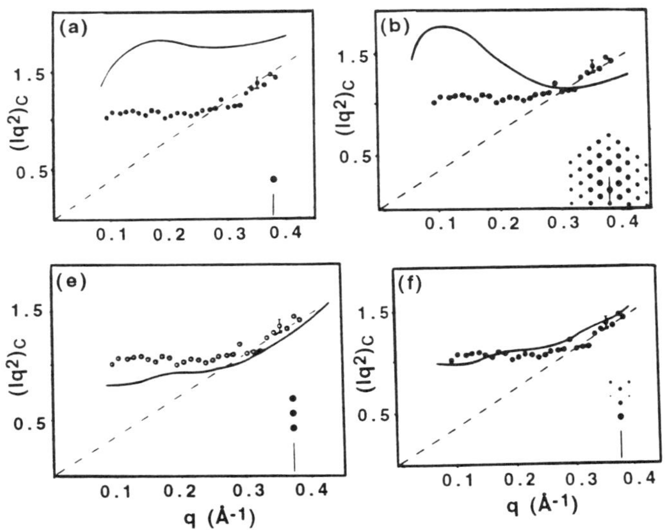

Figure III.10. — Comparison of the neutron scattreing results in the Q-range 0.1 - 0.4 $Å^{-1}$ with the structure factor of a labelled chains in different configurations of re-entry : a- single re-entry, b- two dimensional adjacent re-entry, e- triple re-entry along a row, f- 70% probability of adjacent re-entry along a row (Sadler, 1984).

Is is worth noting here that from steric considerations complete adjacent re-entry may be hindered if the surface of the lamella is perpendicular to the chain axis, while by forming oblique planes of folding (such as (111) or (201)) one increases the surface available for each chain fold (cf. III.3).

III.5. High resolution diffraction using SynchrotronRadiation

Information on the size of the crystallites or more specifically on the coherence length of the crystalline regions can be obtained from the linewidth $\Delta(2\theta)$ of the Bragg reflections. The Scherrer expression :

$$\Delta(2\theta) = 0.9\,\lambda / (L_c \cos\theta)$$

permits evaluations of the coherence lengths of the crystals (perpendicularly to the Bragg planes under consideration) if the angular resolution of the diffractometer is sufficient. Figure III.11 shows an example of lamellar crystals whose large coherence length perpendicular to the chain axis required the use of synchrotron radiation to be resolved accurately (Macchi et al. 1990).

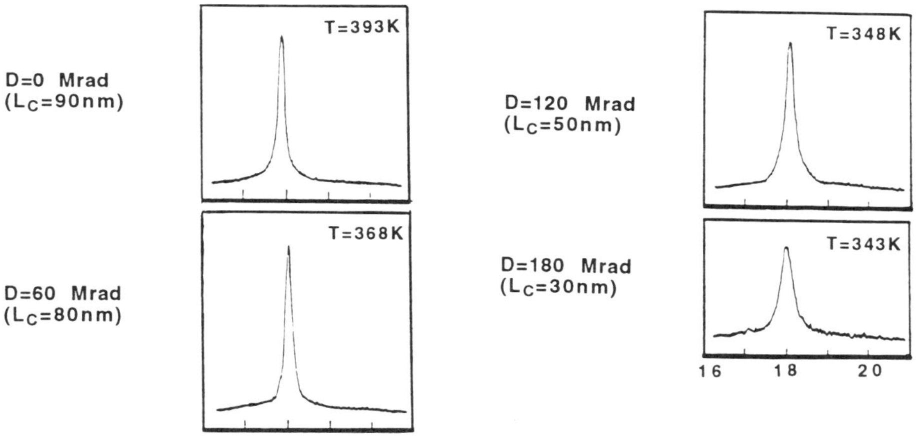

Figure III.11. — Evolution of the Bragg diffraction linewidth showing the fragmentation of lamellar crystals produced by different doses of electron irradiation in the high temperature phase of P(VDF-TrFE) 70/30 copolymer thin films (10 μm)

III.6. Time resolved studies using Synchrotron Radiation

As already shown in Figure III.4, annealing of a semi-crystalline polymer sample close to its melting point produces an increase of the long period. Real time studies of this phenomenon have been performed in combination with calorimetric measurements (Koberstein et al., 1985). Figure III.12 shows the evolution of the SAXS pattern of a polyethylene sample upon heating up to the melting point. The observed increase of the long period which is accompanied by an endothermic signal demonstrate that the smaller the lamellar crystals the lower is their melting

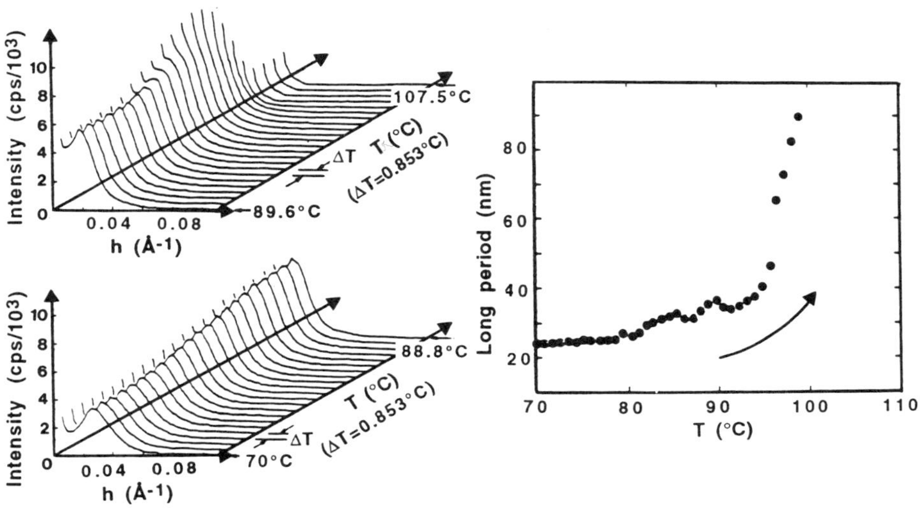

Figure III.12. — Evolution of the SAXS profile and of the corresponding long period during continuous heating of a polyethylene sample

temperature. As a result of this melting, the long period, which is the average distance between crystallites, increases. Moreover, if the temperature is stabilized a few degrees below the melting point, one observes a thickening of the remaining large lamellae. Owing to the chain folded structure of the lamellae this requires a longitudinal diffusion of the chains inside the crystallites; such a process can be efficient only at this high temperature (Bourgaux-Leonard et al., 1991).

III.7. Appendix : Chemical formulae of a few polymers quoted in the chapter

polyethylene	$- CH_2 - CH_2 - CH_2 - CH_2 - CH_2 - CH_2 -$
PVC	$- CHCl - CH_2 - CHCl - CH_2 - CHCl - CH_2 -$
PTFE (Teflon)	$- CF_2 - CF_2 - CF_2 - CF_2 - CF_2 - CF_2 -$
PVDF	$- CF_2 - CH_2 - CF_2 - CH_2 - CF_2 - CH_2 -$
P(VDF-TrFE)	$- CF_2 - CH_2 - CF_2 - CHF - CF_2 - CH_2 -$

polyesters

polyamides (Nylons) $[- NH - (CH_2)_m - NH - CO - (CH_2)_p - CO -]_n$

PMMA

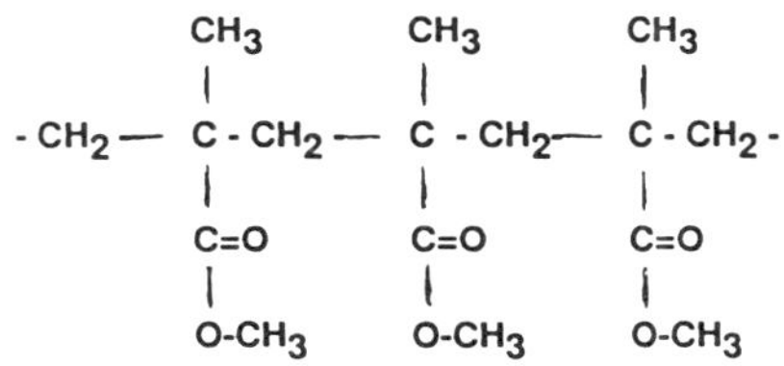

PDMS

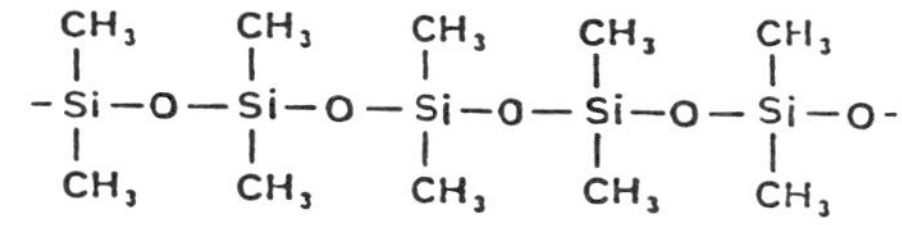

polystyrene

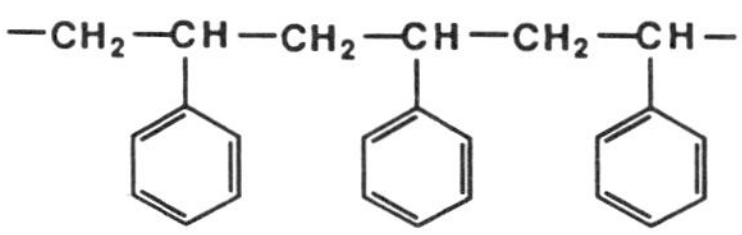

PMDA-ODA (Kapton)

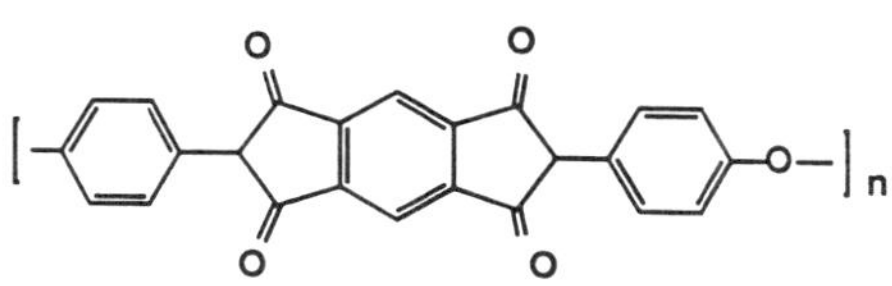

General Reading

TASHIRO K., TADOKORO H., (1989) in "Encyclopedia of Polymer Science and Engineering - Supplement Volume", J. Wiley, New York, p.187-231

HALL I.H., editor (1984) "Structure of Crystalline Polymers" , Elsevier, London

"Organisation of Macromolecules in the Condensed Phase", (1979), Faraday Disc. of the Chem. Soc. 68

WARD I.M., editor (1975) "Structure and properties of oriented polymers"- Applied Science, London

References

BALLARD D.G.H. et al., 1979, Polymer 20, 399

BALLARD D.G.H. et al., 1982, Polymer 23, 1875

BARHAM P.J., SADLER D.M., 1991, Polymer 32, 393

BELLET-AMALRIC E., LEGRAND J.F., STOCK-SCHWEYER M., MEURER B., 1993 - Polymer (in press)

BOURGAUX-LEONARD C., LEGRAND J.F., RENAULT A., DELZENNE P., 1991, Polymer 32, 597

DELZENNE P.,1986, PhD Thesis, Univ,. J. Fourier Grenoble (unpublished)

FACTOR B.J., RUSSEL T.P., TONEY M.F., 1991- Phys. Rev. Lett. 66, 234

FLORY P.J., YOON D.Y., 1978, Nature 272, 236

FORSYTH V.T., 1993 - in "Neutron and Synchrotron Radiation for Condensed Matter Studies", Vol 1, Editions de Physique, les Ullis and Springer Verlag, Berlin, p. 247-260

GRUBB D.T., DLUGOSZ J., KELLER A., 1975, J. Mater. Sci. 10, 1753

KELLER A., POPE D.P., 1971, J. Mater. Sci. 6, 453

KELLER A., 1979, Faraday Disc. of the Chem. Soc. 68, 145

KOBERSTEIN J.T., RUSSEL T.P., 1985, J. Polym. Sci. Polym. Phys. Ed. 23, 1109

LEGRAND J.F., LAJZEROWICZ J., 1983, Ferroelectrics 51, 129

LEGRAND J.F., 1989, Ferroelectrics, 91, 303

LEGRAND J.F., 1993, IEEE Trans. on Electrical Insulation, 28, 336

MACCHI F., DAUDIN B., LEGRAND J.F., 1990, Nucl. Intsr. and Meth. B 46, 334

SADLER D.M., HARRIS R., 1982, J. Polym. Sci. Polym. Phys. Ed. 20, 561

SADLER D., 1984 - in "Structure of Crystalline Polymers" , Elsevier, London p. 125-180

SERVET B., RAULT J., 1979, J. Physique 40, 1145

STEIN R.S., WILKES G.L., 1975 - in "Structure and properties of oriented polymers"- Applied Science (London) p.57

WITTMANN J.C., LOTZ B., 1985, J. Polym. Sci. Polym. Phys. Ed. 23, 205

WITTMANN J.C., SMITH P., 1991 - Nature 352, 414

CHAPTER IV

TWO-DIMENSIONAL CRYSTALLOGRAPHY OF AMPHIPHILIC MOLECULES AT THE AIR-WATER INTERFACE

K. KJAER, J. ALS-NIELSEN, M. LAHAV AND L. LEISEROWITZ

IV.1. Introduction

Ever since Langmuir (1917) discovered that amphiphilic molecules, composed of a hydrophobic tail and a hydrophilic head group, form films at the air-water interface, they have elicited wide interest because of their importance in the pure and applied sciences. For example, Langmuir films may be transferred to solid supports for the formation of multilayer Langmuir-Blodgett films (Kuhn et al, 1971), for potential use as piezoelectric sensors (Blinov et al, 1984), soft X-ray monochromators, non-linear optical devices (Girling et al, 1985), and in molecular electronics (Roberts et al, 1978; Morizumu, 1988). Langmuir monolayers serve as useful models for elucidating structure and function of biological membranes in their interactions with lipids, steroids, sugars, proteins and ions (Swalen, 1987). They may also be engineered to induce nucleation of three-dimensional (3-D) crystals from molecules or ions present in the subphase by a form of epitaxial growth (Landau et al, 1985). In this respect they may be used as models for the study of biomineralization in which two-dimensional (2-D) surfaces of biological macromolecules induce the mineralization (Addadi et al, 1987). The design of monolayers with hydrophilic head group arrangements which mimic the surfaces of 3-D crystals also provides a means of examining crystal surface-solvent interactions, for clarifying the effect of the solvent on crystal growth and polymorphism. Furthermore, a knowledge of the characteristics of Langmuir monolayers of water-insoluble amphiphiles should help to throw light on the packing characteristics at the air-water interface of water-soluble amphiphiles, where even less is known about their 2-D arrangements (Weissbuch et al, 1984).

Many of the classical methods used to characterize the properties of Langmuir films do not give direct information on their crystalline packing arrangements. These methods include measurements of surface pressure versus average molecular area, surface potential data (Gaines, 1966), non-linear optics from surfaces (Heinz et al, 1982) and IR spectroscopy (Allara et al, 1985; Gun et al. 1984). Epifluorescence microscopy does furnish some information on crystallinity and morphology on the micron level, but is limited to films containing dye probes (Weiss et al, 1984; Losche et al, 1983). This limitation does not hold for Brewster angle microscopy and is now widely used (Henon and Meunier, 1991; Hönig and Möbius, 1991). Thus, on the molecular level, knowledge of 2-D monolayers had been achieved, in the main, from 3-D crystals. Only recently, with the advent of intense and well collimated X-ray beams of variable wavelength from synchrotron sources, has it became possible to obtain direct structural information on the crystalline properties of these films by X-ray surface diffraction methods.

Here we shall describe X-ray diffraction techniques, which allow one to study films on liquid surfaces *in situ* and review some of the structural results obtained. Emphasis will be placed on monolayer packing arrangements, particularly in terms of the polar head groups. We shall examine crystalline self-aggregation, the effect of pH, ions and solute molecules in the aqueous subphase on molecular packing and crystallinity, and growth and domain size of 2-D crystallites. Finally, we shall discuss the role played by structural complementarity for the induction of 3-D crystal nucleation at the monolayer-solution interface.

For simplicity and ease of presentation we restricted the topics and systems described in the main to our own work. Nevertheless we do wish to cite the pioneering grazing incidence diffraction (GID) studies by other groups. Major contributions were the first GID study of a monolayer of an amphiphile, dipalmitoyl phosphalidylcholine, but on solid support (Seul, Eisenberger, mcConnell, 1983), the detection of a diffraction peak from a phospholipid monolayer on water in a study performed by Möhwald, Kjaer, Als-Nielsen and coworkers (Kjaer et al., 1987) and a GID study of lead-arachidate on water (Dutta et al., 1987). A state of the art review on techniques and properties of organic thin films, including GID results of monolayers on liquid surfaces, has been compiled in a book by Ulman (1991).

IV.2. Surface Sensitive X-ray Methods

The intensity measured in a conventional X-ray scattering experiment is proportional to the irradiated sample volume. This, in turn, is proportional to the penetration depth of the radiation in a sample. For X-rays of about 1 Å wavelength, this penetration ranges from a few micrometers for highly absorbing material to a few millimeters for low absorbing materials. Consequently, the scattering from the surface region is weak compared to that from the bulk. Therefore, a method restricting the penetration depth to the surface region is a

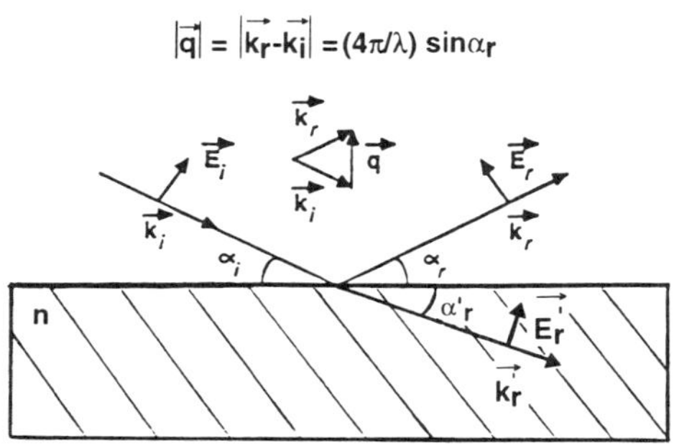

Fig IV.1. — Refracted ($\mathbf{E'_r}$) and reflected ($\mathbf{E_r}$) waves resulting from an incident plane wave with amplitude $|\mathbf{E}_i|$ upon an interface between air and a material of refraction index n. Electric fields $\mathbf{E}$ are illustrated for one polarization only. The symbol $\mathbf{q}$ is the scattering vector, with $\mathbf{k}_i$ and $\mathbf{k}_r$ the incident and reflected wave vectors of magnitude $2\pi/\lambda$. Note that as $n \lesssim 1$, $\alpha'_r < \alpha_i = \alpha_r$. This provides for total reflection at angles $\alpha_r < \alpha_c$, where α_c is the critical angle for total external reflection.

prerequisite for surface diffraction experiments. This restriction can be achieved by using grazing angles of incidence and employing the phenomenon of total external reflection from the surface.

The refractive index n of matter for X-rays in the 1 Å wavelength range is given by (James, 1982):

$$n = 1 - \delta - i\beta \qquad \text{(IV.1)}$$

where $\delta = \rho r_o \lambda^2 / 2\pi$, λ is the wavelength, ρ is the electron density and $r_o = 2.82 \cdot 10^{-13}$ cm is the classical electron radius. Typically δ is of the order of 10^{-5}. The term β is equal to $\eta\lambda/(4\pi)$ where η is the linear absorption coefficient. For X-rays of wavelength $\lambda = 1$Å, absorption is small and $\beta << \delta$.

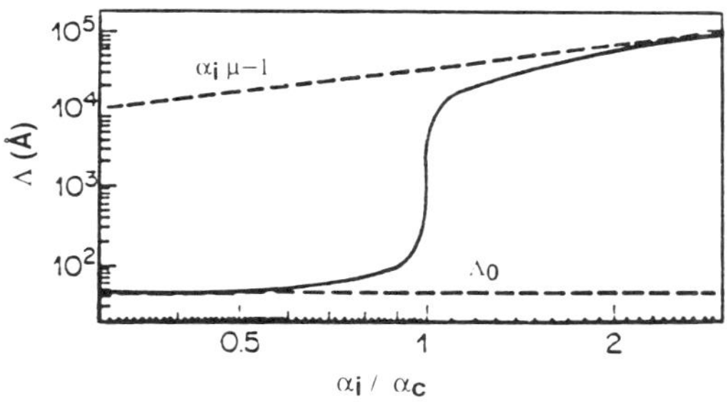

Fig. IV.2. — X-ray beam penetration depth Λ in water at λ = 1.38 Å versus incidence angle α_i/α_c, where α_c is the critical angle for total external reflection. The limiting value Λ_o of the penetration depth for $\alpha_i/\alpha_c << 1$ is a characteristic of the material, $= (\pi r_o \rho)^{-1/2}/4$, independent of the X-ray wavelength λ. For $\alpha_i/\alpha_c >> 1$ the penetration depth is $\alpha_i \mu^{-1}$ where the linear absorption coefficient μ *does* depend on λ.

Consider a plane wave with wavevector $\mathbf{k}_i$ ($k=2\pi/\lambda$) striking at an angle α_i on a planar interface separating a homogeneous medium from the vacuum, as shown in fig. IV.1. The wave will be partially reflected into the vacuum in the direction given by $\alpha_r = \alpha_i$ and partially refracted into the lower medium in the direction given by α'_r. For this case Snell's law yields:

$$n \cos\alpha'_r = \cos\alpha_i \qquad \text{(IV.2)}$$

As $n \cdot < 1$, for angles of incidence $\alpha_i \leq \alpha_c = \cos^{-1}(n) = (2\delta)^{1/2}$, the phenomenon of total reflection occurs (Born & Wolf, 1959): the incident wave is totally reflected, while the refracted wave becomes evanescent travelling along the surface. The amplitude of the evanescent wave decays exponentially with depth. The penetration depth of the X-rays in water versus angle of incidence α_i is shown in fig. IV.2. For $\alpha_i < 0.5\ \alpha_c$, the penetration depth is ~50Å. The evanescent wave may therefore be diffracted by crystalline material in a surface layer of that thickness and provide information on its structure. Such diffraction from a surface layer is called grazing incidence diffraction (GID).

In 3-D crystals, diffraction from a set of crystal planes with an interplanar spacing d occurs only when Bragg's law is obeyed; namely when the scattering vector length $|\mathbf{q}|$ given by $|\mathbf{k}_f - \mathbf{k}_i| = 4\pi\sin\theta/\lambda$ is equal to $2\pi d^*$, where d^* is the reciprocal of d, and the normal to the planes bisects the angle between the incident and outgoing beam. This condition may be expressed mathematically in terms of the reciprocal lattice vectors $\mathbf{d}^* = h\mathbf{a}^* + k\mathbf{b}^* + l\mathbf{c}^*$ where $\mathbf{a}^*, \mathbf{b}^*, \mathbf{c}^*$ are the reciprocal vectors of the unit cell vectors **a, b, c** and h, k, l are integers that represent the Miller

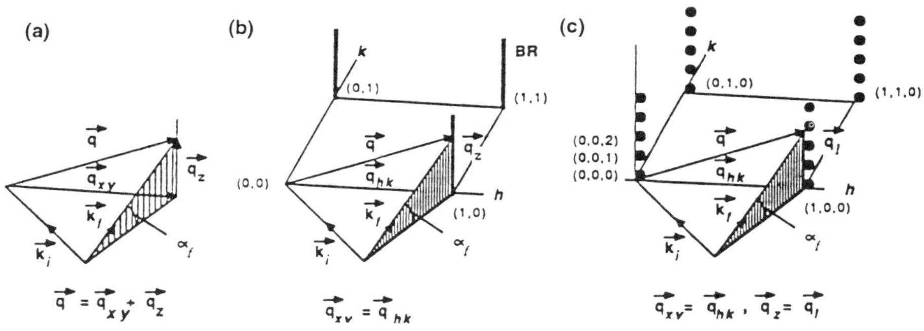

Fig IV.3. — (a) General scattering geometry. $\mathbf{k}_i$ and $\mathbf{k}_f$ are the wavevectors of the incident and diffracted beams respectively. The scattering vector $\mathbf{q} = \mathbf{k}_f - \mathbf{k}_i$ has components $\mathbf{q}_{xy}$ parallel to the monolayer plane and $\mathbf{q}_z$ perpendicular to it. (b) Scattering from a 2-D crystal, depicted in reciprocal space, extends in Bragg rods in the $\mathbf{q}_z$ direction, perpendicular to the plane of the monolayer and of its reciprocal 2-D net. The scattering vector $\mathbf{q}$ must end on an (h,k) Bragg rod for Bragg diffraction to occur. (c) Scattering from a 3-D crystal in a given orientation lies in reciprocal space at the reciprocal lattice points (h,k,l). Diffraction takes place when the scattering vector $\mathbf{q}$ coincides with a specific reciprocal lattice point.

indices of planes with a spacing d_{hkl}. Diffraction only takes place in 3-D crystals when the scattering vector $\mathbf{q}$ coincides with (h,k,l) points of the reciprocal 3-D lattice (Fig. IV. 3c).

For a 2-D crystal, there is no restriction on the scattering vector component $\mathbf{q}_z$ along the film normal; the Bragg scattering extends as continuous rods (BR) through the 2-D reciprocal lattice points (Vineyard, 1982, Feidenhans'l, 1989) (Fig. IV. 3b). The finite thickness of the 2-D crystal causes the BR to extend over finite q_z intervals. The intensity distribution along these intervals is determined by the vertical electron distribution in the molecules and is expressed as the Fourier transform of the resulting electron density along the film normal. This intensity modulation can now be analysed to obtain information, for example, on the molecular packing arrangement in the crystalline part of Langmuir monolayers (Als-Nielsen & Kjaer, 1989, Jacquemain et al, 1989). In addition, detailed information on the electron density distribution in the vertical direction, laterally averaged over both the *ordered* and *disordered* parts of the monolayer can be obtained from the deviation of the measured X-ray reflectivity (Als-Nielsen & Kjaer, 1989) (XR) from the Fresnel law (Born & Wolf, 1959). Finally, grazing incidence X-ray excited fluorescence (Bloch & Eisenberger, 1988) and X-ray standing wave (Batterman, 1969; Bedzyk et al, 1988, 1990) measurements yield information on the location and concentration of ions bound to the monolayer at the liquid interface. We now discuss the methods of specular X-ray reflectivity (XR) and grazing incidence X-ray diffraction (GID) in some detail.

IV.2.1. Specular Reflectivity (XR)

Specular reflection means that the reflected ray measured is in the plane spanned by $\mathbf{k}_i$ and the vector normal to the surface and that $\alpha_i = \alpha_r$ (Fig. IV. 1). The specular reflectivity of an ideal surface is given by the Fresnel law of optics. In the limit of small angle of incidence, it reduces to:

$$R_F(q_z) = \left| \frac{q_z - [q_z^2 - q_c^2 + i(4\pi\mu/\lambda)]^{1/2}}{q_z + [q_z^2 - q_c^2 + (4\pi\mu/\lambda)]^{1/2}} \right|^2 \quad \text{(IV.3)}$$

where $q_z = (4\pi/\lambda)\sin\alpha_i$ is the *z*-component of the vertical scattering vector and $q_c = (4\pi/\lambda)\sin\alpha_c$ is the critical value of q_z for total external reflection. The complex term arises from absorption effects. Neglecting absorption, for $q_z < q_c$, equation (IV.3) yields $R_F = 1$, *i.e.*, total reflection. As q_z is increased beyond q_c, however, R_F decreases and, for $q_z \gtrsim 4q_c$, approaches $R_F(q_z) \approx (q_c/2q_z)^4$. As the range of interest may extend to $q_z/q_c \approx 30$ or more, reflectivities down to 10^{-8} have to be measured. Hence the importance of high X-ray intensity from a rotating anode generator or a synchrotron source Equation (IV.3) is valid only for an ideally flat surface across which the electron density $\rho(z)$ varies step-like between two constant values. If $\rho(z)$ varies continuously in the surface region, the reflectivity is modified, yielding:

$$R(q_z) = R_F(q_z)\ |\phi(q_z)|^2 \quad \text{(IV.4)}$$

where

$$\phi(q_z) = (1/\rho_\infty) \int [d\rho(z)/dz] \cdot \exp(iq_z z) dz \quad \text{(IV.5)}$$

and ρ_∞ is the constant electron density in the subphase bulk. Thus, by measuring $R(q_z)$ it is possible in principle to determine $\rho(z)$, the variation of the electron density normal to the surface, and, in particular, that of a monolayer floating on it. ϕ is a complex function and only its modulus can be derived from the measured reflectivity but not its phase. Thus, one is faced with the usual phase problem of X-ray crystallography. To date virtually all specular reflectivity data have been analysed by fitting a model density $\rho(z)$ to the measured data via equations (IV.3), (IV.4) and (IV.5).

A computationally simple strategy is to represent the monolayer as a stack of slabs (Parrat, 1954), each with a constant density ρ_i and thickness L_i. Two such slabs would be needed for representing, say, a simple fatty acid monolayer. When refining such a *slab* model, a possible choice of fit parameters, then, would be the densities ρ_i and thicknesses L_i of the slabs.

The constant density of the subphase has to be added below the interface. Finally, the model density must be *smeared* out in the *z*-direction to account for the vertical roughness of the interface. The root-mean-square roughness, σ, is of order 3 Å and stems mainly from thermally excited capillary waves on the water

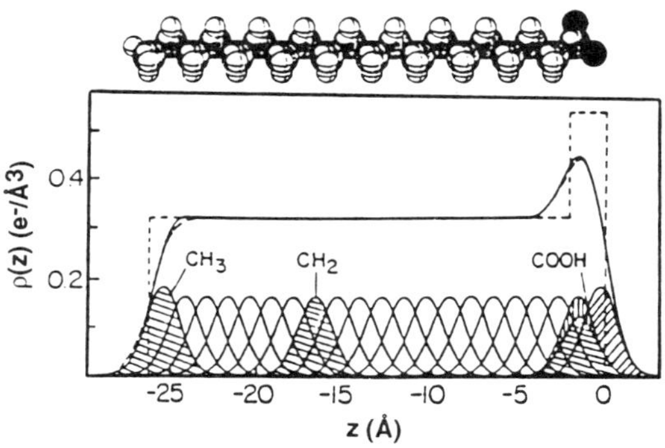

Fig . IV.4. — *Top.* Space filling model of a $C_{19}CO_2H$ molecule.

Bottom. Model electron densities ρ(z). The dashed lines refer to a slab model and a smeared slab model. The full lines refer to an atomic model using CH_3 and CH_2 pseudo-atoms. The contribution of each pseudo-atom (or atom at the CO_2H head group) to the electron density is indicated by Gaussian curves. The two models agree nearly exactly.

surface (Braslau et al, 1988; Als-Nielsen, Kjaer, 1989) It leads to a Debye-Waller factor exp $(-q_z^2\sigma^2)$ in $|\phi(q_z)|^2$ (equation (IV.5)). The rather large value of $\sigma \approx 3$ Å is also the reason that the monolayer may be adequately repesented by the *slab* model. This is illustrated in fig. IV. 4, which shows that the atomic density ρ(z) of a fatty acid monolayer may be represented by a slab model.

IV.2.2. Grazing Incidence Diffraction (GID) and Bragg Rods

In the GID geometry, shown in fig. IV. 5, the angle of incidence α_i of the X-ray beam is kept below the critical angle α_c. This limits the penetration depth of the beam to that of the evanescent wave, so that the background scattering from the subphase is efficiently eliminated. This allows an accurate measurement of the weak diffraction signal originating from the crystalline monolayer. For all the diffraction patterns measured to date, the monolayers were found to be composed of 2-D crystallites randomly oriented on the water surface, namely a 2-D "powder". Hence only a 2θ scan in the monolayer plane is required to measure the diffraction pattern. The analysis is similar to that of a conventional 3-D powder pattern; the reflections can be indexed by two Miller indices, *hk*. Their angular positions $2\theta_{hk}$ corresponding to $q_{hk} = (4\pi/\lambda)\sin\theta_{hk}$ yield the repeat distances $d_{hk} = 2\pi/q_{hk}$ of the 2-D crystal lattice. Their resolution-corrected 2θ-line width (FWHM), Δ, yields the 2-D crystalline coherence length *L* through the Scherrer formula (Guinier, 1968).

$$L = (0.9\lambda)/(\cos\theta_{hk}\Delta) \qquad \text{(IV.6)}$$

The square of the molecular structure factor $|F_{hk}|^2$ integrated along the Bragg rod over the window of q_z seen by the detector, determines the integrated intensity in the peak (James, 1982) . The structure factor F_{hk} is given by:

$$F_{hk}(q_z) = \sum_j f_j \exp i\,(\mathbf{q}_{hk} \cdot \mathbf{r}_j + q_z \cdot z_j) \qquad \text{(IV.7)}$$

where f_j is the scattering factor of the atom j, $\mathbf{r}_j$ is the vector specifying the (x,y) position of the atom j in the unit cell and z_j is the atomic coordinate along the vertical direction.

More information on the monolayer structure can be obtained by resolving the variation with q_z along the Bragg rod of the intensity $I_{hk}(q_z)$, which is given by:

$$I_{hk}(q_z) = KV(q_z)\,|F_{hk}(q_z)|^2\,DW_{hk}(q_z) \qquad \text{(IV.8)}$$

The observed BR intensity $I_{hk}(q_z)$ is actually a sum over those (h,k) reflections whose Bragg rods coincide at a particular horizontal 2θ-angle or q_{xy} position. In equation (IV.8), the most important variation is due to the molecular structure factor $|F_{hk}(q_z)|$. For chain-like surfactant molecules, the square of the structure factor $|F(q)|^2$ is a bell-shaped function which reaches its maximum when the scattering vector $\mathbf{q} = (q_{hk}, q_z)$ is orthogonal to the molecular axis. Thus, when the molecules are vertical or tilted in a plane perpendicular to q_{hk}, the maximum intensity along the Bragg rod will occur at the horizon, for $q_z = 0$ Å^{-1}. For molecules tilted otherwise, the Bragg rod maximum will occur at a finite q_z, dependent upon both the magnitude and direction of the tilt relative to the in-plane scattering vector $\mathbf{q}_{hk}$. The width of the bell-shaped Bragg rod profile measured in a direction parallel to the molecular axis is inversely proportional to the length of the molecule. This description of $F(\mathbf{q})$, may be improved on by calculating $F(\mathbf{q})$ according to an atomic model of the molecule as shown by equations (IV.7)and (IV.8). The exponential (Debye-Waller) factor $DW_{hk}(\mathbf{q}_z) = \exp[-(q^2_{hk} U_{xy} + q^2_z U_z)]$ where U_{xy} and U_z are respectively the mean square displacements in each horizontal direction x, y and along the vertical direction z. They account for the thermal motion of the atoms in the molecule, as well as for ripples on the water surface, which lead to diffuseness of the interface. The factor $V(q_z)$ describes the interference of rays diffracted upwards with rays diffracted down and subsequently reflected back up by the interface (Vineyard, 1982). $V(q_z)$ differs from unity in the vicinity of $q_z = q_c/2$ where it contributes a sharp peak. The factor K includes various corrections, such as Lorentz and polarization factors.

There are thus fundamental differences between the BR and XR methods. The BR originate solely from crystalline material and provide information on unit cell dimensions, plane group symmetry, molecular coordinates (indeed at near atomic resolution in favorable cases) and in plane and out-of-plane molecular motion. This is in contrast with XR, which provides information only on the vertical variation of the electron density, averaged over the illuminated area.

An experimental set-up (Fig. IV 6) for the GID geometry (Als-Nielsen et al, 1982) is shown in fig. IV.5. A monochromatic beam is extracted from the incident synchrotron beam by Bragg reflection from a monochromator crystal. By tilting the normal to the reflecting planes out of horizontal plane, the monochromatic beam can be bent down to yield a grazing angle with the liquid surface. Slit S_2, in front of the sample defines a suitable illuminated beam area on the liquid surface. Slit S_3 excludes diffuse scattered background around the reflected beam.

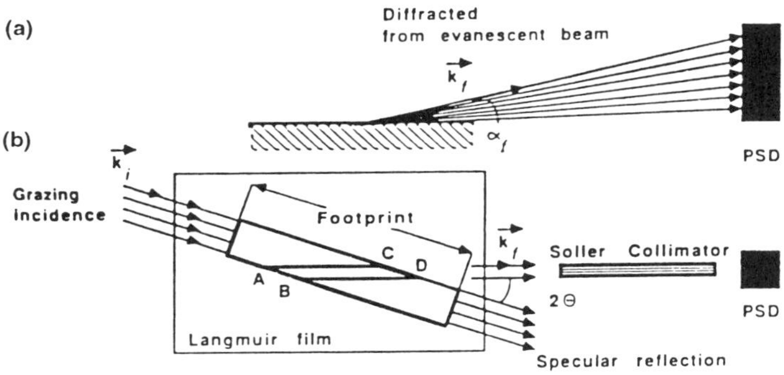

Fig IV.5. — Top and side views of the grazing incidence X-ray diffraction (GID) geometry. The footprint of the grazing incident beam is indicated. The position sensitive detector (PSD) has its axis along the vertical. Only the cross-beam area ABCD contributes to the measured scattering. The Soller collimator consists of thin vertical absorbing foils stacked together to define the horizontal resolution of the detector.

This slit together with a scintillation detector is mounted on an elevator situated on a diffractometer arm which is pivoted around a vertical axis through the sample center. For Bragg rod measurements (Fig. IV. 5), slit S3 is replaced by a Soller collimator giving 1 or 2 mrad collimation and the scintillation counter by a position sensitive detector

IV.3. General Packing Characteristics of Langmuir Monolayers on Water

Before the development of surface X-ray diffraction techniques, molecular arrangements in the three-dimensional state were used to increase understanding of the structure and behaviour of molecules in the monolayer state. Analogous packing is not always possible however, because of environmental differences between the two- and three-dimensional states. Phase diagrams constructed from surface pressure-area (π-A) measurements of Langmuir monolayers revealed phase boundaries, depending upon the pressure, temperature and chain length, but provided little information on their structure. Now, with the recently acquired information from surface-sensitive X-ray techniques, molecular packing within Langmuir films can be determined almost at the atomic level. We will not discuss in any detail Langmuir monolayer structures in terms of their π-A phase diagrams; rather we will focus on their properties in the uncompressed, namely relaxed, state on the solution surface. We will describe examples from different chemical classes studied to date: alcohols, amides, carboxylic acids, salts of the latter, and of α-amino acids (see Table 1 for their chemical formulae and abbreviated names).

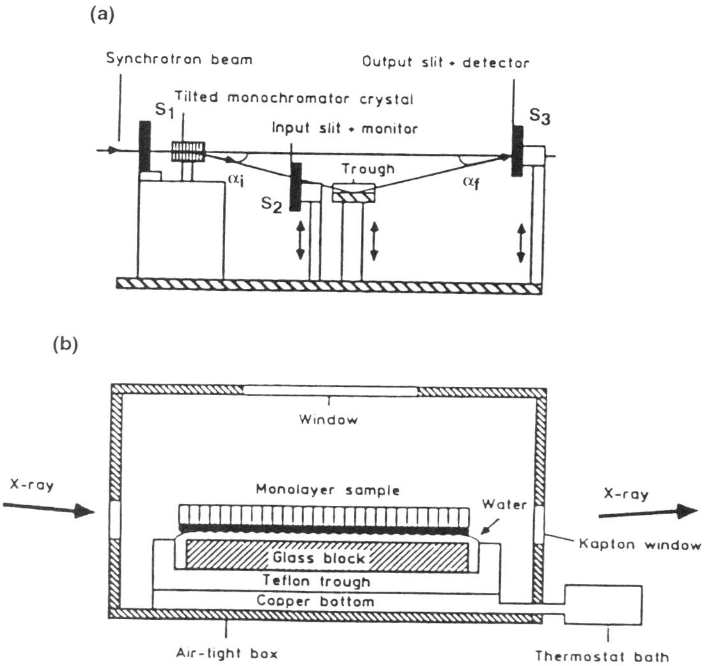

Fig IV.6. — Liquid surface diffractometer (beamline D4, X-ray synchrotron source at Hasylab, Hamburg). *a)* Side view of the vertical, scattering plane.

b) Expanded view of the trough showing the monolayer sample spread on a thin film of water. The glass block provides for a thin liquid film (about 0.3 mm thick) to reduce surface capillary waves.

Table 1 : Abbreviations for monolayer compounds discussed

Abbreviation	Full name and chemical formula
C_nOH	Aliphatic primary alcohols $C_nH_{2n+1}OH$
C_nCO_2H	Aliphatic primary acids $C_nH_{2n+1}COOH$
C_nCONH_2	Aliphatic primary amides $C_nH_{2n+1}CONH_2$
PFA	1H,1H,2H,2H-perfluorododecyl-aspartate $C_{10}F_{21}C_2H_4OCOCH_2CHNH_3{}^+CO_2{}^-$

It is useful at this stage to describe how the unit cell dimensions and molecular orientation may be determined from the GID data for simple cases. We limit the analysis to chain-like molecules with assumed cylindrical symmetry and which pack in three different cells: hexagonal ($a = b, \gamma = 120°$), distorted hexagonal ($a = b, \gamma \neq 120°$) which may be more properly described as centered rectangular ($a' = |\mathbf{a}+\mathbf{b}|$, $b' = |\mathbf{a}-\mathbf{b}|$), and oblique ($a \neq b \neq |\mathbf{a}+\mathbf{b}|$). They are shown in fig. IV.7. In a hexagonal cell the molecules are aligned vertically; in the distorted hexagonal cell the molecules are tilted along a symmetry direction and in the oblique cell they are tilted along an arbitrary direction. As mentioned earlier, all Langmuir monolayers studied to date by GID consist of "powders" of randomly ordered 2-D crystallites on the water surface. Thus the GID measurements are not made on single crystals so that the diffraction spectra consist of coinciding Bragg rods of reflections (h,k,q_z) and $(\bar{h}, \bar{k},q_z)$; the notation $\{h,k,q_z\}$ designating both reflections. In general, the intensity distribution along their two coinciding

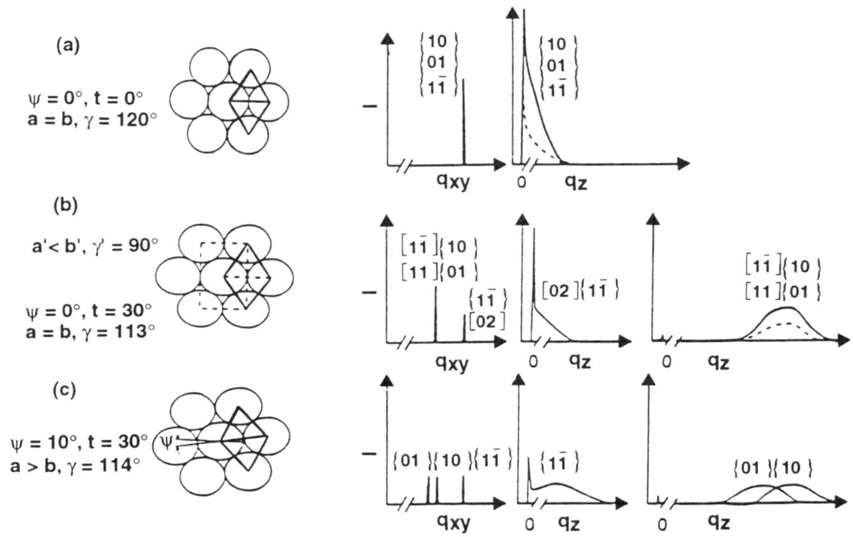

Fig IV.7. — Three structures and diffraction from a monolayer of close-packed cylindrical chains. The left colum shows schematic top views of molecular packing in hexagonal, distorted hexagonal and oblique unit cells. The second column shows the q_{xy}-pattern integrated over the vertical scattering vector component q_z. Last two columns show the Bragg rods integrated over the corresponding q_{xy}. When more than one reflection coincide, the dashed lines represent the intensity from individual reflections as indicated by the $\{h, k\}$ indices. The solid line represents the intensity from a 2-D powder sample; that is the sum of the intensities indicated by dashed lines. When the molecular axis is perpendicular to the water surface the lattice is hexagonal (a), and the $\{1,0\}$,$\{0,1\}$ and $\{1,\bar{1}\}$ reflections are equivalent. The Bragg rod obtained from the single in-plane diffraction peak has its maximum at $q_z = 0$ Å^{-1}. When the molecular axis is tilted towards nearest neighbors (b), the hexagonal cell is distorted to a centered rectangular cell (rectangular notation at top, distorted hexagonal notation at bottom). The degeneracy of the three reflections is partially lifted resulting in two peaks. The Bragg rod of the two equivalent reflections $\{1,0\}$ and $\{0,1\}$($[1,1]$, $\{1,\bar{1}\}$ in the rectangular notation) is centered at $q_z \geq 0$ Å^{-1}, the position of the center depending upon the extent of molecular tilt. The Bragg rod of the $\{1,\bar{1}\}$ reflection is still centered at $q_z = 0$ Å^{-1}. Finally, a tilt of the molecular axis in a non-symmetry direction, yields an oblique 2-D cell (c). The degeneracy is completely lifted and the peak positions of the three Bragg rods determine the tilt angle as well as the azimuthal angle ψ.

Bragg rods will be different unless the monolayer structure has a twofold symmetry along the vertical axis.

The hexagonal cell has three equivalent lattice spacings d_{hk}: d_{10}, d_{01} and $d_{1,\bar{1}}$ The GID spectrum from these three reflections will appear as a single peak with all corresponding Bragg rods peaking at $q_z = 0$ Å^{-1} (Fig. IV. 7a). The corresponding GID pattern for the distorted hexagonal cell will comprise two peaks; one a doublet arising from the coinciding {1,0} and {0,1} reflections, and the remaining {1,$\bar{1}$} peak (Fig. IV 7b). The separation between the two peaks depends upon the deviation of the angle γ from 120°. In the oblique cell the extent to which the three peaks {1,0}, {0,1} and {1,$\bar{1}$} are resolved (Fig. IV. 7c), depends upon the deviation from centered rectangular cell symmetry.

Thus the 2-D crystal lattice parameters (a, b, γ) can be extracted, from the horizontal position of the maxima of the Bragg peaks. Precise information on the molecular orientation can be easily extracted from the BR data assuming the aliphatic chains are uniformly and rigidly tiled in the monolayer (Als-Nielsen & Kjaer, 1989; Jacquemain et al, 1989, 1990). The angle t between the molecular axis and the surface normal is given by:

$$\cos\psi_{hk} \cdot \tan t = q_z^o / |q_{hk}| \quad \text{(IV.9)}$$

where q_z is the position of the maximum along the Bragg rod and ψ_{hk} is the azimuthal angle between the molecular tilt direction projected onto the xy plane and the reciprocal lattice vector q_{hk}. The molecules will therefore be perpendicular to the surface when, for all reflections (h,k), q_z tends to 0 Å^{-1}.

IV.4. Crystalline Self-Assembly of Non-compressed Monolayers

GID has proven a powerful tool for the detection of crystallite self-aggregates formed by insoluble amphiphilic molecules and for the determination of their size and structure, almost at the atomic level. Indeed GID has demonstrated that this crystalline self-assembly is a widespread phenomenon for fluorocarbon (Jacquemain et al, 1990; Barton et al, 1992) and hydrocarbon amphiphiles (Jacquemain et al, 1991; Leveiller et al, 1991). We will focus on hydrocarbon systems, although the fluorocarbons exhibit a greater extent of crystallinity, presumably because of a higher stiffness of their carbon backbone and a more favorable van der Waals interaction between their chains (Jacquemain et al, 1992).

IV.4.1. Structure determination of self-assembled crystallites

Experiments have shown that lowering the temperature considerably enhances the tendency for crystalline self-aggregation of hydrocarbon chain monolayers. Thus GID studies were performed at low temperature on several monolayers C_n-X with different head groups X, i.e. C_nOH, (n=16, 20,23, 30, 31), $C_{19}OCOC_nOH$ (n=9,10), $C_{29}CO_2H$, $C_{19}CONH_2$, $C_{29}CONH_2$ and $C_{19}CONHC_2H_4CO_2H$. The probable packing characteristics of the spontaneously formed crystallites could be extracted from the GID data (Jacquemain, et al 1991; Weissbuch et al, 1992; Leveiller, et al 1992; Wang et al, 1992), as well as making use of atom-atom potential energy calculations to reduce ambiguities . We present here a brief description of the deduced packing arrangements of the acid

($X = CO_2H$) the alcohol ($X = OH$), and the amide ($X = CONH_2$) monolayer.

The GID pattern for the monolayer of $C_{29}CO_2H$ in the uncompressed state at +5°C (Fig. IV.8) comprises two low-order reflections which indicates a rectangular cell (Table 2) for which the symmetry-related {1,1} and {1,$\bar{1}$} reflections coincide and the {0,2} reflection is separate and distinct. The intensity profile of the Bragg rod for the {0,2} reflection has its maximum at $\mathbf{q}_z = 0Å^{-1}$; hence the molecular axis lies in a plane perpendicular to the long *b* axis of the unit cell. The angle of tilt of the hydrocarbon chain within this plane can be extracted from the intensity profile of the {1,1} and {1,$\bar{1}$} Bragg rod making use of eqn. (9). The chain is tilted at 27° towards nearest neighbors along the short *a* axis. This angle is compatible with the length of 5.53 Å. These two lengths yield a molecular tilt angle of 26°(=arcos 4.94/5.53) in the uncompressed state, in the close agreement with the value derived from the Bragg rod profile.

In order to establish the cell symmetry and molecular packing, we now divert to review layered 3-D crystal structures of long chain acids and alcohols (Small, 1986). In each layer the molecules are related by a glide so that neighbouring hydrocarbon chains have parallel mean axes, but are in a mutually perpendicular orientation when viewed down the mean axis; the motif is described as orthogonal $O_\perp$ (Small, 1986). Common to all these structures is a rectangular subcell, which is a projection of the unit cell onto a plane perpendicular to the molecular axis and dimensions a_p = 5.0 Å, b_p = 7.4 Å, corresponding to a cross-sectional area of 18.5 $Å^2$. The calculated 2-D cells of the monolayers projected onto a plane perpendicular to the molecular axis perfectly match the rectangular subcell of the 3-D crystal structures. Thus it was deduced that the $C_{29}CO_2H$ acid monolayer molecules appear in the orthogonal $O_\perp$ packing, *i.e.*, they are related by a glide along the *a* axis in a primitive rectangular cell in plane group symmetry *p*11*g* as shown in figure IV. 9. (The symbol *p* specifies a primitive unit cell and *g* a glide plane perpendicular to the *b* axis, so naturally the glide is parallel to the *a* axis. Note that for plane group *p*1*g*1 , the glide would be perpendicular to the *a* axis and so parallel to *b*).

An *ab initio* demonstration of the orthogonal $O_\perp$ packing of the chains of $C_{29}CO_2H$ in plane group *p*11*g* was provided by an analysis of the BR data making use of eqns. (7) and (8), and a reliability index R_w,* complemented by lattice energy calculation. The molecular arrangement was varied, about its deduced position by rotation about three Eulerian angles and by molecular translation along the *y* axis. The model in which the molecular chain axes are centered parallel between the glide planes, tilted at an angle of 26° and oriented so as to form the orthogonal $O_\perp$ packing gave the lowest R_w index. Although the calculated Bragg rod profiles compared well with the observed ones (Fig. IV. 9), it was not possible to distinguish between two structures which differed only in rotation by 180° about the molecular axis. This ambiguity was eliminated, and the correctness of the model structure established through lattice energy calculations.

---

*The weighted reliability index R_w is defined by $\sum_i \omega_o^{1/2}|I_o - I_c| / \sum_i \omega_o^{1/2} I_o$. The summation is over all the *i* individual Bragg rod points, where I_o is the observed intensity of the ith point, ω, the square root of its weight = $1/\sigma(I_o)$, and I_c, the intensity calculated from the structure factor formula (Eqn. 7) making use of the atomic *xyz* coordinates of the molecular. R_w is analogous to the reliability index used for 3-D crystal structure refinement.

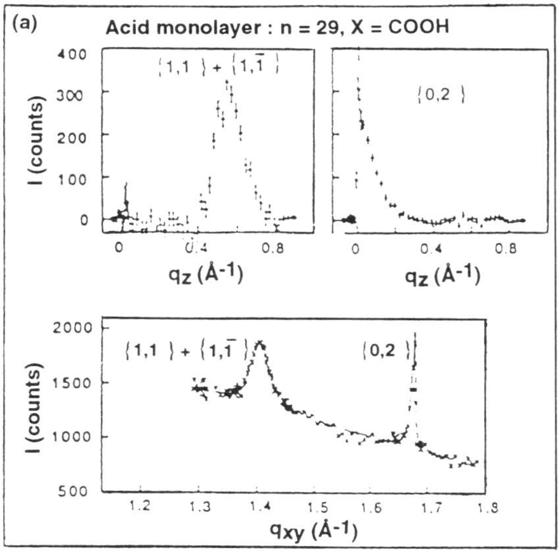
(a)
Acid monolayer : n = 29, X = COOH
I (counts)
{1,1} + {1,1̄}
{0,2}
qz (Å-1)
qxy (Å-1)

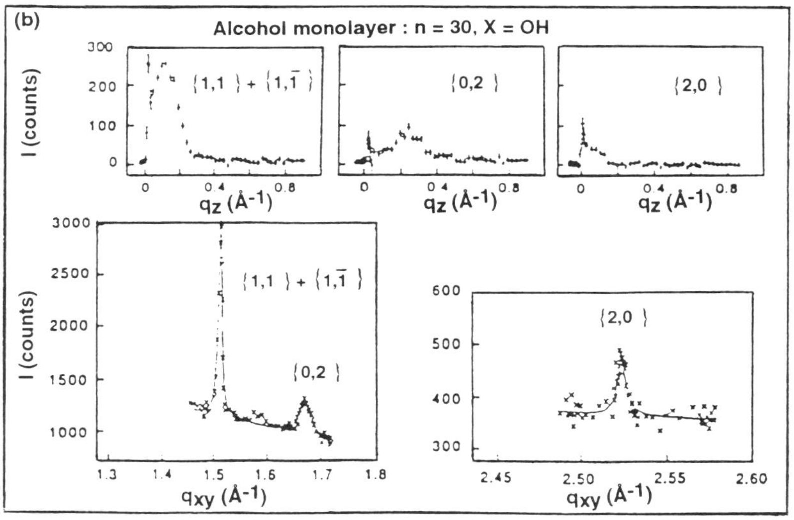
(b)
Alcohol monolayer : n = 30, X = OH
I (counts)
{1,1} + {1,1̄}
{0,2}
{2,0}
qz (Å-1)
qxy (Å-1)

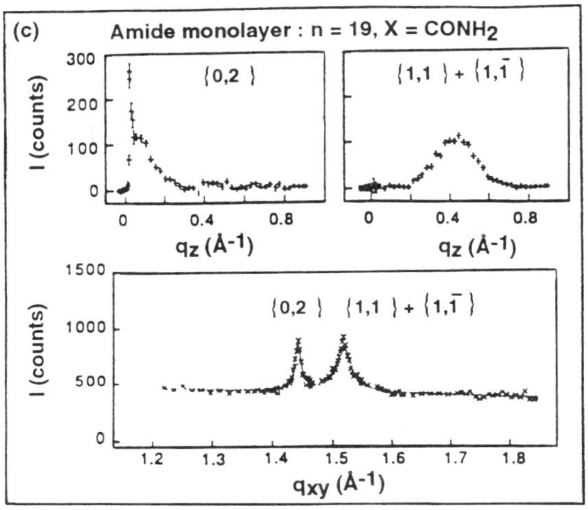

Fig. IV.8. — GID spectra of monolayers at zero surface pressure and at $5^{o}C$ over pure water. *Top.* q_z resolved Bragg rod profiles. *Bottom:* q_z integrated Bragg peaks. The {h, k} reflection assignment is indicated on the two data sets: (a) $C_{29}CO_2H$; (b) $C_{30}OH$; (c) $C_{19}CONH_2$.

The Bragg rod data of uncompressed monolayers of alcohols $C_{30}OH$ and $C_{31}OH$ were analyzed (Wang et al, 1993) in a manner similar to that adopted for $C_{29}CO_2H$. The molecules crystallize in a rectangular cell (Table 2) in space group *p*1*g*1 in the orthogonal $O_\perp$ motif (Fig. IV. 9). Unlike the $C_{29}CO_2H$ structure, the alcohol molecular chains are tilted at a low angle of ~ 10° and in the direction of the long *b* axis. In these systems the presence of the {2,0} reflection enabled a reliable determination of the molecular mean squared motion U_{xy} parallel to the water surface. yielding a value of ~ 0.08Å^2, in keeping with the molecular motion found in layers of 3-D crystal structures containing hydrocarbon chains.

The proposed packing arrangement of the 2-D crystal structure of $C_{19}CONH_2$ (Fig. IV. 9) was based on the GID data and 3-D crystal strucures of long chain diamides. (Jacquemain, 1991). The molecular chains of C_{19} $CONH_2$ are stacked

Table 2. Bragg Peak Analysis of the Different Monolayers $C_nH_{2n+1}X^a$

compound n, x	d_{11} and $d_{1\bar{1}}$ (*L*)	d_{02} (*L*)	d_{20} (*L*)	*a*	*b*
30, OH	4.15 (≥1000)	3.75 (23)	2.49 (>1000)	4.99	7.49
29, CO_2H	4.44 (150)	3.72 (>1000)	5.53	7.44	
19, $CONH_2$	4.34 (200)	4.12 (550)	4.69	8.69	

^{a}For each *h, k* reflection the *d*- spacing (in Å) is given and the crystal correlation length *L* (Å). The unit cell axes *a, b* (Å) are for a rectangular cell.

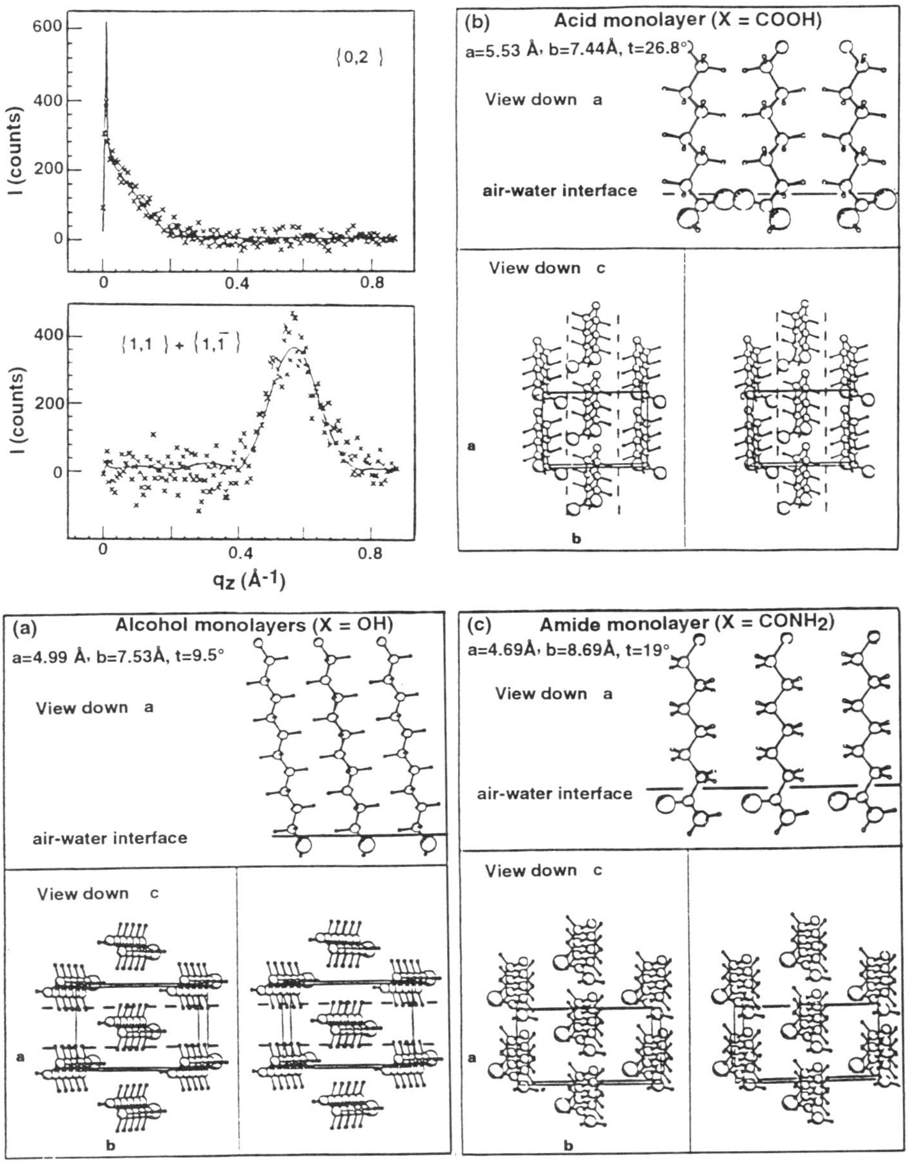

Fig. IV.9. — Side and top (stereoscopic) view of the packing arrangements. Only the head group and maximally 11 CH_2 groups of each molecule are shown. Glide symmetry elements, are shown as dashed lines. *(Top)* Observed and calculated (full line) Bragg rods and packing arrangement of the $C_{29}CO_2H$ acid monolayer. *(Bottom)* Packing arrangements of the $C_{30}OH$ and $C_{19}CONH_2$ monolayers into rows by translation along the 4.7Å *a* axis with the molecular axis tilted at an angle of 18° to the vertical. These molecules are interlinked by N-H...O hydrogen bonds by translation symmetry along the diagonal (**a**+ **b**)/2.

IV.4.2. Correlation between crystalline self-assembly and calculated lattice energies

Positional correlation lengths of crystallites of monolayers such as $C_{29}CO_2H$, $C_{19}OCOC_nOH$ (n=9,10) and C_nOH (n=16,20) where the molecules exhibit a pronounced tilt in the range 20-30° are very anisotropic (e.g. see Table 2 for the $C_{29}CO_2H$ monolayer); these lengths are decidedly shorter in the direction parallel to the molecular tilt than perpendicular to it. Lattice energy calculations were used also to help throw light on coherence length anisotropy. The idea implemented involved calculation of the attachment energy of different (h,k) lines. Such energy computations performed on $C_{29}CO_2H$ and $C_{19}OCOC_nOH$ n=9,10 (Leveiller et al, 1992), showed that the attachment energy was found to be weaker for the molecular row corresponding to the broad {1,1} reflection than to the sharp {0,2} reflection, as exemplified in Fig. IV. 9 and Table 2 for $C_{29}CO_2$.

These results are a corollary of the observation that the stronger the interactions between the amphiphilic molecules on the liquid surface, the more crystalline the material in the self-aggregated state (Jacquemain et al, 1992b, Leveiller et al, 1992). This observation was shown to hold for the fatty acids C_nCO_2H (n=13,19,21,29) over pure water; the longer the chain the more crystalline the material. This increased crystallinity with chain length in the acids may be correlated with the contribution to the lattice energy of ~2Kcal/mole per CH_2 group in the chain. This simple correlation is in keeping with the observation, alluded to earlier, that fluorocarbon amphiphiles are more crystalline than their hydrocarbon counterparts because of more favorable interchain van der Waals interactions.

IV.5. Influence of Ions and Solutes on Monolayer Organization

IV.5.1. Solute Binding and Growth and Dissolution of Monolayers

The binding of solute molecules to hydrophilic groups of monolayers allows us to probe not only its effect on the growth, stability and dissolution of 2-D crystals, but to simulate interactions between the surface of a 3-D crystal and solvent or solute molecules. It has been observed that addition of solute species in the subphase can deter crystalline self-aggregation of monolayers or can bring about dissolution of the crystallites, processes which can be monitored as a function of time (Jacquemain et al, 1990, 1992a).

IV.5.2. Ion Binding from Solution

The interfacial region between a charged surface and an electrolyte is central to many processes such as those occurring during electrodeposition, ion transport through biological membranes, preparation of LB films (Kuhn & Mohius, 1971) biomineralization (Addadi et al, 1987) and induced oriented nucleation of inorganic 3-D crystals under Langmuir monolayers (Landau et al 1986; 1989;

Mann et al, 1988). X-ray reflectivity measurements, demonstrated that metal ions, when present in solution, interact closely with the charged monolayer head groups at the interface (Kjaer et al, 1989; Richardson et al, 1987; Daillant et al, 1989). However, the question was still open as to whether the ion distribution near such charged ordered surfaces is crystalline.

A weak indication of lateral ionic order is provided by GID data of uncompressed PFA monolayers spread over highly basic KOH subphases (pH = 11.5). According to an analysis of the GID data, the K^+ ions occupy ordered sites bridging oxygen atoms of neighbouring PFA molecules (Jacquemain et al, 1990).

An unambiguous demonstration of ordered binding of counterions, with a lateral coherence length of ~ 1000Å was provided by uncompressed cadmium arachidate monolayers over a subphase with an adjusted pH of 8.8 using ammonia as a base, at temperatures lower than 9°C (Leveiller et al, 1991). The diffraction spectrum showed ten distinct peaks (Fig. IV. 10a) , seven of which

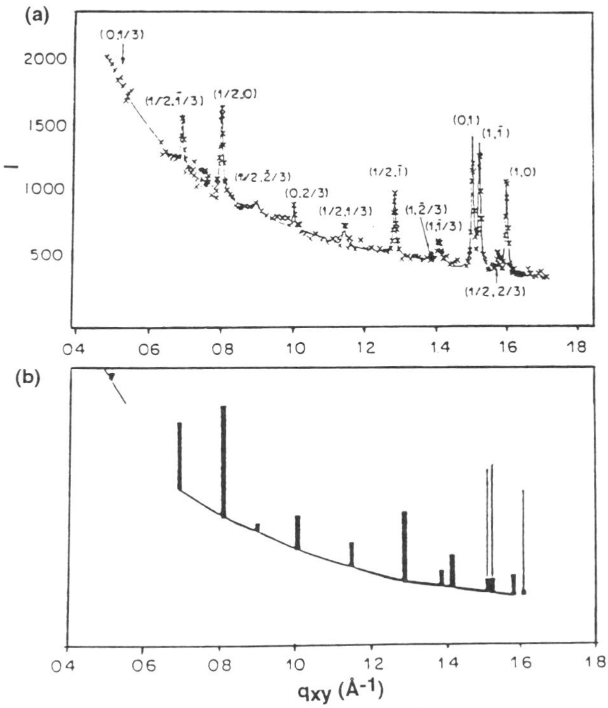

Fig. IV.10 — Arachidic acid monolayer ($C_{19}CO_2H$) over a 10^{-3} M $CdCl_2$ subphase (pH = 8.8) in the uncompressed state and at low temperature (9° C).

(a) GID peaks and reflections assignment. The reflections with integer indices correspond to the scattering arising from the arachidate moieties. The reflections with fractional indices correspond to the scattering from a 2-D layer of $CdOH^+$ ions. (b) Calculated powder pattern for the cadmium net, superimposed on background scattering. The three peaks labeled with thin lines are the arachidate "triplet".

were attributed to the cadmium layer. The remaining three peaks were deduced to stem primarily from the arachidate layer, on the basis of their Bragg rod profiles and q_{xy} positions and so denoted as the arachidate "triplet". The diffraction peaks of the cadmium lattice whose BR profiles were flat and extended, indicative of a thin monoatomic layer, could be indexed according to a 2x3 supercell of the arachidate cell. The packing arrangement of the arachidate molecules could be established from a fit to BR profiles of the arachidate triplet (Leveiller et al, 1993). Specular X-ray reflectivity measurements have shown that the molar ratio of cadmium to arachidate is 5:6 or 1:1, so that the counter layer must consist essentially of $CdOH^+$ ions. The refined Cd positions in the 2 x 3 supercell, containing either 5 or 6 cadmium atoms, gave a very good fit (Fig. IV. 10b) to the observed seven diffraction peaks arising from the counterionic layer.

IV.6. Crystallization of Molecules under Monolayers with a Similar Head Group.

Langmuir monolayers have been shown to induce oriented crystallization from solution of protein(Uzgiris and Kornberg, 1983) and other organic and inorganic compounds (Landau et al, 1985; Mann et al 1988, Weissbuch et al, 1990; Heywood et al, 1991; Gavish et al, 1990). These studies were motivated by their revelance to crystal nucleation, to two-dimensional ordering at the water surface of amphiphilic molecules and to biomineralization. Mechanisms for the oriented crystallization include structural fits and electrostatic attractions. Structures of some monolayers were determined by GID to establish the epitaxial relation, e.g. that of an α-amino acid surfactant (Grayer Wolf et al, 1987) which induces crystallization of glycine (Landau et al, 1985). Here we describe induced nucleation of ice by alcohol monolayers, which involves a structural fit, as shown by GID.

IV.6.1. Nucleation of Ice by Monolayers

Pure water can be supercooled to temperatures as low as -40° C. Therefore the induction or inhibition of the nucleation of ice, in particular through the role of auxiliaries such as membranes or proteins, has far-reaching ramifications for both the living and non-living world.

Here we review experiments conducted in order to design two-dimensional surfaces which raise the ice nucleation temperature (Gavish et al, 1990; Popovitz-Biro et al, 1993). These experiments involved the freezing of drops of water covered with monolayers of aliphatic alcohols (C_nX, X = OH). The structure of hexagonal ice may be described for the oxygen atoms in terms of layers parallel to the *ab* plane where $a = b = 4.5$ Å, $\gamma = 120°$ For purpose of comparison we may describe this *ab* net in terms of a *c* - centered rectangular net $a' = |\boldsymbol{a} + \boldsymbol{b}| = 4.5$Å and $b' = |\boldsymbol{a} - \boldsymbol{b}| = 7.8$Å. Advantage was taken of the observation already discussed in section 4.1 that the uncompressed monolayers of $C_{30}OH$ and $C_{31}OH$ over water at 5°C are crystalline and pack in a rectangular net $a = 5.0$Å, b = 7.5Å (Jacquemain et al 1991; Wang et al, 1993) which is similar in dimension to the *a'b'* rectangular representation of hexagonal ice at 0° C. We may, to a first approximation, regard the alcohol 2-D structures of $C_{30}OH$, and $C_{31}OH$ as *c* - centered, according to Fig. IV. 9 for $C_{30}OH$. The uncompressed monolayer of the carboxylic acid $C_{29}CO_2H$ has axial dimensions a = 5.5 Å, b=7.4 Å (Table 2), and so exhibits a large lattice mismatch to ice. Thus a structural similarity can be proposed for the OH head groups of the alcohol monolayer and the ***ab*** layer structure of ice, but no such similarity exists for the corresponding carboxyl head groups of the fatty acid monolayer (see Fig. IV. 9).

The results of the ice nucleation temperature ("freezing point") measurements (Fig.IV. 11) show that aliphatic chain alcohols are more efficient ice nucleators than the corresponding carboxylic acids (Gavish et al, 1990). The freezing temperature depends not only on chain length but also on chain parity (*i.e.*, n odd or even) of the alcohol $C_nH_{2n+1}OH$ (Fig. IV. 11). This trend suggests that, at the temperature of ice nucleation, the orientation of the alcohol OH groups in the odd and even analogues are not the same. Studies attempting to elucidate these questions are underway.

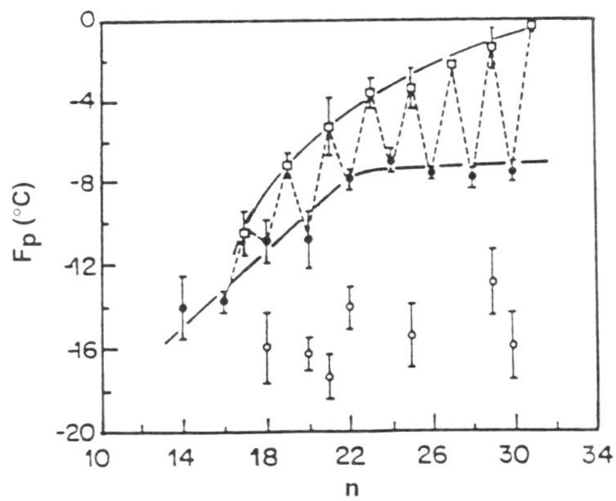

Fig. IV.11 — Freezing points (Fp) of drops of supercooled water covered by monolayers of alcohols C_nOH (n even • and n odd ∘) and carboxylic acids C_nCO_2H (o). Freezing point curves are drawn separately for alcohols with n odd and n even.

REFERENCES

ADDADI L. , MORADIAN J. , SHAY E., MAROUDAS N.G., WEINER S. , 1987- Proc. Natl. Acad. Sci. U.S.A. 84 , 2732.

ALLARA D.L. , NUZZO R.G. , 1985 - Langmuir 1, 52.

ALS-NIELSEN J. , CHRISTENSEN F., PERSHAN P.S., 1982 - Phys. Rev. Lett. 48, 1107.

ALS-NIELSEN J., PERSHAN P. , 1983 - Nucl. Instrum. Meth. 208, 545.

ALS-NIELSEN J. , in S. Ebashi, E. Rubenstein and M. Koch (Eds.): Handbook of Synchrotron Radiation Vol. III, North Holland Amsterdam, in press.

ALS-NIELSEN J. , KJAER K. in T. Riste and D. Sherrington (Eds.): 1989 -The Proceedings of the NATO Advanced Study Institute, Phase Transitions in Soft Condensed Matter, Series B Vol. 211, Plenum Press New York p. 113.

ALS-NIELSEN, J. , MOHWALD H. in S. Ebashi, E. Rubenstein and M. Koch (Eds.): Handbook of Synchrotron Radiation Vol IV , North Holland Amsterdam, in press.

BARTON S.W. , GOUDOT A., BOULOUSSA O., RONDELEZ F., LIN B., NOVAK F., ACERO A., RICE S.A., 1992 - J. Chem. Phys. 96, 1343.

BATTERMAN B.W. , 1964 - Phys. Rev. 133, 1759; BATTERMAN B.W., 1969, Phys. Rev. Letts. 22, 703.

BEDZYK M.J. , BILDERBACK D.H., BOMMARITO G.M., CAFFREY M., SCHILDKRAUT J.S. 1988 - Science 241, 1788; BEDZYK M.J. , BOMMARITO G.M. , CAFFREY M. , PENNER T.L. , 1990 - Science 248, 52.

BERGE B. , FAUCHEUX L. , SCHWAB K. , LIBSHABER A. , 1991 - Nature 350, 322.

BLINOV L.M. , DUBININ N.V. , MIKHNEV L.V. , YUDIN S.G. , 1984- Thin Solid Films 120, 161;

BLOCH J.M. , EISENBERGER P. , 1988 - Nucl. Instrum. Meth. B31, 468; BLOCH J.M. , 1988 -Phys. Rev. Lett. 61, 2941.

BORN, M. , WOLF E. , 1959 - Principles of Optics, MacMillan, New York.

BRASLAU A. , PERSHAN P.S., SWISLOW G. , OCKO B.M. , ALS-NIELSEN J. , 1988 - Phys. Rev. A 38, 2457.

DAILLANT J. , BOSIO L. , BENATTAR J.J., MEUNIER J. , 1989 - Europhys. Lett. 8, 453.

DUTTA, P., PENG, J.B., LIN, B., KETTERSON, J.B., PRAKASH, M., 1987 - Phys. Rev. Lett. 58, 2228.

FEIDENHANS'L R., 1989 - Surf. Sci. Rep. 10, 105.

GAINES G.,1966 - Insoluble monolayers at the Liquid-Gas Interface, Interscience, New York .

GAVISH M. , POPOVITZ-BIRO R. , LAHAV, M. , LEISEROWITZ L. , 1990 - Science 250, 973.

GIRLING I.R. , KOLINSKY P.V. , CADE N.A. , EARLS J.D. , PETERSON I.R. , 1985- Optics Commun. 55, 289.

GRAYER WOLF S. , LEISEROWITZ L. , LAHAV M. , DEUTSCH M. , KJAER K. , ALS-NIELSEN J. , 1987 - Nature 328, 63.

GUINIER A. , 1968 - X-Ray Diffraction, Freeman San Francisco ch. 5.1.

GUN J. , ISCOVICI R. , SAGIV J 1984 - J. Colloid Interface Sci. 101, 201.

HEINZ T.F. , CHEN C.K. , RICARD D. , SHEN Y.R. , 1982 - Phys. Rev. Lett. 48, 478.

HENON, S., MEUNIER, J., 1991 - Rev. Sci. Instrum. 62, 936.

HÖNIG, D., MÖBIUS, D., 1991 - J. Phys. Chem. 95, 4590.

HEYWOOD B.R. , RAJAM S. , MANN S. , 1991- J. Chem. Soc. Faraday Trans. 87, 735.

JACQUEMAIN D. , GRAYER WOLF S., LEVEILLER , F. , LAHAV M. , LEISEROWITZ L. , DEUTSCH M. , KJAER K. , ALS-NIELSEN J., 1989 - Coll. de Phys. 50, C7 p.29.

JACQUEMAIN D. , GRAYER WOLF S. , LEVEILLER F., LAHAV M., LEISEROWITZ L. , DEUTSCH M. , KJAER K., ALS-NIELSEN J., 1990- J. Am. Chem. Soc. 112, 7724.

JACQUEMAIN D. , LEVEILLER F. , WEINBACH S.P. , LAHAV M., LEISEROWITZ L. , KJAER K. , ALS-NIELSEN J. , 1991- J.Am.Chem.Soc. 113, 7684.

JACQUEMAIN D., GRAYER WOLF S., LEVEILLER F., DEUTSCH M., KJAER K., ALS-NIELSEN J., LAHAV M., LEISEROWITZ L., 1992A - Angew. Chem. 31, 130.

JACQUEMAIN D., GRAYER WOLF S., LEVEILLER F., FROLOW F., EISENSTEIN M., LAHAV M., LEISEROWITZ L. 1992B - J. Am. Chem. Soc. 114, 9983.

JAMES R.W. , 1982 -The Optical Principles of the Diffraction of X-Rays, Ox Bow, Woodbridge CT.

KJAER, K., ALS-NIELSEN, J., HELM, C.A., LAXHUBER, L.A., MÖHWALD, H., 1987 - Phys. Rev. Lett. 58, 2224.

KJAER K. , ALS-NIELSEN J., HELM C.A., TIPPMAN-KRAYER P., MOHWALD H. , 1989 -J. Phys. Chem. 93 , 3200.

KUHN H. , MOBIUS D., 1971- Angew. Chem. 83, 672.

LANDAU E.M. , GRAYER WOLF S., SAGIV J. , DEUTSCH M., KJAER K. , ALS-NIELSEN J. , LEISEROWITZ L., LAHAV M., 1989- Pure & Appl. Chem. 61, 673.

LANDAU E.M., POPOVITZ-BIRO R., LEVANON M. , LEISEROWITZ L., LAHAV M. , 1986- Mol. Cryst. Liq. Cryst. 134, 323.

LANDAU E.M., LEVANON, M. , LEISEROWITZ, L., LAHAV M., SAGIV J. 1985- Nature 318, 353; LANDAU E.M., GRAYER WOLF, S., LEVANON, M. , LEISEROWITZ L., LAHAV M. , SAGIV J. , 1989-J. Am. Chem. Soc. 111 , 1436.

LANGMUIR I., 1917 - J. Am. Chem. Soc. 39 , 1848.

LEVEILLER F. , JACQUEMAIN D. , LAHAV M. , LEISEROWITZ L. , DEUTSCH M. , KJAER K. , ALS-NIELSEN J. , 1991 - Science 252, 1532.

LEVEILLER F. , JACQUEMAIN D., LEISEROWITZ L., KJAER K., ALS-NIELSEN J., 1992 - J. Phys. Chem. 96, 10380.

LEVEILLER F., BOHM C., JACQUEMAIN F., MOHWALD D.H., LEISEROWITZ L., KJAER K., ALS-NIELSEN J. Submitted for publication 1993.

LOSCHE M. , SACHMANN E. , MOHWALD H., 1983 - Ber Bunsenges. Phys. Chem. 87, 848.

MANN S., HEYWOOD B.R. , RAJAM S. , BIRCHALL J.D. , 1988- Nature 334, 692,

MORIZUMI T. , 1988 - Thin Solid Films 160, 413.

PARRAT L.G. 1954 - Phys. Rev. 95, 359.

POPOVITZ-BIRO R., WANG J.L., MAJEWSKI J., LEISEROWITZ L., LAHAV M., 1993. Submitted for publication

RICHARDSON R.M. , ROSER S.J. , 1987 - Liquid crystals 2, 797. RICHARDSON Grundy, R.M. , ROSER S.J. , PENFOLD J. , WARD R.C. , 1988 - Thin Solid Films 159, 43.

ROBERTS G.G. , PAUDE K.P. , BARLOW W.A. , 1978 - Proc. Inst. Electr. Eng. Part I, 125, 169.

SEUL, M., EISENBERG, P., McCONNELL, H.M., 1983 - Proc. Natl. Acad. Sci. U.S.A. 80, 5795.

SMALL D.M. , 1986 - The Physical Chemistry of Lipids, Handbook of Lipid Research Vol.4, Plenum Press.

SWALEN J.D. , ALLARA D.L. , ANDRADE J.D., CHANDROSS, E.A. GARROF S. , ISRAELACHVILI, J., McCARTHY, T.J. , MUNRAY R. PEASE, R.F., RABOLT, J.F. , WYNNE, K.J. , YU H. , 1987- Langmuir 3, 932.

ULMAN, A. In Introduction to Ultrathin Organic Films. Academic Press, N.Y. 1991.

UZGIRIS E.E. , KORNBERG R.D. , 1983 - Nature 301, 125.

VINEYARD G. , 1982 - Phys. Rev. B 26, 4146. FEIDENHANS R. , 1989, Surf. Sci. Rep. 10, 105.

WANG J.L., JACQUEMAIN, D., LEVEILLER F., KJAER K., ALS-NIELSEN J., LAHAV M., LEISEROWITZ L., 1993 - Submitted for publication.

WEIS R.M., McCONNELL, H.M., 1984 - Nature 310, 47.

WEISSBUCH I., BERKOVIC G., LAHAV M., LEISEROWITZ L., 1990 - J. Am. Chem. Soc. 112, 5874.

WEISSBUCH I., et al., to be published 1993.

CHAPTER V

STRUCTURE OF LIQUID CRYSTALS

A.M. LEVELUT

V.1. Introduction

Usually, melting brings matter from a solid crystalline state into an isotropic liquid state. In some organic materials this melting process occurs through several well defined steps inducing new phases which simultaneously exhibit some characteristics of a crystalline phase, i.e. anisotropy, and some characteristics of a liquid, such as viscoelastic behaviour. These phases have been called liquid crystals but are also known under the more proper name of mesophases.

In a crystalline structure the content of a unit cell is repeated by periodic translations in three directions, in such a way that a complete description of the structure can be made through a finite number of parameters that define the periodic network and the content of one unit cell. In a liquid, the distribution of atoms is known only from a statistical point of view : the conformation and the environment of a molecule vary continuously in time and space. Starting from a molecular crystal, there are two ways of going stepwise to the liquid state. In one case, the centres of mass of the molecules are still distributed on a three-dimensional (3D) periodic network, but the molecular orientation varies in time and space. In the second case, the positional order can be partially or completely destroyed, while some orientational order is kept over a long range. The first case corresponds to the class of orientationally disordered crystals (ODIC) or plastic crystals, whereas in the second case one obtains liquid crystals.

Liquid crystalline phases are called smectic mesophases when the centres of mass of the molecules form a periodic array in one dimension, building stacks of identical layers with a liquid-like intralayer ordering. In columnar mesophases the molecules form columns organized on a two-dimensional lattice. In the most disordered phase, the nematic phase, the molecules keep a preferred orientation on a large scale, but the centres of mass are not regularly distributed on a lattice (Fig. V.1.).

Mesophases with the same symmetry properties are also obtained by dilution of amphiphilic molecules in solvents. An amphiphilic molecule (or a surfactant) has two amphipatic moieties of opposite affinities: one has a strong affinity with the solvent whereas the second part is not miscible with the same solvent. Soaps in which a hydrophilic carboxylic head group is attached to a hydrophobic paraffinic tail (Fig. V.2.a), are a classic example of amphiphilic molecules. Natural lipids and detergents also form lyotropic mesophases. These molecules, when dissolved in water, form clusters that reduce paraffin-water interactions, the head groups forming the water-hydrophobic interface. At relatively low amphiphile content the clusters adopt a spherical shape, with the chains inside; as the surfactant concentration is increased the clusters transform into cylinders of infinite length located on a two-dimensional planar lattice; for equivalent amounts of water and surfactant, a periodic array of fluid lipid bilayers and thin water layers, alternately stacked over one another, ensure micro segregation between hydrophobic and hydrophilic media. A further increase of the lipid concentration will induce inverse mesophases in which cylinders, and then spheres of water are embedded in a continuous medium of paraffin (fig V.2.b).

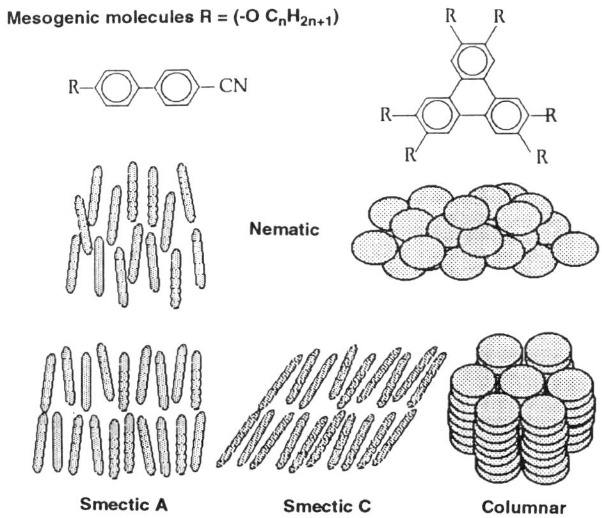

Figure V.1. — Examples of mesogenic molecules and schematic representation of their mesophases.

Depending on the external parameter that induces the mesomorphism the mesophases are called "thermotropic" if they occur under the action of temperature, and "lyotropic" if they are obtained by dilution. In fact many soaps or natural lipids exhibit thermotropic polymorphism (without water) as well as lyotropic. Moreover, temperature influences the stability of lyotropic mesophases. Microsegregation between hydrophobic and hydrophilic media is the major factor that drives lyotropic polymorphism, and the phase diagram of Fig. 2b is typical of what can be encountered with small variations in any lyotropic system. In contrast, it seems at first glance that the molecular organization in thermotropic mesophases is governed by the orientational order of organic molecules of anisometric shape. Thus, rod-shaped or disc-shaped molecules exhibit nematic phases while increasing the positional order, disc-like molecules form only columnar phases and rod-like molecules only smectic phases. In fact, periodic ordering is still a matter of microsegregation between two amphipatic moieties of the molecules. Indeed, a typical smectogenic molecule is made of an elongated aromatic core with two aliphatic chains grafted at each end, whereas a discogenic molecule has a relatively flat core surrounded by a sufficiently large number of paraffinic chains. Recently polymorphism with both columnar and smectic phases has been observed with molecules that are neither rod-like nor disc-like. It is noticeable that amphiphilic molecules do not generally form non-spherical aggregates of finite size. However, there are few examples of lyotropic nematic mesophases formed by anisometric clusters. Rigid rods like DNA PBLG or TMV, while not surfactants, also form lyotropic nematic and columnar mesophases, the symmetry of which is closely related to their shape.

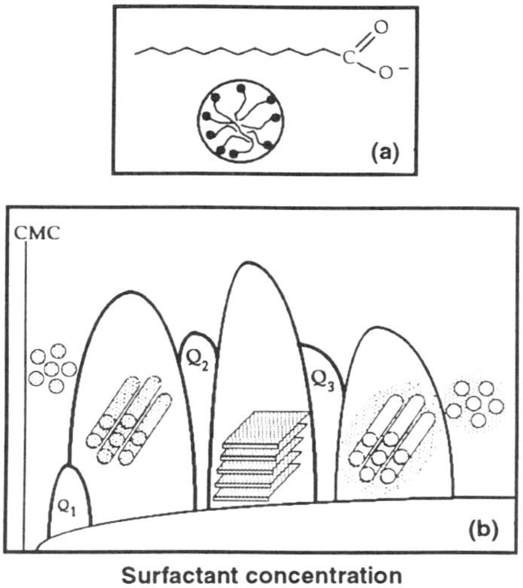

Figure V.2. — a) Typical amphiphilic molecule and mesogenic aggregate; b) Model phase diagram with the main lyotropic mesophases; CMC is the critical micellar concentration; Q_1 is a cubic mesophase of closed direct micelles, Q_2 and Q_3 are cubic phases made up of two direct or two inverse interwoven labyrinths, respectively; a fourth inverse micellar cubic phase (not shown here), can also be observed.

Therefore from the point of view of the microscopic interactions the nematic state stands apart because of the preeminence of orientational order. In the other mesophases, segregation between the two amphipatic components of the system produces interfaces, and the large scale organization is characterized by its periodicity, the interface curvature and the topology of the two media on either side of the interface. The existence of parallel equidistant flat interfaces between an infinite number of lamellae of the two media describes the smectic state. In the case of columnar phases, cylindrical interfaces on a two-dimensional network confine one medium in an assembly of an infinite number of unconnected cylinders embedded in the continuous second medium. In these two cases long-range orientational order can be added inside the lamellae or the cylinders.

In the frontier area between two mesophases, a frustration occurs between two characteristic geometrical parameters (natural periodicities and spontaneous curvature or twist). This frustration induces a new class of mesophases which can be described as liquids modulated by a periodic array of defects. For example the competing effects of spontaneous curvature of the interface and of periodic stacking induce lattices of defects with a cubic symmetry. The medium surface of one of the two media is a periodic minimal surface separating two infinite interwoven labyrinths of the second medium. In smectics the existence of two "natural" periodicities induces a 2D network of walls, while the occurrence of a spontaneous twist induces a twisted array of equidistant walls each one of which is made up of a periodic array of twist dislocations. This is the twist grain boundary smectic phase (TGB).

In summary, all liquid crystalline phases, except the nematic phase, can be described as modulated liquids. The modulation is characterized by the long-range correlation function and a characteristic periodicity spanning the range 1 - 1000 nm. The liquid state could be described by a correlation function that vanishes at large distances.

Therefore we will discuss successively the long-range and short-range order properties through the corresponding spatial and temporal correlation functions.

V.2. Long-range order in liquid crystals

In liquid crystals we have long-range orientational order. If no translational order is added to this orientational order, diffraction methods give access to the atomic pair correlation function The latter takes into account the distribution in length and orientation of segments that join two atoms in the liquid. One cannot simply deduce the orientional distribution function of the symmetry axes of the molecules from a knowledge of the scattering function. For these reasons we restrict ourselves to the question of long-range positional order. Mesophases can be classified according to their degree of periodicity : 1D, 2D or 3D periodicity corresponding to smectic, columnar mesophases and to crystals of defects respectively. Each kind of mesophase has a structure factor S(q) or S(q,ω) in relation with the dimensionality of the periodic ordering. We will discuss in details the case of smectics, the only one in which true long-range positional order does not exist, and which is also the most extensively studied experimentally.

A smectic or its lyotropic equivalent is a stack of identical layers of infinite extension in the layer plane. The first question concerns the symmetry properties and the characteristic wavelength of such systems. Since the periodicity acts only along one direction (say the z direction), it is compatible with any kind of symmetry element, mirror plane, rotation axis, glide mirror or screw axis parallel to z, while the only symmetry elements parallel to the layer planes are mirror and twofold axis. However the only symmetry elements parallel to z compatible with the absence of long-range in-plane ordering are C_∞ or C_∞ / M, which correspond to the smectic A phase (S_A). If there is some long-range orientational order inside the layer the symmetry elements parallel to z will be different depending on the symmetry of this orientational order.

If this symmetry is that of a nematic, the point group symmetry is 2/m with the two-fold axis parallel to the layer plane. This situation is encountered in smectic C (S_C) where the molecular director is tilted with respect to the layer normal and its projection on the layer plane is nematically ordered. The symmetry 2, which is found only with chiral molecules, corresponds to a ferroelectric S_C, but in fact this phase is not stable in the absence of an applied electric field.

The in-plane orientational order is compatible with any Cn axis; the environment of each molecule has the symmetry Cn. This is why it is called a bond orientational order. This kind of ordering has been theoretically predicted as an intermediate step in the melting process of 2D crystals. It is a step in which unbound dislocations destroy the long-range positional order, whereas the bond orientational order is still long-range. Experimentally, this kind of phase corresponds to the melting of a hexagonal 2D lattice, and it is called hexatic smectic B (hex S_B). The tilted equivalent, (melting of a rectangular lattice of monoclinic symmetry), is equivalent to S_C, which obviously presents a bond orientational ordering with the same symmetry properties.

In thermotropic smectics the layer thickness is typically between 1.5 and 5 nm, lying in between once or twice the molecular length. The layer thickness is equal to the lamellar periodicity unless a screw axis or a glide mirror parallel to z is added. A screw axis is found in chiral C smectics (S_{C*}), in which the azimuthal projection of the director is twisted with a pitch that can vary between few nm and few µm. However, the density modulation is still that of one molecular layer and there is no means of revealing this helicoidal array by diffraction methods, unless the corresponding susceptibility is non-scalar. For pitches of the order of the wavelength of visible light, one can see Bragg reflections corresponding to the pitch periodicity, with a structure factor that depends of the state of polarization of the incident light. In the X-ray range, similar experiments can be done only in anomalous scattering conditions, and consequently the experiment will be easier with a mesogenic molecule containing heavy atoms.

In lyotropic lamellar phases the periodicity is equal to the sum of the lipid bilayer and water sublayer thickness. It is also in the range of few nm in most cases. However it is possible to increase considerably the layer thickness of either the lipid sublayer by adding oil, or the water sublayer by adding brine. In both cases the layer thickness could be increased continuously until a few hundreds nm are reached.

In the absence of thermal fluctuations the layer structure can be described by a set of surfaces given by: $\mathbf{z}_n = n\,\mathbf{d} + \mathbf{z}_0$ where $\mathbf{d}$ is the layer periodicity and n an integer. If $F_0(\mathbf{Q})$ is the Fourier transform of the electronic (or diffusion length) density profile of each layer the diffracted intensity $I(\mathbf{Q})$ is:

$$I(\mathbf{Q}) = F_0(\mathbf{Q})\,\delta(\mathbf{Q} - l\mathbf{d}^*) \tag{V.1}$$

where l is an integer and $\mathbf{d}^*$ the unit vector of the reciprocal lattice.

At finite temperatures the scattered intensity is modified by thermal fluctuations. These fluctuations correspond to displacements parallel to z : $z = z_n + u_n(\mathbf{x})$ where $\mathbf{x}$ is the in-plane coordinate. These displacements can be considered as a sum of acoustic modes of wave vector $\mathbf{q}$. If $\mathbf{q}$ is parallel to $\mathbf{z}$, the acoustic mode is a compression mode, and if B is the compressibility, the dispersion law is $\omega^2 \propto B\,q^2_{//}$ and the energy $(1/2)\,B\,q^2_{//}\,u^2(q_{//})$. If $\mathbf{q}$ is perpendicular to $\mathbf{z}$, a displacement in the z direction corresponds to an undulation of the layers that does not change the period: the elastic energy is then due to splay distortion of the director field. The energy associated with the undulation mode is $(1/2)\,K\,q_\perp^4\,u^2\,(q_\perp)$, where K is the nematic splay elastic constant. The corresponding dispersion law is $\omega^2 \alpha\; K\,q_\perp^4$ (fig.V. 3.).

Starting from the total energy :

$$E(\mathbf{q}) = (1/2)\,B\,(\mathbf{q}_{//}^2 + \lambda\,\mathbf{q}_\perp^4) \tag{V.2}$$

where $\lambda = \left(\frac{K}{B}\right)^{1/2}$ measures the penetration depth for undulations, one can estimate the mean square fluctuations using the equipartition theorem :

$$\langle u^2 \rangle = \frac{k_B T}{4\pi\lambda B}\,\mathrm{Ln}\,(L/a) \tag{V.3}$$

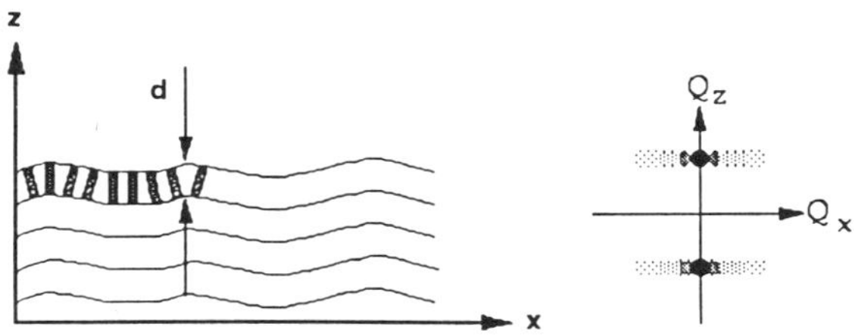

Figure V.3. — Undulating smectic layers and the corresponding reciprocal image.

Roughly speaking, the undulation wavelengths are necessarily between a molecular size, a, and a length, L, which could be either the size of the sample or a coherence length associated with an aligning field. The mean fluctuations diverge logarithmically with L. For $L \Rightarrow \infty$ i.e. $q >> 2\pi/L$, and with a harmonic approximation, the prediction of A. Caillé for the X-ray quasi-elastic scattering function gives :

$$S\,(0,Q_{//})\;\alpha\;(Q_{//} - Q_0)^{-2+\eta} \tag{V.4a}$$

$$S(Q_\perp,Q_0)\;\alpha\;Q_\perp^{-4+2\eta} \tag{V.4b}$$

where $\eta = k_B T\, Q_0^2 / (8\pi\, B\lambda)$

A true Bragg scattering function is observed only for $(\mathbf{Q} - \mathbf{Q_0}) < 2\pi/L$, but in fact this condition is not easily fulfilled, given the resolution of the X-ray diffraction experiment.

For a crystal, in the vicinity of the reciprocal vector $\mathbf{Q_0}$, the quasi-elastic scattering function scales as $|\mathbf{Q_0} - \mathbf{Q}|^{-2}$ in all directions. For a smectic, in a direction perpendicular to the layers, the quasi-elastic scattered intensity comes from longitudinal modes, and the correction to the inverse square law increases with η. In thermotropic smectic phases η is of the order of 0.2 for $Q_0 = 2\pi/d$. Indeed, the value of η increases with $\mathbf{Q_0}$ ($\mathbf{Q_0} = \mathbf{l d^*}$). The structure factor of a thermotropic liquid crystal is, however, very weak for $l \geq 2$. Consequently, in order to check the Caillé relation, an experimental resolution function with sharp edges is needed (a single reflection on a monochromator will introduce $1/q^2$ wings in the resolution function; a channel-cut monochromator would be better). Moreover, measurable effects occur with small B, i.e. close to the S_A-nematic transition. The first experimental verification of Caillé's prediction was obtained by Als Nielsen and co-workers, using a rotating anode X-ray source. Experiments performed by Safinya and co-workers on lyotropic swollen lamellar phases, using synchrotron radiation, have confirmed the l^2 dependence. The dependence of the Caillé exponent upon the layer thickness fits the theoretical Helfrich predictions for the elastic behaviour of swollen lamellar phases.

The dynamic behaviour of the thermal fluctuations is also anisotropic : longitudinal modes propagating in the z direction behave like acoustic waves propagating in a crystal, whereas the undulation modes have a dispersion law of magnon type. Experimentally, the undulation mode is easily detected by light scattering experiments, since an undulation corresponds to a local change of the birefringence. Coherent inelastic neutron scattering experiments have not been conclusive: defects could have added an elastic scattering contribution around the quasi-Bragg singularity.

It is also interesting to look at the large scattering vector limit ($2\pi/a$). In themotropic smectics the diameter of the molecule is about 0.5nm while the layer thickness is of the order of 2nm. Consequently the typical lengths of the long-range order are fairly large compared to the short-range order scale. Some lyotropic systems also exhibit a nematic-lamellar transition, but the nematic phase is an assembly of disc-shaped aggregates which could be considered as finite size fragments of a surfactant bilayer. As the temperature decreases the aggregates form extended layers, while still keeping their individual shape. In this case the diameter of the disc is larger than its thickness and larger than the lamellar period, so that the short-range and long-range scales are mixed together in the diffraction pattern.

In a columnar phase infinite columns with a liquid order inside each column are packed into a 2D network. In the absence of thermal fluctuations, the diffracted intensity will be localized at points on a planar reciprocal lattice perpendicular to the column axis (oz). The thermal fluctuations correspond to displacements perpendicular to z; in fact modes with q perpendicular to the column axis are the usual acoustic modes and for each $\mathbf{q}_\perp$ the two modes are :

$$\omega_1{}^2 \propto C_{11}\, q_\perp{}^2$$

$$\omega_2{}^2 \propto C_{66}\, q_\perp{}^2$$

where C_{11} and C_{66} are the compression and shear elastic constants respectively.

For q parallel to z the columns bend easily and the mode is equivalent to the smectic undulation mode :

$$\omega_3{}^2 \propto K_3\, (q_{//})^4$$

where K_3 is the bending elastic constant of the columns.

In columnar phases, the thermal fluctuations do not diverge, as in thin 2D crystals. The scattering law is then a sum of δ functions. The thermal fluctuations induce around each Bragg peak a scattering intensity which decreases as $C_{ij}{}^{-1}Q_\perp{}^{-2}$ when $(Q_0 - Q)$ is perpendicular to z, and as $K_3{}^{-1}(Q - Q_0)^{-4}$ when $(\mathbf{Q} - \mathbf{Q}_0)$ is parallel to z. X-ray diffuse scattering experiments performed on a single domain of a lyotropic hexagonal phase allow an estimation of $\lambda = \sqrt{\frac{K_3}{B}}$ which measures the rigidity length of a column: $\lambda \sim 2$ nm.

There are other kinds of modulated fluids. The simplest corresponds to the cholesteric phase when there is only orientational modulation. Other systems correspond to 1D modulation of a smectic with a wave vector parallel to the layer plane. They have the symmetry of a columnar phase. Modulations can also develop in three independent directions, and generally the symmetry is described by one of the classical crystallographic space groups, but the lattice constants are in the range of 10 nm to 1μm. Finally, more exotic structures can be found, as the twist grain boundary smectic phase which can have a quasi-crystalline symmetry. However, in all liquid crystalline phases, the common structural problem is the characterization of the density distribution function of the liquid state.

V.3. Short-range order in liquid crystals

The short-range order is related to the liquid state of the mesophase. In thermotropic liquid crystals the short-range order is characterized by an atomic pair correlation function describing the organization of molecules inside a column, a layer, or in the whole sample, depending on the nature of mesophase, i.e. columnar, smectic or nematic. In contrast, in lyotropic systems there are two levels of short-range order, corresponding to the molecular scale and to the aggregate scale. The latter corresponds to lengths of few nm, i.e. to the small angle scattering range, whereas the molecular organization yields an intensity distribution similar to that observed in any liquid, if one neglects the anisotropy. Since liquid crystals are composed of rather complex molecules it is not easy to derive precise information from the behaviour of the structure factor S(q) at large q (> 0.6 $Å^{-1}$). Generally the structure factor S(q) is modelled using an orientational distribution function of identical molecules, while the mean molecular form factor is calculated using crystal structure data, if these are available. Mesogenic molecules contain a large number of atoms, and in the liquid crystalline state, they are highly flexible. The approach described above is therefore very crude. However it allows us to assign the origin of the different diffuse maxima : intra- or inter-molecular interferences. Two methods provide more specific information.

First, the incoherent scattering cross section gives access to the self-correlation function of individual atoms (mainly hydrogen atoms). Analysis of the time-space distribution of the incoherent inelastic scattering intensity gives information about the rotational and translational diffusion motions. Moreover the incoherent quasi-elastic structure factor is the Fourier transform of the spatial auto-correlation function of the hydrogen atoms. Using this method, it was established that elongated mesogenic molecules undergo fast uniaxial rotational motions around their long axis in the nematic and Sm A phases. Going down into the biaxial smectic C phase this rotational motion displays an anisotropy that is barely measurable. The weakness of this anisotropy is responsible of the small value of spontaneous polarization measured in many ferroelectric liquid crystals.

The second method is a labelling technique. This technique is particularly interesting in the study of the conformation of liquid crystalline polymers. Polymeric liquid crystals made of mesogenic units linked one to another can have smectic and nematic mesophases. The smectic or the nematic order describes the monomer organization, but the relation between macromolecular conformation and mesomorphic properties cannot be completely understood without the use of labelling methods.

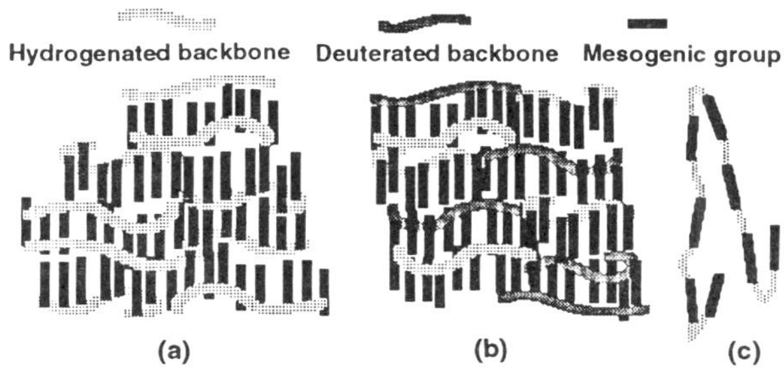

Figure V.4. — Liquid crystal polymers; a) pure side chain polymers; b) H/D mixture of side chain polymers; c) main chain polymer with hairpin defects.

Small angle neutron scattering experiments, performed on mixtures of randomly distributed deuterated and hydrogenated polymers give the shape factor of a macromolecule. Within this assumption of random distribution the shape of the scattering curve is independent of the composition, while a fifty percent composition yields the best contrast. In side chain polymers (i. e. when the mesogenic units are grafted as side groups to a flexible polymer backbone) the polymer conformation changes according to the nature of the mesophase. In a S_A phase the polymer is disc-shaped, with the plane of the disc parallel to the layer planes. In contrast, in the nematic phase the polymer is usually elongated along the director ($R_{G//} > R_{G\perp}$); however, if we approach a smectic phase even in the nematic range, $R_{G//}$ could be smaller than $R_{G\perp}$ owing to the influence of pretransitional smectic fluctuations. Moreover the structure factor of the smectic layer gives the layer density profile. A comparison of the two profiles obtained for a fully hydrogenated polymer and for the same molecule deuterated on the backbone provides information about the localisation of the backbone in the layer (fig.V.4.).

In main chain polymers, where the mesogenic units are linked end-to-end, the temperature dependence of the small angle scattering intensity in the nematic phase is related to conformational changes of the polymer. At low temperature the macromolecule takes on a fully elongated shape, and at high temperature, a folded conformation with hairpin defects.

The same labelling techniques could be used in anomalous X-ray scattering, but this is restricted to molecules with heavy atoms. It will be probably useful to solve some structural problems in metallomesogens. EXAFS experiments have been also used to answer questions about the environment of the metal.

Contrast variation can be used in lyotropic systems, by virtue of the presence of water in the system. Neutron scattering experiments are very useful for discriminating between the intra- and inter-aggregate distribution functions, for example in the case of lyotropic nematic phases. Lyotropic nematic mesophases are seen mostly in ternary systems (two surfactants and water). In a very narrow range of temperature and concentration three nematic phases are seen; two of them are uniaxial, corresponding to rod-shaped and disc-shaped micelles, and one is biaxial. The simplest description of these micelles corresponds to short cylinders ended by two hemispheric cups in the first case, and to circular or elliptical fragments of lamellae surrounded by a semi-toroidal ring, for the others. In fact, these model shapes are averaged by the orientational disorder of more anisotropic aggregates. Let us assume that the micelles have an ellipsoidal shape with three principal axes of lengths $l > i > s$. Then the rod-like uniaxial phase consists of ellipsoids with $i \sim s < l$, where l is parallel to the director. Uniaxiality is ensured if $i = s$, i.e. if the aggregate is intrinsically uniaxial, or if $i \neq s$, but the aggregate rotates around the l axis. The diffusion length profile has cylindrical symmetry in both cases, but it decreases more abruptly in the first case than in the second. In fact, when $i/s \approx 1$, it is not easy to discriminate between the two solutions. The problem was solved by taking advantage of the two kinds of surfactant in the micelles. Each surfactant can be characterized by its factor (V/Al), where V is the molar volume, A the hydrophilic head group area and l the chain length. If the hydrophobic moiety is inside the aggregate, then $(V/Al) < 1$ and the local curvature of the interface increases as (V/Al) decreases. Therefore the two components will be inhomogeneously distributed in the elliptical sections of the aggregate. This experiment has been carried out by Y. Hendrikx and J. Charvolin on ternary mixtures of potassium laurate-decanol-water. The contrast between

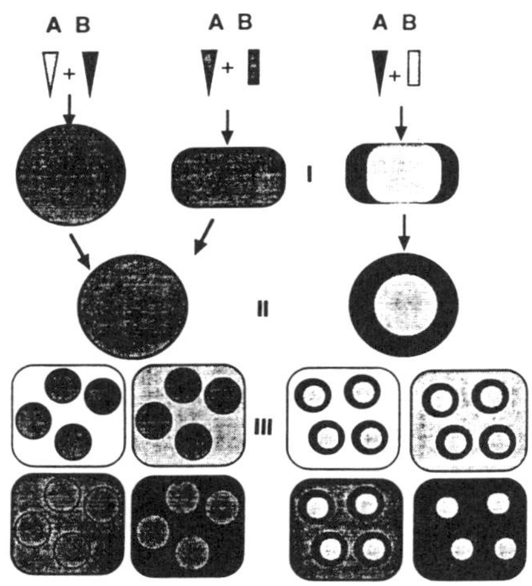

Figure V.5. — Structure of two component aggregates; the two surfactants A and B have (from left to right), the same shape and different densities , the same density and different shapes, different shapes and densities; I structure of the aggregates; II the two resulting mean structures of aggregates in uniaxial rotation; III application of the contrast variation method to suspensions of the two kinds of aggregates.

the two surfactants is obtained by using deuterated laurate and hydrogenated decanol. The scattering function is measured for different H_2O/D_2O ratios. In a first approximation if i = s, the two surfactants will not segregate inside each aggregate and S(q) will keep the same behaviour as H_2O/D_2O varies. The fact that the experimental intensity profiles change with the H_2O/D_2O ratio shows an inhomogeneous distribution of surfactants in the cross-section (i,s) of the aggregates (fig. V.5.).

This method is especially useful in solving the problem of the organization in the nematic phase, since there is no long-range order. However it has also been used in 2D ordered phases surrounding the nematic areas in similar ternary phase diagrams. It has been demonstrated that the cylinders depart from uniaxial shape as the alcohol content increases, even though the overall symmetry is still hexagonal. Complementary information could be obtained in a X-ray diffraction experiment performed near the absorption edge of the counter-ion. Actually this method was also used in the determination of the charge distribution around a spherical micelle.

V.4. Conclusion

In liquid crystals long-range order corresponds to long periods, especially in the case of mesophases with a periodic network of defects, which are presently being investigated. A structural study of these phases taking into account their main features, namely large lattice size, shallow density modulation and a large Debye-Waller factor, needs instruments with especially high resolution and good signal to noise ratio, particularly if we wish to characterize the short-range fluctuations. X-ray diffraction methods using synchrotron sources provide a high resolution beam that can be used to investigate the long-range order. Moreover, owing to its brilliance, it is also possible to follow phase transformations in time-resolved experiments. Neutron diffraction is efficient for short-range order studies and for investigations into the dynamics of individual molecules. Until now the study of the coherent inelastic scattering cross section has yielded very little information on the collective modes. However, use of spin-echo spectrometers in the investigation of the 2D columnar phases and of the 3D lattices of defects seems to be promising. Nowadays there is an increasing number of mesogenic organometallic complexes: the role of the metal can be ascertained using diffraction near the absorption edge. It will be also interesting to extend these methods into the soft X-ray range in order to reach the absorption threshold of lighter atoms.

GENERAL READING

P.G. DE GENNES : The physics of liquid crystals, Clarendon Press, Oxford, (1974).

J. PROST and P.G.DE GENNES : The physics of liquid crystals, 2nd edition, Clarendon Press, Oxford, to appear.

P.S. PERSHAN : Structure of liquid crystals phases,World Scientific, Singapore, New Jersey, Hong Kong, (1988).

Papers in honour of the 100th anniversary of liquid crystal research, Molecular Crystals and Liquid Crystals, **165**, (1988).

P. DAVIDSON and A-M. LEVELUT, Liquid crystals, **11**,469, (1992).

L. NOIREZ, P. KELLER, J-P. COTTON, submitted to Liquid Crystals.

CHAPTER VI

LARGE-AMPLITUDE MOTIONS IN MOLECULAR CRYSTALS AND POLYMERS

M. BEE

VI.1. Introduction

This lecture aims at giving some illustrations of the use of Incoherent Quasielastic Neutron Scattering (IQNS) in the investigation of motions of molecules in phases with dynamical disorder. This technique provides information about both the dynamical and the geometrical aspects of the motions. The term "Incoherent" neutron scattering refers to a scattering process in which the neutron interacts with the same scatterer at two successive times t_0 and t_0+t. "Quasielastic" scattering is mainly interested in small energy-transfers with the atoms of the sample, typically ± 2 meV (± 16 cm^{-1}). These originate from interactions of the neutrons with particles moving over time-scale c.a. 10^{-10} - 10^{-12} s. Such phenomena introduce a broadened component under the elastic line originating itself from the neutrons scattered without energy-change. An accurate investigation of the exact shape of this quasielastic component and especially a comparison of its relative amount with respect to the purely elastically scattered intensity, yield the Elastic Incoherent Structure Factor (EISF), i.e. the long-time limit of the intermediate scattering function, which is directly related to the region of space accessible to a scatterer and contains precise information about the geometry of their motions. Its determination permits to build a particular dynamical model and to write a scattering law to be compared to the data. The parameters are the characteristic times associated to the different motions. They are determined from the width of the broadened part and their temperature-dependence gives an idea of the potential barriers against the reorientations. A flourishing development of theoretical models has made the IQNS technique complementary to infrared or Raman light-spectroscopy, nuclear magnetic resonance, dielectric relaxation and X–rays.

The incoherent scattering function $S(\mathbf{Q},\omega)$, where $\mathbf{Q}$ is the momentum transfer vector and $\hbar\omega$ the energy gain for the neutron, is the time-Fourier transform

$$S(\mathbf{Q},\omega) = \frac{1}{2\pi}\int_{-\infty}^{\infty} I(\mathbf{Q},t)e^{-i\omega t}dt \qquad \text{(VI.1)}$$

of the intermediate scattering function $I(\mathbf{Q},t)$. itself given, for a single scatterer, by the correlation function (ensemble average) (Bée, 1988):

$$I(\mathbf{Q},t) = \left\langle e^{i\mathbf{Q}.\mathbf{r}(t)}e^{-i\mathbf{Q}.\mathbf{r}(0)}\right\rangle \qquad \text{(VI.2)}$$

$\mathbf{r}(t)$ and $\mathbf{r}(0)$ are the positions vectors of the scatterer at time t and at time origin, respectively. Thus the dynamical structure factor $S(\mathbf{Q},\omega)$ is readily expressed in terms of the spectrum of spontaneous fluctuations in the variable that describes the neutron-matter interaction. Splitting the intermediate scattering function $I(\mathbf{Q},t)$ into its constant limit at infinite time $I(\mathbf{Q},\infty)$ and its time-dependent part, we obtain

$$S(\mathbf{Q},\omega) = I(\mathbf{Q},\infty)\delta(\omega) + S^{qe}(\mathbf{Q},\omega) \tag{VI.3}$$

where

$$S^{qe}(\mathbf{Q},\omega) = \frac{1}{2\pi}\int_{-\infty}^{\infty} [I(\mathbf{Q},t) - I(\mathbf{Q},\infty)]\, e^{-i\omega t}\, dt \tag{VI.4}$$

The scattering function appears composed of a purely elastic component superposed on another, $S^{qe}(\mathbf{Q},\omega)$ which, because it is the time-Fourier transform of a time-dependent term, exhibits a non-vanishing broadening. The overall width of this quasielastic term is directly related to the characteristic times of the motions

When the system is in equilibrium, there is clearly no correlation between the positions of the scatterer at initial and infinite time, i.e. the system relaxes to the same equilibrium distribution of the scatterers after an external perturbation has been switched off. Therefore,

$$I(\mathbf{Q},\infty) = \langle e^{i\mathbf{Q}.\mathbf{r}(\infty)} e^{-i\mathbf{Q}.\mathbf{r}(0)}\rangle = \langle e^{i\mathbf{Q}.\mathbf{r}(\infty)}\rangle \langle e^{-i\mathbf{Q}.\mathbf{r}(0)}\rangle$$
$$I(\mathbf{Q},\infty) = |\langle e^{\,i\mathbf{Q}.\mathbf{r}(\infty)}\rangle|^2 = |\langle e^{-i\mathbf{Q}.\mathbf{r}(0)}\rangle|^2 \tag{VI.5}$$

Thus the coefficient of the delta-function which appears in the scattering law provides direct information about the region of space accessible to the scatterers i.e. about the equilibrium distribution of their positions. More precisely, in the case of a scatterer moving freely inside a volume V, the probability to find the scatterer in the unit volume around any given point $\mathbf{r}(\infty)$ at infinite time is simply $1/V \rightarrow 0$ for sufficiently large volume V, so that $I(\mathbf{Q},\infty) = 0$. Therefore a characteristic feature of the scattered intensity from a liquid or any material with dynamical translational disorder, like hydrogen in metals, superionic or protonic conductors, etc, is the absence of elastic peak (if $Q \neq 0$). Conversely, the existence of an elastic component in the scattered intensity indicates the presence in the sample of a scatterer, the motion of which is essentially located in space.

By integrating the incoherent scattering law over energy transfer $\hbar\omega$ at constant Q, we get, from the definition of $S(\mathbf{Q},\omega)$ above:

$$\int_{-\infty}^{\infty} S(\mathbf{Q},\omega)\, d\omega = \int_{-\infty}^{\infty} I(\mathbf{Q},t)\, \delta(t)\, dt = I(\mathbf{Q},0) = 1 \tag{VI.6}$$

$I(\mathbf{Q},\infty)$ is the fraction of the total quasielastic intensity contained in the purely elastic peak. It has the dimension of a structure factor and it is called the Elastic Incoherent Structure Factor (EISF). The direct importance consequence is that, providing that the separation between the sharp, purely elastic, component and the wider, quasielastic, contribution can be performed, by natural extrapolation or from computer techniques, the EISF is a measurable quantity, evaluated for each module and direction of the momentum transfer, **Q**, from the ratio

$$A(\mathbf{Q}) = \frac{I^{el}(\mathbf{Q})}{I^{el}(\mathbf{Q}) + I^{qe}(\mathbf{Q})} \quad \text{(VI.7)}$$

where $I^{el}(\mathbf{Q})$ and $I^{qe}(\mathbf{Q})$ the integrated intensities corresponding to the elastic and quasielastic parts of the spectra, respectively.

It must be noticed that the value of the EISF is directly related to the instrument resolution and to the time-scale accessible to the experiment. Depending on the instrument, different aspects of the motions will be revealed, and different models will be considered.

VI.2. Reorientations about a single axis

VI.2.1. Two-site jump model: Caesium Hydroxyde Monohydrate $CsOH.H_2O$

Above 340 K, an hexagonal structure (P6/mmm) alternates layers of Cs^+ and $H_3O_2^+$. Within the latter, all the protons appear to be equivalent, with no distinct OH^- and H_2O groups. Each layer can be described as a two-dimensionnal net, with pairs of neighbouring O atoms linked by hydrogen bonds. The P6/mmm symmetry is fulfilled by a dynamical disorder of the protons among the minima of a double-well potential between the atoms. Three mechanisms can be put forwards, i.e. a simple jump model between two sites, a 120°-rotation around the O atoms, or a combination of both yielding an exchange of H atoms and a translational diffusion. The EISF of the two-site model was found to coincide with the data. A characteristic time $\tau = 1.25 \times 10^{-11}$ s and a mean-square amplitude of vibration $\langle u^2 \rangle = 0.07$ Å^2 at 402 K were found from refinement. The jump distance is relatively large (1.03 Å) because of a non-linear H bond (Stahn et al., 1983)

A similar double proton-exchange mechanism in a double-well potential between oxygen atoms was evidenced in terephtalic acid at 220 K (Meier et al., 1984). The same mechanism was supported by NMR measurements of p-toluic acid, benzoic acid and other carboxylic acids (Meier et al., 1982, 1983; Graf et al., 1981; Nagaoka et al., 1983).

VI.2.2. Three sites over a circle: 120°-jumps of methyl groups

Two solid-solid phase transitions at about 200 K and 250 K separate the three solid phases of trimethyloxosulphonium iodide. The low-temperature (III) phase is fully ordered. Methyl group reorientations are mainly involved in the phase (III) → phase (II) transition, while whole-cation motions are partly involved in the phase(II) → phase (I) transition. According to NMR (Jurga et al., 1981), the barrier height against CH_3 rotation was estimated equal to 10.9 kJmol^{-1}, in agreement with the barrier against CH_3 torsion of 13.3 kJmol^{-1} derived from light-scattering study (Sourisseau et al., 1985), and with a correlation time $\tau = 9.6 \times 10^{-11}$ s at 300 K. 120°-jumps of the whole cation were evidenced over a much slower rate ($\tau = 1.75 \times 10^{-5}$ s at 300 K). From the comparison of the experimental values of the EISF with the theoretical curves based upon several models, the methyl dynamics alone was found responsible for the quasielastic broadening, with a unique Arrhenius law, without any discontinuity at the transitions: $\tau = 1.05 \times 10^{-12} \exp(\Delta H/RT)$ s, with an activation energy $\Delta H = 10.6$ kJmol^{-1} (Bée et al., 1985a).

Methyl torsion is certainly the motion which has been the most extensively investigated in polymers: polyacetaldehyde (Longer and White 1969), polydimethylsiloxane (Henry and Safford 1969, Allen et al. 1975, Amaral et al. 1976), polymethylmethacrylate (PMMA, Allen et al. 1975), polypropylene (Yasukawa et al. 1971). Measurements on stretch-oriented polypropylene (Takeuchi et al. 1982) evidence a large frequency-dispersion for the CH_3-torsion mode around 230 cm^{-1} as a result of a coupling between chains by methyl groups on neighbouring molecules. Biopolymers have also stimulated neutron studies (see review by Middendorf 1984). Drexel and Peticolas (1975) in the case of poly-L-alanine have shown conclusively that methyl torsion, not α-helical backbone vibration was responsible for the strong band at 230 cm^{-1}. Because the main chain motions broaden the quasielastic spectra, side-chains rotations are preferentially observed in samples at around or below their glass-transition temperature. PMMA has two methyl groups in the monomer unit, with barriers for the torsional frequencies different enough to separate the two contributions to the quasielastic broadening. As a matter of fact, it was found that at room-temperature, the α-methyl group is rotating at less than 10^9 Hz while the ester methyl moves much faster at 10^{11} Hz (Ma 1981). Gabrys et al. (1985) have observed a non-Arrhenius behaviour of the widths of the quasielastic peaks and a difference in the values of the reorientational rates between the two stereospecific forms of PMMA (Gabrys 1984). This effect was explained by inclusion of hugher than V_3 terms in the Fourier series expansion of the potential function hindering the motion of the CH_3 group.

VI.2.3. Reorientations over more than three sites

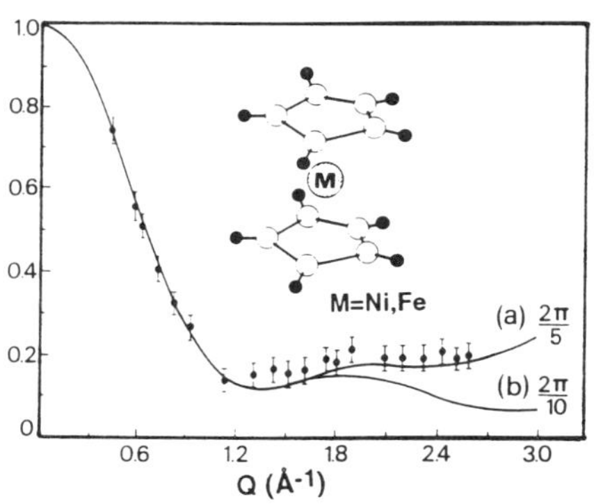

Figure VI.1. — EISF for different reorientations of C_2H_5 rings of nickelocene on a circle of radius 2.15 Å.

Both $2\pi/5$ and $2\pi/10$ - jump models lead to practically the same EISF curve while $Qr \leq 3$ (fig. VI.1). In the case of rotating C_5H_5 rings of nickelocene $(C_5H_5)_2Ni$, the distinction becomes possible only for $Q \geq 1.6$ Å^{-1}: $\tau = 3.3 \times 10^{-13} \exp(\Delta H/RT)$ s, $\Delta H = 6.3$ kJmol^{-1} (Sourisseau et al., 1983).

In the high-temperature, f.c.c. plastic phase of cyanoadamantane $C_{10}H_{15}CN$ (T 228 K) the CN group can occupy six orientations along the [100] lattice axes. In the same time, each molecule has three possible orientations around each [100] axis. IQNS (Bée et al., 1980a) confirmed the fast rotation by 30°-jumps ($\tau = 10^{-12}$ s at 300 K) evidenced by NMR. The tumbling of the CN group amongst the [100] axes was much too slow to be visible ($\tau > 10^{-8}$ s at 300 K).

VI.3. Reorientations about several axes in space: Single-crystal study of adamantane

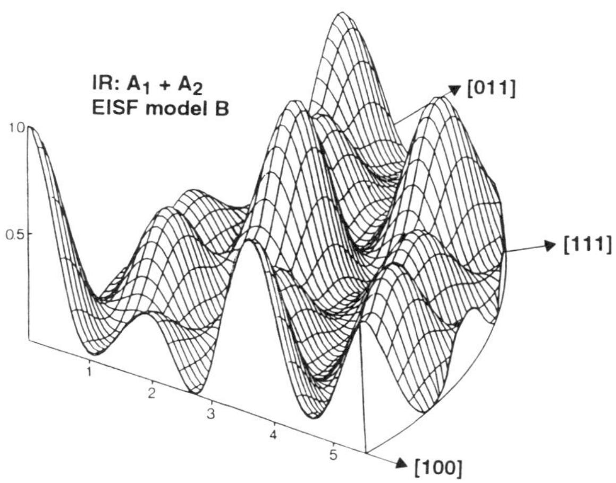

Figure VI.2. — Variation of the EISF as a function of Q in the [-110] plane of the reciprocal lattice containing the three [001],[111] and [110] directions.

Adamantane $C_{10}H_{16}$ is a cage-like molecule with tetrahedral symmetry (Td). At T = 209 K, a transition (ΔS = 16.2 kJmol^{-1}) occurs from a b.c. tetragonal cell to a f.c.c. phase (a = 9.45 Å, Fm3m, Z = 4), stable until the melting point T_m = 540 K (ΔS_m= 20.9 kJmol^{-1}). This phase is disordered, with two distinguishable orientations, related by 90°-rotations about the [100] axes of the cubic cell, the molecule and lattice threefold axes being coincident. Group theory leads to 5 characteristic times, related to each of the irreducible representations of the O group. Three models can be envisaged. With the A-model 90° rotations about [100] axes and/or 180° rotations about [110] axes occur, independently of the simultaneous existence of other rotations. Conversely with the B-model these rotations do not exist: only 120° jumps about [111] axes and 180° jumps about [100] axes) occur(fig. VI.2). In the C-model, only 180°-rotations about the [100] axes exist.

When using a single crystal sample, the structure factors, because of the occurrence of scalar products **Q.r** in the scattering law depend on the orientation of the scattering vector with respect to the specimen. The distinction could be made between the three hypotheses in favor of A-model (Bée et al., 1980b).

VI.4. Reorientations about molecular and crystalline axes

The application of the group theory permits to calculate the scattering function when motions about both fixed (crystal) and mobile (molecular) axes are involved. Bicyclo-octane (BCO) and triethylenediamine (TEDA) are both cage-like molecules, with the same symmetry. They show a phase transition from an hexagonal ordered phase to a high-temperature, cubic plastic phase (Fm3m, Z=4) in which they reorient themselves between eight, equally weighted orientations with coincident threefold molecular and crystal axes. Around each [111] direction, there are two orientations separated by a 60°-rotation. The evolution of the characteristic times associated to each motion versus temperature suggests a stronger steric hindrance in TEDA than in BCO, in accordance with the lattice parameters at T = 300 K (a = 8.86 Å and 9.10 Å for TEDA and BCO respectively). The dominant role of the interatomic potential between nitrogen atom and other atoms was also evidenced (Bée et al., 1982, 1985b; Leadbetter et al., 1982).

Following the original study of adamantane, adamantanone $C_{10}H_{14}O$ and adamantane halides $C_{10}H_{15}X$ (X = F, Cl, Br) were extensively investigated in their plastic phase: according to the nature of the substituting atom on the adamantyl cage, the departure from globular symmetry is more or less important. The effects of steric hindrance were shown and all the experimental data could be interpreted in terms of jump-models based upon two types of reorientations.

VI.5. Molecular Isotropic Rotational Diffusion (IRD)

Norbornane (NBA) exhibits two plastic phases: an h.c.p. phase (131 K < T < 306 K, a = 6.17 Å, c = 10.03 Å) and a f.c.c. phase (a = 8.73 Å) which subsists up to melting (T_m = 360 K). The IRD model was found well-adapted. No discontinuity was observed for D_R at the h.c.p.-f.c.c. phase transition; $D_R = 2.03 \times 10^{12} \exp(-\Delta H/RT)\ s^{-1}$ with the activation energy: $\Delta H = 6.14\ kJmol^{-1}$, in agreement with NMR (Bée et al., 1986). At low temperature, the EISF deviates from the IRD model, but does not reduce into a uniaxial rotation. An explanation was proposed with the formation of ordered domains of close-packed molecules, the other remaining free to rotate. Similar effects were observed with hexamethylethane.

Octaphenylcyclotetrasiloxane (OPCTS) is formed by a central, almost planar tetrasiloxane ring, on which eight phenyl groups are attached, roughly pointing towards the vertex of a 6.5 x 6.5 x 4.5 $Å^3$ parallepiped. Between 461 K and 478 K, a highly disordered phase exists, with a very weak melting entropy ($\Delta S_m/R = 0.49$). The experimental values of the EISF obtained on the 10^{-10} s time-scale show a good agreement with the IRD model(fig. VI.3). Furthermore, no discontinuity was found for D_R at the melting point, with a unique Arrhenius law: $D_R = 0.86 \times 10^{13} \exp(-\Delta H/RT)\ s^{-1}$ with the activation energy: $\Delta H = 33\ kJmol^{-1}$ (Bée et al., 1984).

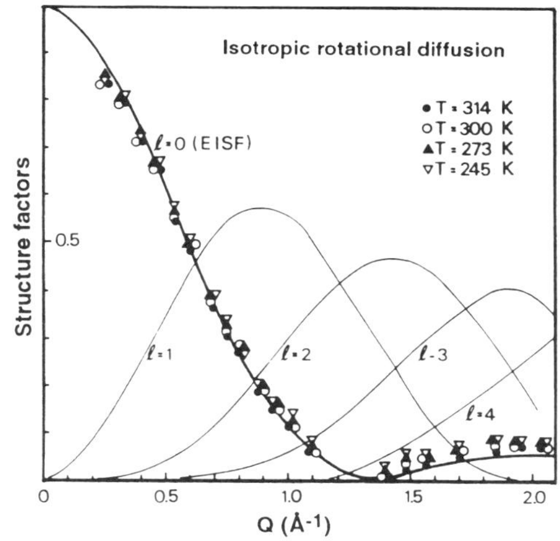

Figure VI.3. — Experimental EISF for NBA and comparison with the theoretical curve predicted by the IRD model.

VI.6. Rotation-Translation coupling in Norbornadiene

Norbornadiene (NBE) differs from NBA only by the presence of two double bonds. The liquid crystallises at 254 K either in a stable h.c.p. (a = 5.89 Å, c = 9.51 Å) or in a metastable f.c.c. structure (a = 8.80 Å), depending on the conditions of cooling. The EISF is identical in both phases but the motions appear much less isotropic, even at the highest temperature, as compared with observed with NBA. in particular, no IRD is observed, because the EISF does not decrease to zero. A correct description requires a combination of uniaxial reorientations with a simultaneous displacement of the centre of mass (c.o.m.) of c.a. 0.6 Å (Bée et al., 1990). The characteristic time of the motion follows an Arrhenius law: $\tau = 2.5\text{x } 10^{12} \exp(\Delta H/RT)$ s with the activation energy: $\Delta H = 4.62$ kJmol^{-1}.

In the liquid phase of NBE, a long-range displacement of the c.o.m. of the molecules is responsible for an additional broadening of the spectra which can be described by the Chudley-Elliott model.

VI.7. Alkyl chain motions in columnar liquid-crystal phases

A phase transition at about 390 K was evidenced for binuclear copper(II), rhodium(II) and ruthenium(II) complexes of fatty acids, from a lamellar crystalline lattice to a liquid crystalline mesophase. The former structure is made of layers of polar cores separated by layers of paraffinic chains. The latter is characterized by a two-dimensional hexagonal lattice where binuclear units are stacked and surrounded by alkyl chains pointing away from the Cu-Cu axis.

Five dicopper complexes with various chain-lengths were compared by IQNS (time-scale 10-10 s) (fig. VI.4). From the EISF analysis large-amplitude motions were evidenced in the columnar and crystalline phases. Their amplitude continuously increases when increasing the length of the chain, and, together with the characteristic times they are temperature-dependent in all phases. It also appears that compounds with eight external chains are more disordered in their crystalline phase and less disordered in their columnar phase, with a more continuous change in the EISF at the transition.

An analysis with selectively deuterated samples was focused on the columnar phase of the palmitate compound. On the 10^{-11} s time-scale it revealed no particular behaviour of the methyl groups at the chain ends, but conversely, a different dynamics of the hydrogen atoms depending whether they are bound to carbons close to the core or far from it and finally a similar behaviour of the part of the chain close to the core of the molecule, irrespective to the chain length.

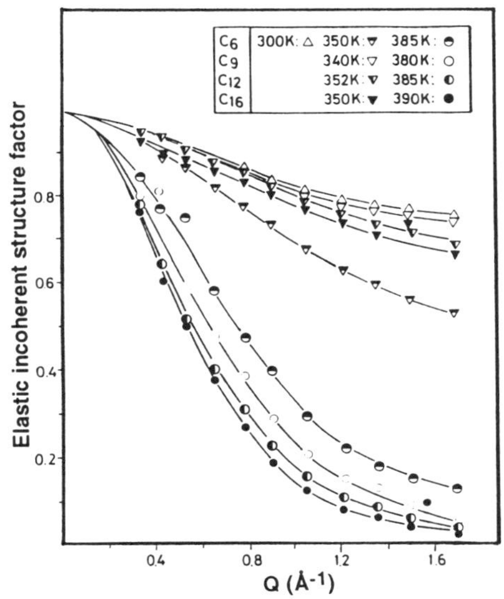

Figure VI.4. — Experimental EISF for different alkanoates complexes with different chain-lengths.

A model, allowing for a diffusion of each hydrogen inside a sphere, and averaged over the different radii, was used. The radius of the sphere is a linear function of the distance from the core between the first and the fifth carbons, and then reaches a limiting value in relation to the spacing between the stacked molecules (Carpentier et al., 1989; Giroud-Godquin et al., 1989).

VI.8. Decylammonium chains and alkylenediammonium chains in perovskite-type layered compounds

Decylammonium tetrachloride compounds $(C_{10}H_{21}NH_3)_2MnCl_4$ with M^{2+} = Mn, Cd,... are two-dimensional compounds built of corner-sharing MCl_6 octahedra parallel to the (**a**, **b**) plane. In between these planes, double layers of organic chains are intercalated. A dynamical model was elaborated, based on a gradient of disorder on going from the NH_3^+ polar heads to the CH_3 end, the volume visited by each CH_2 unit being approximated by a sphere of radius $R_i = R_o/(1 + \exp[-\alpha(i-\beta)])$. Just above the transition, the occurrence of kink defects ($t_4gtg't$) was confirmed. As the temperature increases, cooperative torsions of the CH_2 groups along the chain axis take place, as suggested from computer simulations (Guillaume et al., 1989).

The dynamics of alkylene diammonium chains in $(NH_3(CH_2)_nNH_3)MnCl_4$ with n = 3, 4, 5 was analysed on the basis of instantaneous jumps of the protons over 4 sites which correspond for the chains to two equiprobable "all-trans" conformations and two equiprobable "twisted" conformations. Two possibilities for the geometry of the twisted conformers were considered i.e. i) a rotation of half of the molecule around the central C-C bond and ii) a progressive and cooperative torsion running along the molecule axis. The EISF indicates a population of twisted (ii) forms, which increases with the temperature. The characteristic times confirm that these forms become more stable at high temperature. In conjunction, 53°-oscillations of the NH_3 groups were found, independently of the nature of the cation (Guillaume et al., 1987).

VI.9. Alkyl Chains in Inclusion Compounds

$C_9H_{16}NO_2$ (TANO) forms stable crystalline inclusion compounds with alkanes C_nH_{2n+2} ($7<n<44$). Alkane molecules are enclosed end to end in parallel channels, 5 Å diameter. IQNS from selectively deuterated samples revealed large-amplitude deformations of the TANO molecules, in relation with local changes in the conformation of the chains with restricted displacements along the channels (Bée et al., 1992b). Similar experiments carried out with hexane molecules adsorbed inside two different zeolites, ZSM5 and mordenite, namely, could be described on the basis of the same hypotheses for the alkyl chain motion.

VI.10. Disordered metastable phases and critical phenomena

With some molecular compounds, it is possible by very fast cooling, to avoid the normal transition from the orientationally disordered phase into the ordered low-temperature phase. Then the sample passes into a glassy state, where the molecules are fixed at random among their preferential orientations.

VI.10.1. Cyanoadamantane

The quasielastic broadening observed in the glassy phase on the 10^{-10} s time-scale confirms the persistence of the uniaxial rotation, but the EISF does not agree with any uniaxial jump model. Moreover it depends on the temperature, indicating an evolution in the geometry of the motions. A direct Fourier transformation of the scattering function reveals several relaxation times. A possible interpretation is the occurrence of antiferroelectric domains, responsible for noticeable differences in the intermediate surrounding of the molecules, with a direct influence on their dynamics. Annealings of 20 to 60 hours yield a strong increase of the elastic intensity and a change of the shape of $S(Q,\omega)$. The rate of these effects increases with the temperature of annealing. However, the analysis of the EISF leads to an interpretation in terms of an increase of the correlation length between molecules rather than on the basis of nucleation of anti-ferroelectric domains (Bée et al. 1992a).

VI.10.2. Amorphous and crystalline phases of Methoxybenzylidenebutylaniline

MBBA has a rich polymorphism: quenching from the R.T. nematic liquid-crystal (NLC) phase yields the glassy state C_0. Subsequent reheating reveals two smectic phases, C_1 followed by C_2 and then two crystalline phases C_3 and C_4. All of them can be cooled to very low temperature without further transformation. On the other hand, the slow cooling of the NLC phase yields the formation of two crystalline phases C_6 and C_5 (Pepy et al. 1989).

On the 10^{-11} s time-scale, a quasielastic broadening is visible in the C_6 phase (and also in C_5), which is wider than in C_4 at the same temperature. Fixed-window measurements (time-scale 10^{-10} s) monitored the purely elastic scattering in the C_0 phase as a function of the temperature. The intensity decreases strongly between T_g - 30 K and T_g (T_g = 205 K) when approaching the transition towards the C_1 phase (more rapidly than predicted by the usual Debye law). But no quasielastic broadening could be observed in any of the two phases: the decrease of the purely elastic scattering is rather related to the increase of the density of states (d.o.s.) observed with IN6 (Belushkin et al., 1987; Derollez et al., 1989).

VI.10.3. Polybutadiene

Above T = 320 K, a gradually increasing quasielastic broadening occurs. The intermediate scattering function I(Q,t) measured from neutron spin-echo was described by a stretched exponential law ($\exp[-(t/\tau)^{\beta}]$) with β ranging from 0.65 to 0.4 depending on Q but independently of the temperature. In the same time, a faster relaxation component was indirectly observed, revealing itself as a missing fraction in I(Q,t)/I(Q,0) for $t \rightarrow 0$. Time of flight data showed a square root behaviour of the mean square displacement $\langle u^2 \rangle$. The fast component increases strongly in intensity near T_g at low-energies and was interpreted either as the β-relaxation predicted by the mode-coupling theory or from the change in the phonon d.o.s., without definite conclusion (Frick et al., 1988).

VI.11. Conclusion

We tried, through a restricted series of examples, to give some idea of the informations which can be obtained from IQNS. Considerable help in the choice of a model is obtained from the determination of the EISF, which shows the importance of the molecule delocalization. The jump model is the most commonly used. Although it is based on some hypotheses which are strictly not true (instantenous jumps, well-defined orientations...) it is often able to give an adequate description of the motions. The corresponding scattering law is rather simple, with a finite number of terms in its expansion. More complicated cases can be strongly simplified by taking advantage of group theory to evaluate the scattering law. Measurements carried out with single crystals can make easier the distinction between several models. True IRD is rarely encountered, except in some cases of highly disordered crystals. When the molecules are not rigid, or when the phases are not well crystallised it can be difficult to assign precise equilibrium positions to the scatterers and to analyse the motions in terms of jump-distances. Another difficulty arises when the different kinds of motions are no more decoupled e.g. when the displacement of the c.o.m. of the molecule occurs on the same time-scale as the reorientations, or is directly linked to them. Then, the standard models are no more adapted. Nevertheless, the IQNS technique is still able to give precise information: it revealed a translation-rotation coupling mechanism when passing from NBA to NBE. It evidenced the flexibility of alkyl chains in different compounds and precised the relative motions of their parts (here the selective deuteration appears as a powerful tool). Molecular glasses attract much interest: IQNS permitted to follow the evolution of the dynamics of the molecule of cyanoadamantane in relation with the nucleation of the tetragonal domains. It was used to investigate the rich polymorphism of quenched MBBA. It evidenced critical phenomena near the glass-transition in polybutadiene.

REFERENCES

ALLEN G., WRIGHT C.J. and HIGGINS J.S., 1975 - Polymer 15, 319

AMARAL L.W., VINHAS L.A. and HERDADE S.B., 1976 - J. Pol. Sci. Pol. Phys. 14, 1077

BEE M., 1985 - Journal de Chimie Physique 82, 2/3, 205

BEE M., 1988 - Quasielastic Neutron Scattering: Principles and Applications in Solid State Chemistry, Biology and Materials Science - Adam Hilger Ltd (Bristol)

BEE M., 1992 - Spectrochemica Acta 48A, 3, 429

BEE M., 1992 - Physica B 182, 323

BEE M., AMOUREUX J.P. and DIANOUX A.J., 1980a - Mol. Phys. 41, 325

BEE M., AMOUREUX J.P. and LECHNER R.E., 1980b - Mol. Phys. 39, 945

BEE M., SAUVAJOL J.L. and AMOUREUX, J.P., 1982 - J. Physique 43, 1797

BEE M., DIANOUX A.J.. and VOLINO F., 1984 - Mol. Phys. 51, 221

BEE M., JOBIC H. and SOURISSEAU C., 1985a - J. Phys. C: Sol. State Phys. 18, 5771

BEE M., SAUVAJOL J.L., HEDOUX A. and AMOUREUX, J.P., 1985b - Mol. Phys. 55, 637

BEE M., JOBIC H. and CAUCHETEUX C., 1986 - J. Chim. Phys. 83, 10

BEE M., ELKHAMKAMI J. and JOBIC H., 1990 - J. Phys.: Condensed Matter 2, 1945

BEE M., DEROLLEZ P., DESCAMPS M. and FOURET R., 1992a - Physica B, 180/181, 655

BEE M., RENAULT A., LAJZEROWICZ-BONNETEAU J. and LE BARS-COMBE M., 1992b - J. Chem. Phys. 97, 10, 7730

BELUSHKIN A.V., DOLGANOV K., NATKANIEC I. and SIMKIN V.G., 1987 - Molec. Cryst. Liq. Cryst. 150B, 187

CARPENTIER L., BEE M., GIROUD-GODQUIN A.M., MALDIVI P. and MARCHON J.C., 1989 - Mol. Phys. 68, 1367

DEROLLEZ P., BEE M., GORS C., FOURET R. and DIANOUX A.J., 1989, J. Chim. Phys. 86, 2001

DREXEL W. and PETICOLAS W.L., 1975 - Biopolymers 14, 715

FRICK B., RICHTER D. PETRY W. and BUCHENAU U., 1988 - Z. Phys. B70, 73

GABRYS B., HIGGINS J.S. and YOUNG D.A., 1985 - Polymer 26, 355

GABRYS B., HIGGINS J.S., MA K.T. and ROOTS J.E., 1984 - Macromol. 17, 560

GIROUD-GODQUIN A.M., MALDIVI P., MARCHON J.C., BEE, M. and CARPENTIER L., 1989 - Mol. Phys. 68, 1353

GRAF F., MEYER R., HA T.K. and ERNST R.R., 1981 - J. Chem. Phys. 75, 2914

GUILLAUME F., CODDENS G., DIANOUX A.J., PETRY W., REY-LAFON M. and SOURISSEAU C., 1989 - Mol. Phys. 67, 665

GUILLAUME F., SOURISSEAU C., LUCAZEAU G. and DIANOUX A.J., 1987 - in "Dynamics of Molecular Crystals", Ed. J. Lascombe, p. 173, Elsevier (Amsterdam)

HENRY A.W. and SAFFORD G.J., 1969 - J. Polym. Sci. A2, 7, 433

JURGA S., JURGA K. and PAJAK Z., 1981 - J. Phys. C: Solid State Phys. 14, 4433

LEADBETTER A.J., PIPER J., RICHARDSON R.M. and WRIGHTON P.G., 1982 -J. Phys. C: Solid State Phys. 15, 5921

LONGER G.F. and WHITE J.W., 1969 - Mol. Phys. 17, 1

MA K.T., 1981 PhD Thesis Imperial College (London)

MEIER B.H., GRAF F. and ERNST R.R., 1982 - J. Chem. Phys. 76, 767

MEIER B.H., MEYER R., ERNST R.R., ZOLLIKER P., FURRER A., HALG W. and ANDERSON I., 1984 - Chem. Phys. Lett. 108, 522

MEIER B.H., MEYER R., ERNST R.R., STOCKLI A., FURRER A. and HALG W., 1983 - Chem. Phys. Lett. 103, 169

MIDDENDORF H.D., 1984 - Ann. Rev. Biophys. Bioengineering 13, 425

NAGAOKA S., TERAO T., IMASHIRO F., SAIKA A., HIROTA N. and HAYASHI S., 1983 - J. Chem. Phys. 79, 4694

PEPY G., FOURET R., MORE M. and ROSTA L., 1989 - Physica Scripta 39, 485

SOURISSEAU C., LUCAZEAU G., DIANOUX A.J. and POINSIGNON C., 1983 - Mol. Phys. 48,367

SOURISSEAU C., BEE M., JOBIC H. and DWORKIN A., 1985 - J. Ram. Spect. 16, 44

STAHN M., LECHNER R.E., DACHS H. and JACOBS H.E., 1983 - J. Phys. C: Solid State Phys. 16, 5073

TAKEUCHI I., HIGGINS J.S., HILL A., MACCONACHIE A., ALLEN G. and STIRLING G.C., 1982 - Polymers 23, 499

YASUKAWA T., KIMURA M., WATANABE N. and YAMADA Y., 1971 - J. Chem. Phys. 55, 983

CHAPTER VII

DYNAMICS AND DIFFUSION IN MACROMOLECULES, COLLOIDS AND MICROEMULSIONS

B. FARAGO

VII.1. General introduction

The observed quantity in scattering techniques is in general the Fourier transform of the relevant (density-density) correlation function (space $\Rightarrow q$, time$\Rightarrow\omega$). This reads:

$$I(q,t) = \sum_{i,j} b_i b_j \left\langle e^{-iq(r_i(t)-r_j(0))} \right\rangle$$

if we consider the space Fourier transform. When the scattering particles are molecules that consist of many atoms we can separate the centre of the particle and the position of the atoms inside the molecules $r_i = R_n + r_i^n$. $(n,m$ denote different particles, r_i^n and r_j^m the positions i and j inside n and $m)$

$$= \sum_{n,m,i,j} b_i b_j \left\langle e^{-iq(R_n(t)-R_m(0))} e^{-iq(r_i^n(t)-r_j^m(0))} \right\rangle$$

If furthermore the orientation (or shape) of particle n is independent of particle m , the sum can be separated into two independent averages:

$$= \sum_{n,m} \left\langle e^{-iq(R_n(t)-R_m(0))} \right\rangle \sum_{i,j} b_i b_j \left\langle e^{-iq(r_i(t)-r_j(0))} \right\rangle$$

The first will give the interparticle interference factor $S(q,t)$ and the second the particle form factor $P(q,t)$. The time dependence of the first gives the structural (interparticle) relaxation and that of the second the form fluctuations (intraparticle relaxation). If the particles do not interact (eg very dilute solution) the first sum is different from zero only if n=m and we have an 'incoherent' sum of the scattered intensites from the individual particles:

$$= N \left\langle e^{-iq(R(t)-R(0))} \right\rangle \sum_{i,j} b_i b_j \left\langle e^{-iq(r_i(t)-r_j(0))} \right\rangle$$

VII.1.1. Statics

Thumb rule: A measurement performed at a particular q value probes distances ~ $1/q$ (up to factors of 2π).

In the small q region the preaveraging over $1/q$ (if $1/q$>> a typical interatomic distance, ca 2 Å) means that instead of discrete scattering centres we can define scattering-length densities.

Frequently static scattering can already reflect some dynamic properties but a knowledge of it is nearly always necessary in order to extract physical parameters from the dynamic measurements.

Let us consider a solution of large macromolecules (polymers, viruses or surfactant aggregates). The scattering as a function of dilution will give us the following information:

Dilute system => form factor $P(q)$

Denser system => form factor + interference $P(q)*S(q)$

Ordered system => form factor + structural information $P(q)*S(q)$

In a perfectly regular structure (crystal) S(q) consists of Dirac delta functions (Bragg peaks). Imperfect structure will lead to broadening or diffuse scattering between the peaks. This might have its origin in softening - dynamics.

VI.1.2. Dynamics

Thumb rule: the energy of the probe must be close to the energy exchange investigated. When dealing with thermally excited movements the probes that have the proper energy (in the range of kT) are light and neutrons.

In the case of neutrons with a known incoming velocity (energy) v_{in} and a measured velocity after scattering v_{out}, $\omega = 1/2\, m \left(v_{out}^2 - v_{in}^2\right)$ the energy exchange is obtained. In general the measured quantity is S(q,ω), the space and time Fourier transform of the scattering correlation function. Only Neutron Spin Echo (NSE) measures $I(q,t)$ (space but no time Fourier transform) directly as we will see later.

The smaller the ω, the more precisely v_{in} and v_{out} must be determined. The more the beam is monochromatic the less intensity we will have. The following inelastic techniques are in order of increasing resolution: 3-axis, TOF, backscattering. For NSE see later. To see discrete excitations, measurements in ω space are better. To analyse line shapes of relaxations $I(q,t)$ is more advantageous.

VII.1.2.1. Neutron Spin Echo

(Mezei 1980)

When a polarized neutron beam enters a magnetic field region it undergoes Larmor precession (Fig. VII.1.). Let us now use two regions with opposite field. Choose $H_1 l_1 = H_2 l_2$ and if $v_1 = v_2$ (that is to say no energy change during the scattering) then $\Delta\varphi = \varphi_2 - \varphi_1 = 0$. If there is a small energy change $v_1 \neq v_2$ and $\Delta\varphi \neq 0$.

To first order in $(\omega/\frac{1}{2}mv^2)$ $\Delta\varphi$ is proportional to ω . Let us denote it $\Delta\varphi = t\omega$.

To calculate the final polarization we have to take the average over all $\Delta\varphi$ values.

$$P_{final} = P_{init}\langle\cos\Delta\varphi\rangle$$

As shown above $\Delta\varphi$ is proportional to the energy change and the probability of an energy change ω is nothing else than the so called scattering law $S(q,\omega)$.

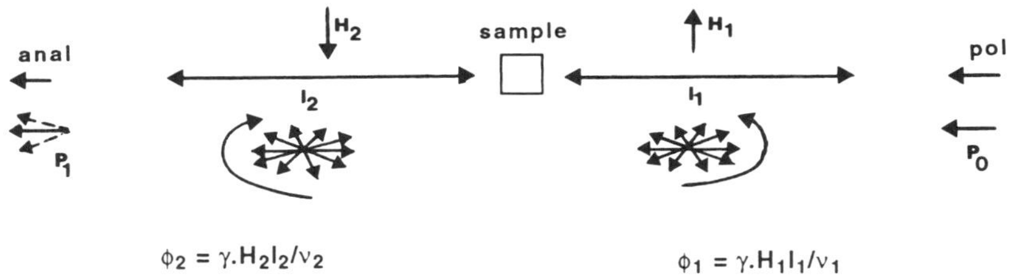

Fig. VII.1. — Schematic outline of the principle of the neutron spin echo technique.

$$P_{final} = P_{init}\frac{\int S(q,\omega)\cdot\cos\omega t\cdot d\omega}{\int S(q,\omega)\cdot d\omega} = P_{init}\frac{I(q,t)}{I(q,0)}$$

By measuring P_{final} and P_{init} with NSE we obtain directly $I(q,t)$.

To give some typical numbers, φ_1 can be as much as 10000 $*2\pi$ and $\Delta\varphi\approx 10$ degrees can be detected. The time range t can go from 1 psec to 30 nsec.

VII.1.2.2. De Gennes narrowing for solutions of dilute particles:

(Pusey 1975)

In soft matter mostly slow, diffusive (overdamped) motions can be expected which will lead to quasielastic scattering. For example in the case of simple Brownian motion of very dilute non-interacting particles the scattering law is:

$$\left\langle e^{-iq(R(t)-R(0))}\right\rangle = e^{-\frac{1}{6}q^2\left\langle (R(t)-R(0))^2\right\rangle} = e^{-Dq^2t} \text{ ---> Fourier---> } \frac{1}{\omega^2+(Dq^2)^2}$$

in time domain $\qquad$ in ω domain

We have seen that finite dilution affects the static scattered intensity (at least in the first approximation) via a multiplying factor $S(q)$. Let us examine what happens to the dynamics. Consider only the interparticle term (see General Introduction) :

$$S(q,\tau) = \sum_{n,m}\left\langle e^{iq(R_n(t)-R_m(t+\tau))}\right\rangle$$

R_n and R_m are the positions of particles n and m, t is the starting time and τ is the correlation time. ($S(q,\tau)$ should not depend on t).

Let us define an effective diffusion coefficient D_{eff} by:

$$\frac{\partial}{\partial\tau}\left[\frac{S(q,\tau)}{S(q,0)}\right]_{\tau\Rightarrow 0} = -D_{eff}q^2$$

We want to find a relation between the zero concentration limit D_0 (or self diffusion) and D_{eff} at finite concentration.

We can readily obtain:

$$\frac{\partial}{\partial\tau}[S(q,\tau)] = -iq\sum_{n,m}\left\langle v_m(t+\tau)e^{iq(R_n(t)-R_m(t+\tau))}\right\rangle$$

As $S(q,\tau)$ should not depend on t, using the relation:

$$\frac{\partial}{\partial t}[S(q,\tau)] = 0 = iq\sum_{n,m}\left\langle v_n(t)e^{iq(R_n(t)-R_m(t+\tau))}\right\rangle - iq\sum_{n,m}\left\langle v_m(t+\tau)e^{iq(R_n(t)-R_m(t+\tau))}\right\rangle$$

we have:

$$\frac{\partial}{\partial\tau}[S(q,\tau)] = -iq\sum_{n,m}\left\langle v_n(t)e^{iq(R_n(t)-R_m(t+\tau))}\right\rangle$$

with the identity:

$$R_m(t+\tau) = R_m(t) + \int_t^{t+\tau} v_m(t')dt'$$

For short times we can approximate:

$$e^{iqR_m(t+\tau)} = e^{iqR_m(t)}\left(1 - iq\int_t^{t+\tau} v_m(t')dt'\right)$$

Substitute it in, and use the plausible assumption that the velocities and the positions are independent. In this case the expectation values can be separated into two. Since we no longer use the t independence, we set it to zero to see more clearly

$$\frac{\partial}{\partial\tau}[S(q,\tau)] = -iq\sum_{n,m}\langle v_n(0)\rangle\left\langle e^{iq(R_n(0)-R_m(0))}\right\rangle - q^2\sum_{n,m}\int_0^{\tau}\langle v_n(0)v_m(t)\rangle dt\left\langle e^{iq(R_n(0)-R_m(0))}\right\rangle$$

The expectation value $\langle v_n(0)\rangle$ will be zero. We need to calculate the integral of the velocity-velocity correlation. In the case of simple diffusion $\left\langle (x(0)-x(t))^2\right\rangle = 2D_0 t$. Applying $\frac{\partial}{\partial t}$ to get the velocity and using the same trick with the starting time independence as above, we obtain:

$$\langle (x(t)-x(0))v(0)\rangle = D_0 = \int_0^t \langle v(0)v(t')\rangle dt'$$

Furthermore we can suppose that $v_n(0)$ and $v_n(t)$ are uncorrelated if $n \neq m$.

With that

$$\frac{\partial}{\partial\tau}[S(q,\tau)] = -q^2 D_0$$

Or finally the desired result

$$\frac{\partial}{\partial\tau}\left[\frac{S(q,\tau)}{S(q,0)}\right]_{\tau\Rightarrow 0} = -\frac{q^2 D_0}{S(q,0)} = -D_{eff}q^2$$

The effective (measured) diffusion coefficient (frequently called collective diffusion) is related to the single particle (or tracer) diffusion by a division of $S(q)$. If we measure D_{eff} as a function of q we find that where $S(q)$ has a peak, the diffusion is apparently slowed down, the line width of the inelastic scattering in energy is 'narrower'. As mentioned already this is only a first approximation. One of the main assumptions, which breaks down very quickly, is that individual particles still move as if they were unperturbed. Our calculation is only about how the interference introduces an extra q dependence. If the dynamics of the individual particles is also influenced (as in the case of the long range hydrodynamic interaction) it might have comparable influence on D_{eff}.

We will see later through examples that whenever interference becomes important a similar effect will show up. Although, as usual, generalisations must be considered with caution, this can be a useful hint for guessing what can be expected from a measurement.

VII.2. Surfactants, Microemulsions

VII.2.1. Cubic Systems

Certain surfactants can form crystalline structure. When e.g. a cubic structure is convoluted with large non-isotropic objects the static scattering can have a complicated and fast decaying form factor.

Let us have a look at the X-ray pattern of a simple crystal made up of $C_{12}E_6$ surfactant and water (Fig. VII.2.). The elongated spots in certain directions might

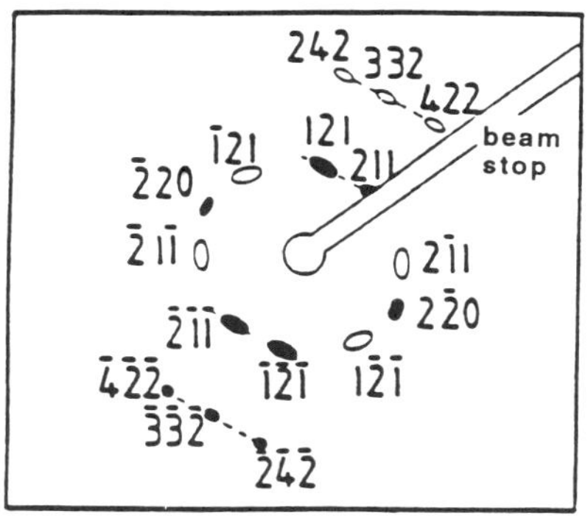

Fig. VII.2. — X-ray scattering on the $C_{12}E_6$ cubic system (Rancon et al. 1987)

indicate loose structure (easy deformation in certain directions) or just badly defined next neighbour distances.

The q resolution of NSE ($\Delta q/q \approx 15\%$) does not allow investigation of different points within these elongations but only the integrated intensity. If, however, we had a solid, but badly defined structure, $S(q,t)$ would be constant as a function of time (elastic scattering). The test measurement performed on this system (Fig. VII.3.) shows clearly that a large fraction of the intensity is in fact quasielastic (decaying $S(q,t)$). We can conclude that the elongation of the static Bragg spots has a dynamic origin.

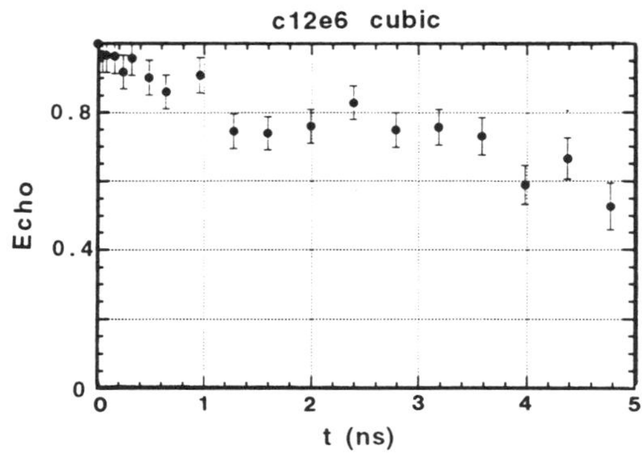

Fig. VII.3. — $S(q,t)$ as measured by NSE . As the sample was not a monocrystal, the intensity integrates the (211) and (220) spots.

VII.2.2. Lamellar Systems

There are many different kinds:

- double layer + water or oil
- single layers and alternating water, oil
- etc.

They give no true Bragg peaks at finite T (Peierls-Landau theory) (Safinya et al. 1986; Safinya 1989; Safinya et al. 1989), but instead exhibit power law singularities.

In the z direction (stacking direction):

$$I(0,0,q_z) \propto (q_z - q_m)^{-2+\eta_m}$$

where q_m is the q vector of the m-th order peak:

In the perpendicular direction:

$$I(q_\perp,0,q_m) \propto q_\perp^{-4+2\eta_m}$$

with

$$\eta_m = \frac{m^2 \eta_1^2 kT}{8\pi\sqrt{BK}}$$

K being K_C/d , d the interlamellar spacing, and K_C the elasticity modulus. Depending on the dominant stabilising mechanism, η_1 has different forms.

In the Helfrich (Helfrich 1978) mechanism the dominant force is due to the 'collisions' between undulating membranes: $\eta_1 = 1.33\left(1 - \frac{\delta}{d}\right)^2$, where δ is the thickness of the lamellae and d is the interlamellar spacing. For thin layers only the first order has a real singularity!

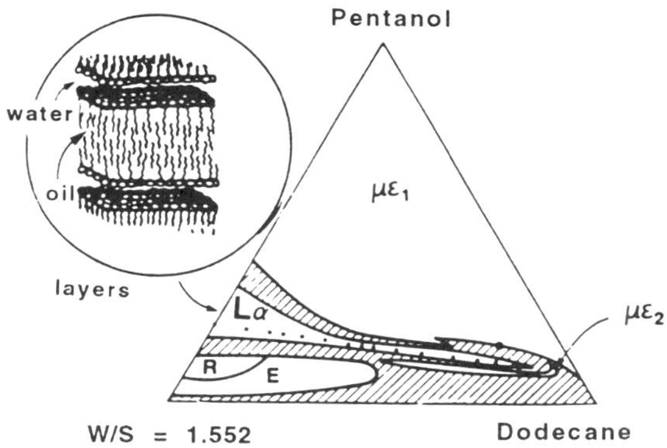

Fig. VII.4. — A section of the phase diagram of the quaternary mixture of SDS pentanol, water and dodecane shown in the plane with a constant water/SDS weight ratio=1.55.

$$\text{Electrostatic long range interaction:}\ \eta_1 = \sqrt{\frac{kTL_e}{2gK_c}}\left[\left(1-\frac{\delta}{d}\right)^{\frac{3}{2}}\sqrt{d}\right]$$

When the displacement is comparable to the separation strong concentration fluctuations will show up (Porte et al. 1988)

We tried a direct dynamic measurement (Farago et al. 1988) to separate undulation and thickness fluctuation mode. It is problematic to calculate the former for big amplitude fluctuations:

How can one separate the two experimentally ? Consider the undulation as the motion of the contour. By preparing with two different contrasts:

- full sheet (essentially contour motion) H-D
- double layer (contour motion + thickness fluctuation) D-D

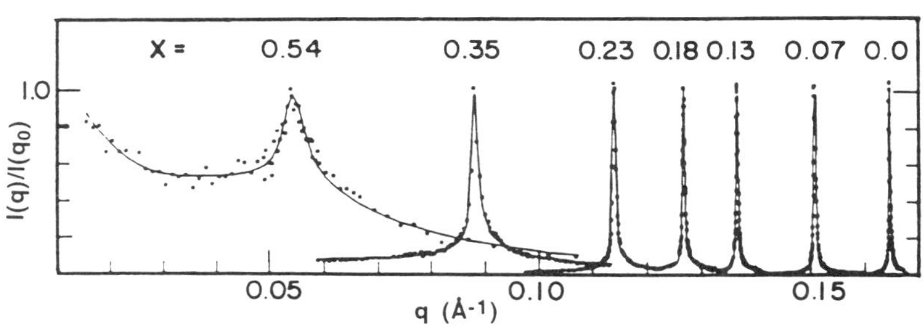

Fig. VII.5. — Typical singularities from references (Safinya et al. 1986; Safinya 1989; Safinya et al. 1989) as studied with synchrotron radiation. Longitudinal profiles of the first harmonic for seven different mixtures along the dodecane dilution path. The solid lines are fits to the Caillé [Caillé 1972] power-law shape.

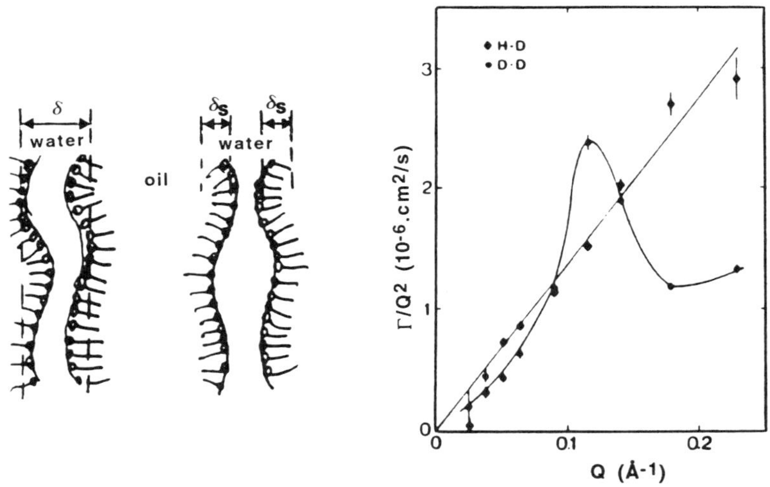

Fig. VII.6. — Left: Schematic representation of the undulation and baroclinic mode. Right: effective diffusion coefficient as measured by NSE under the two different contrast conditions.

One can also study the local motion in a layer [Pfeiffer et al., 1989]. The method used was partial deuteration in conjunction with incoherent neutron scattering on a backscattering instrument. With incoherent scattering the individual motion of the H atoms (bound the lipid) is observed. The EISF (Elastic Incoherent Structure Factor) via the q dependent intensity of the elastic scattering is partly due to the Debye-Waller factor. This gives valuable information about the mean-square amplitude. With an oriented sample, comparing q parallel and perpendicular to the sheets gives further information on this amplitude in these directions. With partial deuteration one can look at different H atoms in the lipid molecule. If they all show the same dynamics then this probably reflects the dynamics of the lipid molecule as a whole. If not, one can draw conclusions about internal motion (twisting, bending) within the molecule.

VII.2.3. Spherical microemulsion

We consider dilute (non-interacting) systems. The AOT (surfactant) water-decane system can form small (R ~ 50 Å) water droplets, coated with the AOT in the solvent decane. By measuring $S(q,t)$ via NSE

- at small q => only the centre of the mass diffusion is observed
- at high q => centre of the mass diffusion + shape fluctuations

We have shown that interference effects can give a q dependent D_{eff} via de Gennes narrowing. To be sure of avoiding $S(q)$ and $H(q)$ (the latter denotes the hyrodynamic interaction) two contrasts were used: 1.) - shell - (deuterated oil+surfactant+deuterated water) where for the neutrons the deuterated liquids look the same and they see only the surfactant molecules (full of H) surrounding the water sphere. 2.) - sphere - (deuterated oil+surfactant+normal water) where now the normal water and the surfactants are indistinguishable for the neutrons and they see the full sphere.

S(q) and *H(q)* do not depend on contrast (only on the outer diameter). Having verified the absence of these factors let us consider more in detail what can be expected:

A rigid thin shell of radius R gives for the form factor:

$$P(q) \propto \left(j_0(qR)\right)^2 \quad \text{and} \quad I(q,t) = e^{-Dq^2t}P(q)$$

A fluctuating shell can be described by an expansion of spherical harmonics as

$$R = R_0 + \sum_{l,m} u_{lm} Y_{lm}(\vartheta,\varphi)$$

which leads to:

$$I(q,t) = e^{-Dq^2t}\left[f_{static}(q,R) + \sum_{l>2} \frac{2l+1}{4\pi} f_l(q,R) u_l^2 e^{-t/\tau_l} \right]$$

For simplicity we do not give explicitly the time independent (static) contribution.

The q dependent static form factor has a minimum at $qR=\pi$, while $f_l(q)$ of the $l=2$ mode has a maximum. This means the observation of an extra contribution is the most favourable in this q region.

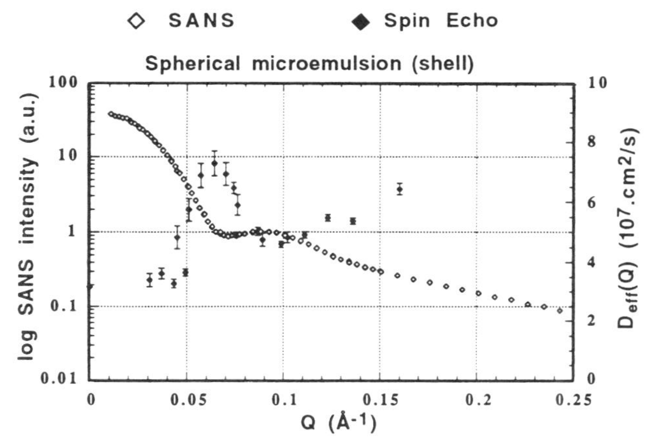

Fig. VII.7. — Log of the SANS intensity (left scale) and the q dependent effective diffusion coefficient as measured by NSE on the AOT-water - Decane ternary water-in-oil microemulsion system.

The physics of interest is the bending elasticity. The bending energy can be given in terms of K and $\overline{K}$, the two splay elastic constants, R_1 and R_2 are the principal curvatures and R_s is the spontaneous curvature. (In the absence of external constraint the surfactant molecules would form a surface with a curvature R_s that corresponds to the most favourable packing) (Helfrich 1978)

$$E = \frac{K}{2}\int dS\left(\frac{1}{R_1}+\frac{1}{R_2}-\frac{2}{R_s}\right)+\overline{K}\int dS\frac{1}{R_1 R_2}$$

A model $I(q,t)$ can be obtained by calculating:

eigenmodes => energies => equipartition => fluctuation amplitudes

energies + hydrodynamics => time constants (overdamped oscillator)

The polydispersity is also a function of the same parameters (Farago et al. 1990) Experimentally, essentially u_0 is measured by small angle neutron scattering and u_2 can be measured by NSE.

VII.2.3. A biological example: pig immunoglobulin [IgG] (Alpert et al. 1985)

The immunoglobulin consists of three loosely connected parts. A central one and two elongated arm-like parts. The question of interest is how flexible are these loose connections, as they might have a role in the biological function of the molecule. We can assume rotational diffusive motion of the arms in a limited angular range but we have to consider that the molecule can rotate as a whole too. The q dependence of D_{eff} is attributed to the flexibility of the "arms" of the IgG. The numerical result strongly depends on D_{tr} at $q \ll 1/Rg$.

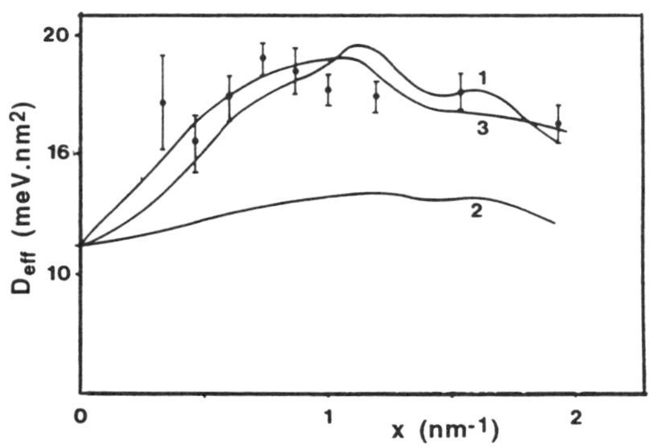

Fig. VII.8. — Computed plots of D_{eff} versus q for three different models of immunoglobulin-G. Curves: 1. rigid model (τ=1600ns) 2. rigid model (τ=6000ns) 3. flexible model (β_{max} = 50, τ=1220 ns)

VII.3. Polymers (Doi et al. 1986)

VII.3.1. The basics of dynamics

The basic concept: beads connected by entropic springs of length b. Spring constant : $K=\frac{3kT}{b^2}$

Rouse dynamics: $F=f(x_{n-1}-x_n, x_n-x_{n+1})=-v_n\xi$ localized, pairwised interaction

Zimm dynamics: as Rouse + backflow of the solvent, which introduces long range hydrodynamic interaction.

The main difference is that a polymer consisting of N beads that is dragged with a constant speed v feels, in the case of Rouse dynamics, a force proportional to N. The individual friction on the beads sums up. In the case of Zimm dynamics the force is proportional to the radius of gyration, and the polymer drags the solvent along with it.

The following is a simple scaling argument on the functional form of the correlation function $I(q,t)$:

When measuring at high q compared to $1/R_g$ we look inside the chains, so $I(q,t)$ must not depend on N. From dimensional analysis we can construct only $I(q,t) = f\left(\frac{tD}{R_g^{\,2}}\left(R_g q\right)^x\right)$ (b is not included as we are interested in the universal range where microscopic properties do not yet make themselves felt). Rouse dynamics can be expected in polymer melts where Rg scales with N as $N^{0.5}$ and Zimm dynamics in dilute solutions where Rg scales as $N^{-\nu}$. From the above arguments the diffusion coefficient (D) should scale as N^{-1} and $N^{-\nu}$ respectively.

	Rouse	Zimm
D_{trans}	N^{-1}	$N^{-\nu}$
Rg	$N^{0.5}$	N^{ν}
this gives	x=4	x=3
	$f(tWr\sigma^4q^4)$	$f(tWr\sigma^3q^3)$

For incoherent scattering from the Gauss approximation we know:

$$I(q,t) = \left\langle e^{-iq\left(R_n(t)-R_m(0)\right)}\right\rangle = e^{-\frac{1}{6}q^2\left\langle\left(R_n(t)-R_m(0)\right)^2\right\rangle}$$

To make this expression agree with the functional form above $I(q,t)$ must have the following forms for Rouse and Zimm dynamics, respectively.

$$I(q,t) = \qquad e^{-K\left(W_r\sigma^4q^4t\right)^{\frac{1}{2}}} \qquad e^{-K'\left(W_r\sigma^3q^3t\right)^{\frac{2}{3}}}$$

where K and K' are numerical constants.

For coherent scattering the functional form is more complicated (we cannot use the Gauss approximation), but its argument will contain the same scaling dependent combination of t and q.

VII.3.2. Dilute - semidilute solutions

The characteristic parameter is ξ, the average distance between crossings of different chains (mesh size)

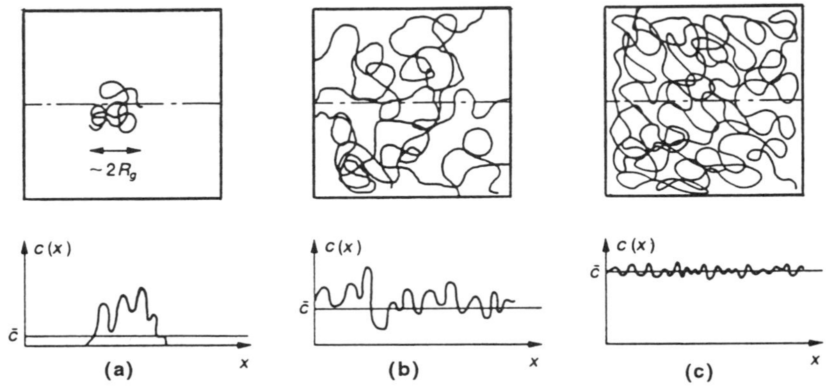

Fig. VII.9. — Three concentration regimes in good solvent: a)dilute, b) semidilute and c) concentrated. c(x) denotes the concentration profile along the dot-dashed lines. (Doi et al. 1986).

Cooperative dynamics:

Unlike compact molecules, polymers can interpenetrate. The cooperative dynamics become more complicated than the simple picture of de Gennes narrowing. At $q\xi \ll 1$ the concentration fluctuations will have a diffusive form e^{-Dq^2t} while at $q\xi \gg 1$ we look inside the chains and we should recover the Zimm-like motion as in the dilute case (Richter et al. 1978)

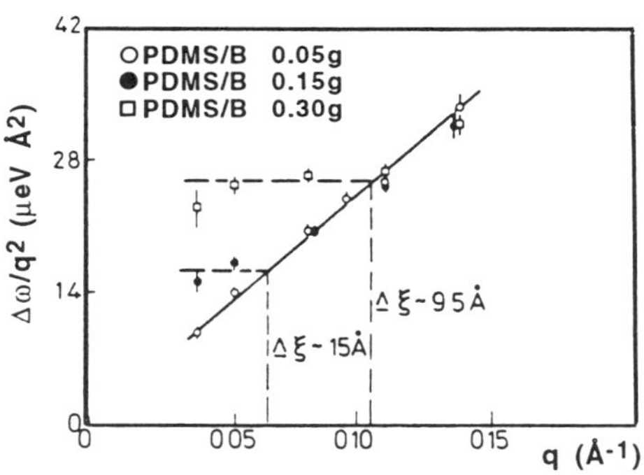

Fig. VII.10. — Linewidth divided by q^2 as a function of q: solid line $q\xi \gg 1$ dashed lines shows the cross over to the $q\xi \ll 1$ regime

This picture corresponds to de Gennes narrowing: coherent scattering shows cooperative behaviour, but locally nothing is different.

In the case of semidilute and concentrated solutions there is an influence of the cooperative motion on the dynamics itself (Richter et al. 1984). Screening of the hydrodynamic interaction gives:

on very local scale, still the original Zimm behaviour;

at larger distances screening: e^{-r/ξ_H}

How can this behaviour be measured :

d-solvent + d-polymer + little h-polymer to see only one polymer among the others. Results were in disagreement with the theory. It seems that the hydrodynamic interactions were incompletely screened

VII.3.2. Melts

(Doi et al. 1986; Richter et al. 1989; Kremer et al. 1990; Richter et al. 1990)

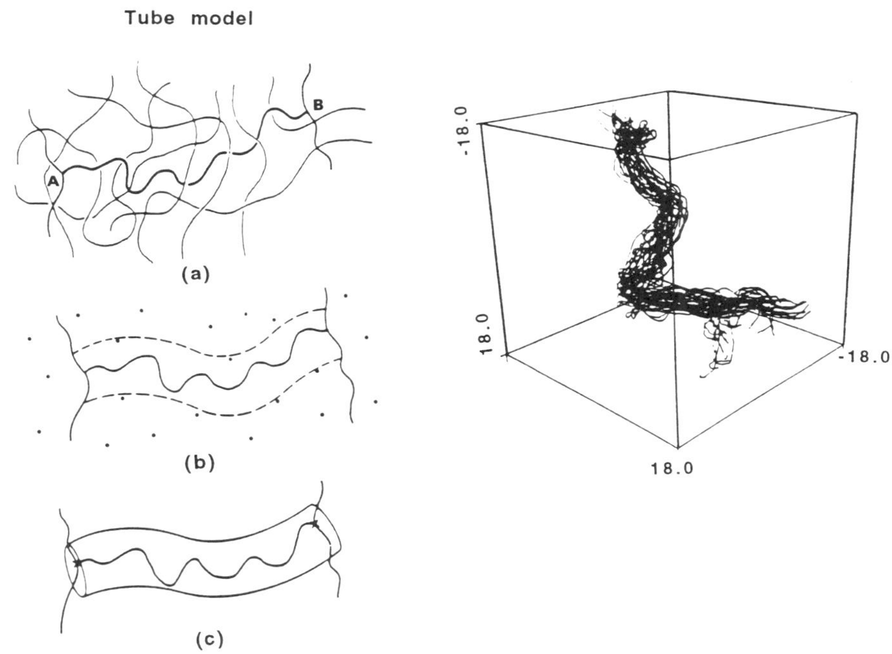

Fig. VII.11. — The reptation concept (Doi et al. 1986) and a computer simulation result (Kremer et al. 1990).

In a melt we deal with "Real gaussian polymers". The hydrodynamic interaction is completely screened => pure Rouse dynamics. The complication comes from the influence of the entanglements. Indeed the polymer chains can interpenetrate in space but cannot cross each other. This has no influence on the static form (static scattering) but dramatically modifies the dynamics. A model proposed by de Gennes treats this problem in the following way: Let us consider a chain among the others. The surrounding chains will form a sort of tube around our chain. The polymer relaxes freely inside the tube (short times) but at long times cannot relax further if the tube has a long life-time. On much longer times it will "reptate" along the tube (which gave the name of reptation theory or tube concept); at even longer times the tube itself might relax.

By measuring $I(q,t)$ we should see a fast relaxation, corresponding to the first relaxation process, which should become flat at a non zero level at longer times, as these "longer times" are still too short to see the reptation itself or the relaxation of the tube. We can look at it also in the scaling representation of $f(tWr\sigma^4q^4)$ presented in VII.3.1. We recall that this scaling function was obtained by the assumption that the only length scale in the system is R_g. If the measured $I(q,t)$ at different q values are plotted versus tq4 they should fall on the same $f(tWr\sigma^4q^4)$ curve. If not, it shows the presence of an additional length scale in the system.

VII.3.3. More complicated polymer systems

VII.3.3.1. Copolymers

Let us put a polystyrene polymer chain, which is made up of a deuterated and a protonated chain connected together, in a 50%-50% mixture of deuterated and protonated solvent (seen by the neutrons as a half black, half white chain in a gray solvent). Two eigenmodes will show up which will reflect a) the relative motion b) the collective motion of the two connected parts In zero average contrast the relative (optical) mode is observed (Borsali et al. 1989).

VII.3.3.2. Star polymers

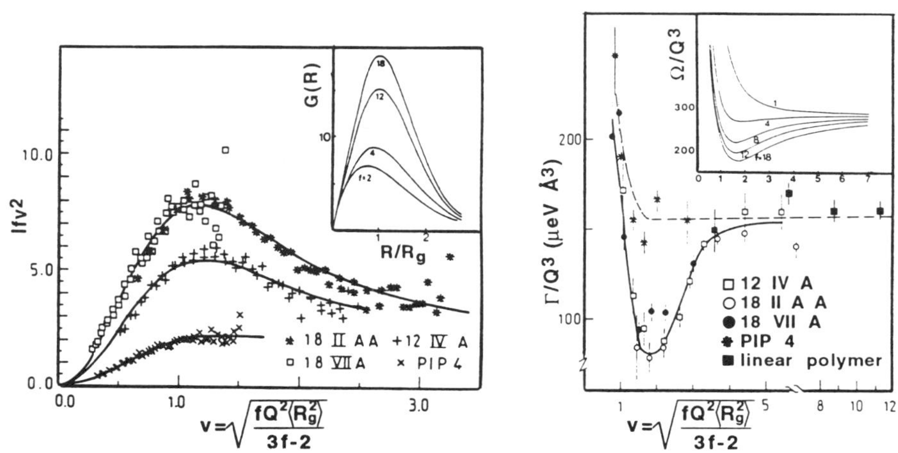

Fig. VII.12 — Left: Small angle scattering results from different molecular weight stars in scaled form (see ref. (Richter et al. 1987; Richter et al. 1989)). The lines are fits. Inset: related radial segment distribution functions.

Right: Reduced relaxation rates $\Gamma/q2$ as a function of the scaling variable. Inset: calculated first cumulants, same representation.

Star polymers consist of chains connected to small central molecules. This imposed configuration leads to a characteristic interference between the chains. Both the statics and dynamics can be well understood within the framework of the so called Random Phase Approximation (Richter et al. 1987; Richter et al. 1989).

Building up the star polymer from H and D copolymer (Richter et al. 1990) will enhance even more this effect and one can study it more in detail.

VII.3.3.2. Networks
(Oeser et al. 1988)

By marking specifically only the crosslink points their spatial movement can be studied. Again on a short time they will follow normal polymer dynamics but their movement is restricted in space, so the correlation function will not decay to zero but to a finite value. The q dependence of this elastic fraction gives information about the length scale of this restricted motion.

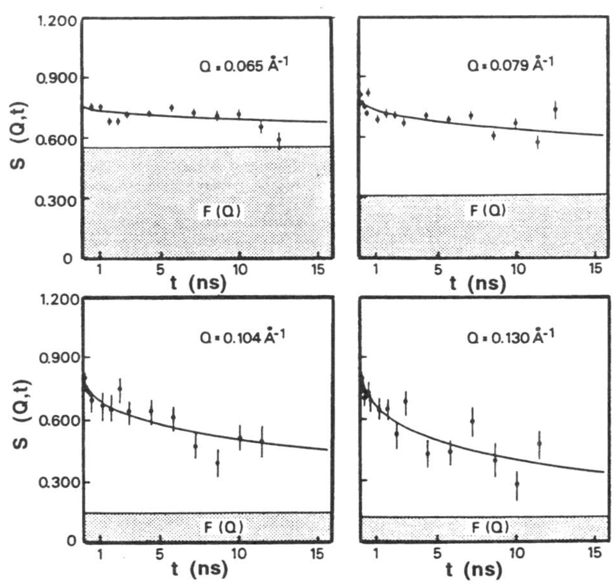

Fig. VII.13. — NSE spectra for labelled crosslinks at four different q values. The shaded area displays the non decaying (EISF) part in the spectra.

VII.4. Conclusion

We tried to make a short survey on the application of large instruments to a dynamic study of macromolecules and colloids. The examples were taken from recent works of many colleagues. The purpose was to give an idea of what can be done with these really powerful instruments rather than to go deeply into one subject or another. The hope of the author is to see some of the readers become active users with new exciting physical problems.

REFERENCES

ALPERT, Y., L. CSER et al. (1985). Biopolymers 24: 1769-1784.

BORSALI, R., H. BENOIT et al. (1989). Macromolecules : 4119-4121.

CAILLE, A. (1972), C.R. Acad. Sci. Ser. B274: 891.

DOI, M. AND S. F. EDWARDS (1986). The Theory of Polymer Dynamics. Oxford, Clarendon Press.

FARAGO, B. AND D. RICHTER (1988). ILL Annual Report : 87-89.

FARAGO, B., D. RICHTER et al. (1990). Phys. Rev. Lett. 65: 3348-3351.

HELFRICH, W. (1978). Z. Naturforsch. 33a: 305-315.

KREMER, K. AND G. S. GREST (1990). J.Chem. Phys. 92: 5057-5086.

MEZEI, F., Ed. (1980). Neutron Spin Echo. Lecture Notes in Physics. Berlin, Springer-Verlag.

OESER, R., B. EWEN et al. (1988). Phys. Rev. Lett. 60: 1041-1044.

PFEIFFER, W., T. HENKEL et al. (1989). Europhys. Lett. 8: 201-206.

PORTE, G., J. MARIGNAN et al. (1988). Europhys. Lett. 7: 713-717.

PUSEY, P. N. (1975). J. Phys. A. 8: 1433-1440.

RANCON, Y. AND J. CHARVOLIN (1987). J. Phys (France) 48: 1067-1073.

RICHTER, D., K. BINDER et al. (1984). J. Phys. Chem. 88: 6618-6633.

RICHTER, D., B. EWEN et al. (1989). Phys. Rev. Lett. 62: 2140-2143.

RICHTER, D., B. FARAGO et al. (1990). Macromolecules 23: 1845-1856.

RICHTER, D., B. FARAGO et al. (1990). Phys. Rev. Lett. 64: 1389-1392.

RICHTER, D., B. FARAGO et al. (1989). Macromolecules 22: 468-472.

RICHTER, D., J. B. HAYTER et al. (1978). Phys. Rev. Lett 41: 1484-1487.

RICHTER, D., B. STÜHN et al. (1987). Phys. Rev. Lett. 58: 2462-2465.

SAFINYA, C. R. (1989). Rigid and fluctuating surfaces: A series of synchrotron X-ray scattering studies of interacting stacked membranes. Phase Transitions in Soft Condensed Matter, Geilo,Norway, Plenum Pres, New York.

SAFINYA, C. R., D. ROUX et al. (1986). Phys. Rev. Lett. 57: 2718-2721.

SAFINYA, C. R., E. B. SIROTA et al. (1989). Phys. Rev. Lett. 62: 1134-1137.

BIOLOGY

BIOLOGY

INTRODUCTION

D. M. Blow

The major scientific achievement of the last 40 years has been the integration of molecular science and biology. Although some of the most spectacular details are common knowledge, presented in high-school textbooks and popular television programmes, it is much more difficult to grasp the huge breadth of knowledge which has been brought together in this process, and the vast range of techniques which have been developed for its achievement. About 40 years ago, at the beginning of the molecular biological revolution, Linus Pauling wrote a wonderful *General Chemistry* which provided a survey of the whole field as it then was, ripe for development. Today, no single undergraduate textbook tries to cover the whole range of achievement, and indeed no undergraduate course could cover it.

There is an integrated range of techniques and discoveries, based in Biochemistry, Microbiology and Genetics, which have come to be known as "Molecular Biology", which are basic knowledge for Life Sciences students, and are well documented in a number of first-class textbooks. There is also a wide range of Physical, Organic and Physical Organic Chemistry, basic for molecular insights into biology, which is recognised as part of the standard equipment of the chemist, and is standard material of Chemistry textbooks.

The part of the subject which is based in Physics, Colloid Science and Statistical Mechanics is less well provided with textbooks, and those which exist tend to be specialised monographs. Thus there are many good texts on Crystallography, Polymer Science, Nuclear Magnetic Resonance, Liquid Crystals ... but few which try to present the impact of all these topics on the broad understanding of biology in molecular terms.

In designing the HERCULES course, the organisers have brought together a wide range of specialised topics in this area, concentrating on those which draw research workers to use large facilities. The authors invited to contribute to this part of HERCULES have provided something approaching a textbook, extending in some cases to a fairly advanced level, and which can be of particular use in the biological exploitation of large facilities.

In my introduction to the biological section of the course, I address the fundamental character of biological systems, and try to provide some essential background within the areas often regarded as Chemistry and "Molecular Biology".

Living cells are self-contained functional units, able to organise themselves, to maintain an energy flow from whatever they can find in the environment, and capable of self-reproduction. To think in detail about the requirements of such a system, we may consider the design of an automatic factory, whose most important function is to build copies of itself.

A bronze statue may be copied by forming a "mould" of it in an appropriate type of cement, into which liquid metal may be poured. Such direct copying occurs also in biology, at the level of DNA replication, where a matching piece is inserted against every specific feature of the nucleic acid, to produce a negative copy or "mould" of it, which can in turn reproduce the original. But for the copying of more complicated structures, such a mechanism breaks down. Try to imagine a machine for copying a mechanical watch by sensing and reproducing its shape, and the types of material which are used. Then imagine doing the same for a quartz crystal watch or a CD player. Long before these complexities are encountered (and these are simple structures compared to you and me!), it is easier to ask for a specification, or blueprint, of the parts to be made, than to copy them by inspection.

In just the same way, the automatic factory's first requirement is for specifications of the total automatic factory which is to be constructed. It is essential to safeguard the originals from the aggressive environment of the construction area, where they will be spattered with oil and swarf or torn up by the machinery.

The automatic factory needs **information**

- stored in a library;
- arrangements to find and copy required information;
- only copies allowed out;
- the library itself needs to be copied for the new factory under construction;
- this copy library needs to be carefully checked, since errors will propagate;
- to eliminate errors use digital rather than analogue storage media.

The factory also needs **functional machines** including

- machine tools - with automatic control
  - including machine-tool-making machines;
- transporters - both passive flow channels
  - and active carriers;
- two types of copiers
  - to copy the whole library of the original factory
  - to copy selected information to control machine tools;
- cleaning machines - to clean up and re-use rubbish and spoilt work
  - including products rejected by quality control;
- energy supplies - using fuel brought in from outside, or rubbish
  - providing energy for tools and transporters.

The factory requires a **structural organisation**, placing boundaries

- between the organised factory and the mass of rubbish outside;
- between the library and the shop floor;
- between parts of the shop floor where incompatible activities take place.

It also requires **temporal organisation** with sequencers and time signals

- to organise tasks in their proper order;
- to make sure no task is begun until a preliminary stage is finished;
- to control the commissioning of the new factory when it is completed, and its separation from the old one.

All of these components are found in living cells. The information stores are *nucleic acids,* long copolymers made of four alternative monomers whose sequence carries the information. The machines are *proteins,* amino-acid polymers with the unique property that they fold up spontaneously into three-dimensional shapes. The structural organisation is mediated by *membranes* which are usually composed of phospholipid glycerides. We do not know the mechanism which specifies the topology and the specific properties of different membranes. The temporal organisation of DNA replication and cell division is mediated by mechanisms for chemical modification of proteins controlled by *cell-cycle proteins.*

INFORMATION	
original library information	DNA
copying original information	DNA polymerase
checking and correction	repair enzymes
copying for issue	DNA-dependent RNA polymerase
issued information	RNA
MACHINES	
machine tools	enzymes
machine-tool-making machines	ribosomes (enzyme-RNA complex)
cleaning machines	proteinases, nucleases
energy supplies	sugar, ATP
energy transduction	metabolic enzymes
ORGANISATION	
spatial boundaries	phospholipid membranes
passive transporters	specific pores in membranes specific binding proteins
active transporters	ATP-driven pumps in membranes motile fibrous proteins
time signals, sequence control	cell cycle proteins

The two fundamental types of nucleic acid, in which the sugar is either deoxyribose (DNA) or ribose (RNA) have significantly different conformational properties. DNA has a strong tendency to form stiff double helices with very regular external structure, ideal for information storage. DNA double helices are stored in chromosomes in supercoiled form, where the coils are stabilised by protein molecules (histones). RNA is more versatile structurally, though a degree of specific base-paring is important in maintaining a particular structure. RNA occurs in a variety of different biochemical contexts, for example

- messenger RNA (mRNA) – an edited DNA transcript;
- ribosomal RNA (rRNA) – part of the machinery of the ribosome;
- transfer RNA (tRNA) – a specific "label" attached to an amino acid to render it recognisable during protein synthesis.

The code which specifies the conversion of a RNA message to an amino-acid sequence has several important consequences. In freshly synthesised DNA there are only four possible bases, which pair specifically:

adenine <-> uracil

guanine <-> cytosine.

There are twenty different types of amino-acid in freshly synthesised protein.

There are thus four possible "letters" in the DNA code, which have to specify one of the 20 possible amino-acids in the target protein. This is achieved by a triplet code, in which each of the 64 possible 3-letter "words" in DNA specifies one of 20 amino-acids (or "stop"). For most amino acids, there are several different DNA codes. Crick's "fundamental dogma" concerns the flow of information from nucleic acid to protein, but not in the reverse direction. It may be written

$$\mathrm{DNA} \rightleftarrows \mathrm{RNA} \longrightarrow \mathrm{protein}$$

and is a natural consequence of this type of code, since although a DNA or RNA sequence uniquely specifies the amino-acid sequence of a protein (once the "reading frame" which separates the triplets is defined), the protein sequence cannot specify a unique nucleic acid sequence.

The machine tools for making machine tools are very special. They are huge complexes of protein and RNA known as *ribosomes*. They provide a system through which the mRNA is fed, and on which an amino-acid polymer is synthesised according to the information encoded in the mRNA. This process has been studied biochemically in great detail, and the steps by which it proceeds are known in detail. But the molecular structure of the ribosome, and the detailed mechanism of its action, are still unknown.

Proteins, as amino-acid polymers, have a fundamental property which has no counterpart in man-made machines. The polymer is specified and assembled as a linear polymer, but the specific sequence of the polymer defines a three-dimensional conformation which can be reached efficiently by a well-defined route, and which is more stable than the extended polymer. The *linear* polymer thus specifies a definite *three-dimensional shape*. Despite all we know about protein structure today, this folding process sets a baffling problem which is beyond the power or accuracy of present conformational modelling techniques. Although the amino-acid sequence specifies a three-dimensional shape, the only way we can recognise it is if the amino-acid sequence has a clear and close resemblance to that of a protein whose structure is already known.

The variety of amino-acid side chains, bulky or small, polar or non-polar, acidic or basic which occur in proteins, confers a versatility on their properties, and their structure is often complex. A nomenclature to classify the various levels of structural organisation is generally used :

Primary structure – the sequence of individual side-chains along the peptide chain which specifies the chemical formula of the protein.

Secondary structure – local regularity in the relationship of amino-acids, creating a structure which is in general helical.

Tertiary structure – The overall organisation of the complete polypeptide chain, including the interactions of different secondary structural elements.

Quaternary structure – The interaction of complete polypeptide chains to form a molecule composed of several chains.

Quaternary structures often show exact or approximate symmetries relating different peptide chains in a molecule. Tertiary structures also often include two or more similar structures within the same polypeptide chain, leading to the suggestion that more complicated protein structures have often evolved by gene duplication from simpler structures.

While the structures of hundreds of soluble proteins are now known, much less is known about proteins which reside in membranes. In a very few cases well-ordered three-dimensional crystals have been prepared, and the structures determined by conventional x-ray diffraction techniques. In a few other cases, ordered two-dimensional arrays of proteins can be isolated in membrane-bound form, and these specimens of monomolecular thickness are feasible for study by electron microscopy and electron diffraction.

The fundamental limitation on the study of very small biological specimens is the problem of radiation damage. The energy transferred to the specimen in observing it is so great, for both x-rays and electrons, that a high resolution view of a single molecule would vaporise it, breaking many covalent bonds in the process. Electron microscopy of single structures is never likely to yield an interpretation at atomic resolution, but where the structure is highly repetitive, processing of the images of molecular arrays improves the quantity of data available. This gives insight into the reasons why crystallography can produce detailed results, where microscopy fails : it is fundamentally a matter of spreading the radiation, and the consequent damage, over a larger number of individual units. Many radiation damage problems can be overcome by the use of neutrons, which give much more scattered energy than the energy absorbed in the sample, but which scatter much more weakly, requiring larger specimens and more expensive detectors.

At large facilities in the future, we should expect to see

- successful structural study of larger and more complex structures;
- the exploitation of absorption effects (MAD and XAS) to give more structural information;

- study of very short-lived structures, such as intermediates in muscle contraction or in enzyme activity;
- a better understanding of the dynamic properties and vibrational modes of biological specimens;
- deeper insight into the properties of specimens in solution, and of the effects of the solvent on stability and kinetic properties;
- the overcoming of radiation damage problems, by completing the data collection before the molecule has had time to fall apart;
- the study of much smaller specimens than has been possible hitherto, widening the range of structures which can be studied.

CHAPTER VIII

MACROMOLECULES IN DILUTE SOLUTIONS

H.B. STUHRMANN

VIII.1. Contrast

"Suppose a region [= volume] v of constant shape to be occupied by the molecule which for the present purposes is defined as the region into which the salt does not penetrate when a salt solution is substituted for water. It is assumed that the density of the liquid outside v is uniform. The effect on the value of F [= structure factor] of increasing density everywhere outside v is equivalent to that of decreasing density by an equal amount everywhere inside v, since a uniform distribution of density throughout the cell has no effect on F. It follows that the values F(water) - F(salt) are the F values for a region v of uniform density equal to the difference between the electron concentration of the salt solution and water."

This argument was introduced by Bragg and Perutz (1952) to explain reversible changes in the intensities of low order x-ray reflections from a haemoglobin crystal and to deduce from these the shape and orientation of the haemoglobin molecule occupying the volume v within the unit cell. It is also a description of what became known later as contrast variation in small-angle scattering (Stuhrmann and Kirste, 1964).

Following this idea, the scattering density $\rho(\mathbf{r})$ of the dissolved macromolecule which effectively contributes to the observed scattered intensity (short: the effective scattering density) is

$$\rho(\mathbf{r}) = \rho(\mathbf{r})_{solute} - \rho(\mathbf{r})_{solvent} \tag{VIII.1}$$

$$= \rho(\mathbf{r})_{solute} - \rho_{solvent}.$$

The quantity F(water) - F(salt) used by Bragg and Perutz is the Fourier transform of $\rho(\mathbf{r})_{solute} - \rho_{solvent}$. In (VIII.1) we have assumed, as did Bragg and Perutz, that the density of the solvent is uniform outside the volume of the dissolved particle. Or

$$\rho(\mathbf{r}) = \Sigma\ b_i\ \delta(\mathbf{r} - \mathbf{r}_i) - \rho_{solvent}\ v(\mathbf{r})_{particle} \tag{VIII.2}$$

where $v(\mathbf{r})_{particle}$ (the index is suppressed in the following) defines the shape of the volume of the dissolved particle with $v(\mathbf{r}) = 1$ inside the volume occupied by the particle and zero elsewhere (a more precise definition is given below). b_i is the scattering length of the i-th atom at the position $\mathbf{r}_i$. The index i runs over all atoms of the particle, e.g. in proteins, from 1 to some thousands.

ρ is the average value of $\rho(\mathbf{r})$. Both ρ and $\rho(\mathbf{r})$ represent contrasts. The latter is the excess scattering density in a volume element of the solute particle at $\mathbf{r}$ whereas the former refers to the whole particle. With this definition of the contrast of the solute with respect to the solvent we can rewrite the effective scattering density

$$\rho(\mathbf{r}) = u(\mathbf{r}) + \rho v(\mathbf{r}) \qquad \text{(VIII.3)}$$

$u(\mathbf{r})$ is the local excess scattering density of the dissolved particle with respect to the solvent at zero average contrast, i.e. if $\rho = 0$ then $\rho(\mathbf{r}) = u(\mathbf{r})$. $v(\mathbf{r})$ describes the shape of the volume which is inaccessible to solvent molecules (or dissociated parts of solvent molecules). The integral over $v(\mathbf{r})$ yields the total volume V of the particle. We call the contrast $\rho(\mathbf{r})$ of the solute molecules with respect to the solvent the **external contrast**. The external contrast can be varied, e.g., by varying the salt concentration within the solvent. One can speak of a **labelling** of the solvent.

We can assign a more general meaning to (VIII.3). Instead of assuming that $\rho v(\mathbf{r})$ coincides with the shape of the whole molecule, we restrict it to a portion of the dissolved particle. In an extreme case, $v(\mathbf{r})$ may be just one atom of a macromolecule. The atoms belonging to $v(\mathbf{r})$ are called labelled atoms. They are earmarked in some way and give rise to an **internal contrast**.

Most importantly, in the case of internal contrast, the scattering length of the labelled atoms can be changed, i.e. the contrast is varied, as well as it can be done with the labelled solvents in the external contrast variation.

VIII.2. Structure factors

The aim of this section is to introduce the reader to two mathematical descriptions of small-angle scattering both of which have their own merits. One of them is derived from the scattering of X-rays by gas molecules (Debye, 1915) and most frequently used nowadays. The other way has been inspired more recently by the theory of angular momentum in quantum mechanics as it was published in excellent textbooks in the fifties (e.g. Edmonds, 1957).

We start with the latter approach. The structure $\rho(\mathbf{r})$ is expanded into a series of multipoles (Harrison 1969, Stuhrmann 1970).

$$\rho(\mathbf{r}) = \sum_{l=0}^{\infty} \sum_{m=-l}^{+l} \rho_{lm}(r)\, Y_{lm}(\omega) \qquad \text{(VIII.4)}$$

The Y_{lm} are *spherical harmonics*. The members of this double sum are also called multipoles. Spherical harmonics are the angular part of the eigenfunctions of the operator L2 , the square of the angular momentum operator. A description of these functions is found in textbooks of quantum mechanics or spectroscopy of atomic excitations, for instance. For l=0, we have $Y_{oo} = 1/\sqrt{4\pi}$. It is associated with the s orbital. There are three Y_{1m} with m = -1, 0, 1. They are related to the orbitals $p_x \pm ip_y$ and p_z. The angular functions of the five d-orbitals are described by Y_{lm} with m = -2, -1, 0, 1, 2.

Given a density distribution $\rho(\mathbf{r})$, the coefficients can be calculated as projections of $\rho(\mathbf{r})$ on $Y_{lm}(\rho)$:

$$\rho_{lm}(r) = \int_{\rho} \rho(\mathbf{r})\, Y^{*}_{lm}\, d\omega \qquad \text{(VIII.5)}$$

This is so because the $Y_{lm}(\rho)$ are a set of normalised, orthogonal functions. The structure factor F(**Q**) is the Fourier transform of $\rho(\mathbf{r})$:

$$\begin{aligned} F(Q) &= \int_{v} \rho(\mathbf{r})\, e^{(i\mathbf{Q}\cdot\mathbf{r})}\, d^3\mathbf{r} \\ &= |F(\mathbf{Q})| \exp[i\Phi(\mathbf{Q})] \end{aligned} \qquad \text{(VIII.6)}$$

where $|F(\mathbf{Q})|$ is called the scattering amplitude and $\Phi(\mathbf{Q})$ the phase angle.

Q is the momentum transfer. It is the difference between the wave vector of the incident photon (or particle) and the wave vector of the scattered photon (or particle). In the case of elastic scattering there is no energy transfer. Only the momentum, i.e. the direction of propagation of the photon (or particle) is changed: $Q = |\mathbf{Q}| = (4\pi/\lambda)\sin(\theta)$. λ is the wavelength. 2θ is the full scattering angle. All possible end-points of the vectors **Q** are situated on a sphere of radius $2\pi/\lambda$, the so-called Ewald sphere.

In terms of multipoles, F(**Q**) is

$$F(Q) = \sum_{l=0}^{\infty} \sum_{m=-l}^{+l} F_{lm}(Q)\, Y_{lm}(\Omega) \qquad \text{(VIII.7)}$$

The $F_{lm}(Q)$ are obtained from the $\rho_{lm}(r)$ by Hankel transforms of order l:

$$F_{lm}(\mathbf{Q}) = \sqrt{(2/\pi)}\; i^{l} \int_{r=0}^{\infty} \rho_{lm}(r)\, j_l(Qr)\, r^2\, dr \qquad \text{(VIII.8)}$$

The j_l are spherical Bessel functions. For l=0 we have $j_0(z) = \sin(z)/z$.

For the inversion of (VIII.6) and (VIII.8) use (-i) instead of i.

The intensity S(**Q**) which is scattered by one single macromolecule, i.e., the coherent scattering cross section of this macromolecule per unit solid angle, is the absolute square of the structure factor:

$$\begin{aligned} S(\mathbf{Q}) &= F(\mathbf{Q})\, F^{*}(\mathbf{Q}) \\ &= |F(\mathbf{Q})| \exp[i\Phi(\mathbf{Q})]\; |F(\mathbf{Q})| \exp[-i\Phi(\mathbf{Q})] \\ &= |F(\mathbf{Q})|^2 \end{aligned} \qquad \text{(VIII.9)}$$

In crystal diffraction, **Q** is a point of the reciprocal lattice. Note the loss of the phase angle $\Phi(\mathbf{Q})$ when the structure factor is multiplied with its complex conjugate.

Macromolecules in dilute solution may have any orientation. The mean scattering intensity I(Q) from an ensemble of randomly oriented macromolecules is obtained as the average of S(**Q**) with respect to the unit vector Ω:

$$\begin{aligned} I(Q) &= 1/4\pi \int S(\mathbf{Q})\, d\Omega \\ &= \int_{V}\int_{V'} \rho(r)\, \rho(r')\; 1/4\pi \int \exp(i\mathbf{Q}\cdot(\mathbf{r}-\mathbf{r}'))\, d\Omega\; d\mathbf{r}\, d\mathbf{r}' \end{aligned}$$

The scalar product $\mathbf{Q}\bullet(\mathbf{r}-\mathbf{r}')$ can be written as $|\mathbf{Q}|\,|\mathbf{r}-\mathbf{r}'|\cos\alpha$, where α denotes the angle between the momentum transfer $\mathbf{Q}$ and the vector connecting the scatterer in $\mathbf{r}$ with that in $\mathbf{r}'$. The scattering of a pair of scatterers situated at $\mathbf{r}$ and at $\mathbf{r}'$ is determined by the projection of $\mathbf{r}-\mathbf{r}'$ onto $\mathbf{Q}$.

The average over the solid angle Ω is then

$$\begin{aligned}\int_\Omega \exp(i\mathbf{Q}\cdot(\mathbf{r}-\mathbf{r}'))\,d\Omega &= 2\pi\int_{\cos\alpha=-1}^{1}\exp(iQ|\mathbf{r}-\mathbf{r}'|\cos\alpha)\,d\cos\alpha\\ &= 2\pi\,[\exp(iQ|r-r'|)-\exp(-iQ|r-r'|)]\,/(iQ|r-r'|)\\ &= 4\pi\sin Q|r-r'|\,/\,Q|r-r'|\end{aligned}$$

Hence

$$I(Q)=\int_V\int_{V'}\rho(\mathbf{r})\,\rho(\mathbf{r}')\,[\,\sin(Q|\mathbf{r}-\mathbf{r}'|)\,/\,Q|\mathbf{r}-\mathbf{r}'|\,]\;d\mathbf{r}\,d\mathbf{r}' \qquad \text{(VIII.10)}$$

The corresponding multipole expansion is

$$I(Q) = 2\pi^2\sum_{l=0}^{\infty}\sum_{m=-l}^{+l}|F_{lm}(Q)|^2 \qquad \text{(VIII.11)}$$

In practice we are limited to the observation of scattering up to a maximum value of Q; thus the series will be truncated at some maximum l = L (the increase of L corresponds to a higher structural resolution in (VIII.4)). The addition of the terms belonging to l+1 has no influence on terms of lower index. (VIII.11) lends itself easily to the discussions of uniqueness problems in structure determination from small-angle scattering (Stuhrmann 1970, Feigin and Svergun 1987).

For a solid sphere ($\rho(\mathbf{r}) = 1$ if $r \le R_s$) we obtain

$$I(Q) = 2\pi^2\,|F_{oo}(Q)|^2 = 2\pi^2\left(\sqrt{(2/\pi)}\int_{r=0}^{R_s}\rho_{oo}(r)\,(\sin(Qr)/(Qr))\,r^2dr\right)^2$$

$$\text{As }\rho_{oo}(r) = \int 1\text{x}\,Y_{oo}(\Omega)\,d\Omega = 4\pi\times 1/\sqrt{(4\pi)} = \sqrt{(4\pi)}$$

$$\begin{aligned}I(Q) &= 2\pi^2\,(2/\pi)\,4\pi\left(1/Q\int_{r=0}^{R_s} r\sin(Qr)\,dr\right)^2\\ &= 16\,\pi^2\,(1/Q\,[\sin(QR_s)/Q^2 - R_s\cos(QR_s)/Q])^2\\ &= 16\,\pi^2\,(\,[\sin(QR_s) - QR_s\cos(QR_s)]/Q^3\,)^2 \qquad \text{(VIII.12)}\end{aligned}$$

For small Q, the power series of Q^2 is:

$$\begin{aligned}I(Q) &= 16\,\pi^2\,(\{[QR_s - (QR_s)^3/6 + (QR_s)^5/120+..]-[QR_s - (QR_s)^3/2 + (QR_s)^5/24]\}/Q^3\,)^2\\ &= 16\,\pi^2\,(\{\,(QR_s)^3/3 - (QR_s)^5/30\}/Q^3\,)^2\\ &= 16\pi^2/9\,(\,1 - (QR_s)^2/10 + ..)^2\\ &= (4\pi/3)^2(\,1 - (QR_s)^2/5 + ..)\quad \text{(cf (VIII.17))} \qquad \text{(VIII.13)}\end{aligned}$$

Remember:

Crystals:	$S(Q) =$	$\mid F(Q) \mid^2$
Solutions:	$I(Q) \approx$	$\sum_{l=0}^{\infty} \sum_{m=-l}^{+l} \mid F_{lm}(Q) \mid^2$

Crystal diffraction and scattering from solutions have in common absolute square terms. - The difference is that scattering from dilute solutions is the sum of many absolute squares and the partial structure factors add up incoherently. In crystal diffraction there is only one single square term.

VIII.3. Forward scattering

Macromolecules give rise to small-angle scattering. For sufficiently small Q ("forward scattering") the cross section σ_{coh} of coherent scattering of a macromolecule *'in vacuo'* is

$$d\,\sigma_{coh}(0) / d\Omega = \mid \Sigma\, b_i \mid^2 \qquad \text{(VIII.14)}$$

b_n is the scattering length of an atom of the macromolecule. For neutron scattering, it is a length of the order of 10^{-12} cm. Hence, the cross section σ is of the order of 10^{-24} cm^2 = 1 b (1 barn).

We remember the definition of the cross section:

cross section σ = number of scattered neutrons or photons/

number of incident neutron or photons = area (VIII.15)

The cross section of one macromolecule times the number of scatterers per unit volume has (as the absorption coefficient μ) the dimension of a length, e.g., cm^{-1}.

To calculate the cross section of forward scattering of a macromolecule in solution we have to subtract from the sum Σb_n of the scattering lengths of the 'free' macromolecule the sum of the scattering lengths of the solvent molecules which are displaced by the solute macromolecule. Thus, we are led to define the volume V of particle inaccessible to solvent molecules. The cross section of coherent forward scattering is

$$d\,\sigma_{coh}(0) / d\Omega = \mid \Sigma\, b_i \; - \; \rho_{solvent} V \mid^2 \qquad \text{(VIII.16)}$$

$\rho_{solvent}V$ *is the total coherent scattering length of solvent molecules displaced by one macromolecule* .

As an example we calculate the neutron forward scattering from a 1% solution of hen egg white lysozyme in heavy water, where all 194 labile hydrogen atoms are replaced, in ^2H_2O, by $^2H_2$, and in light water.

Table VIII.1. Composition, molecular weight and scattering length of lysozyme

atom	number	weight	scattering length $/(10^{-12}$ cm)	
lysozyme excluding labile hydrogens:				
fixed hydrogen (^{1}H)	755	x 1 = 755	755 x (-0.374) =	-282.4
carbon	613	x 12 = 7356	613 x 0.665 =	407.6
nitrogen	192	x 14 = 2688	192 x 0.937 =	179.9
oxygen	186	x 16 = 2976	186 x 0.58 =	107.9
sulphur	10	x 32 = 320	10 x 0.28 =	2.8
Σ :	1845;	M = 14059		Σb_n = 415.8
in ^2H_2O add				
labile deuterium (^{2}H)	194	x 2 = 388	194 x 0.667 =	129.4
Σ :	1950;	M = 14483		Σb_n = 545.2
in ^1H_2O add				
labile hydrogen (^{1}H)	194	x 1 = 194	194 x (-0.374) =	-72.6
Σ :	1950;	M = 14289		Σb_n = 343.2

The 194 labile hydrogens which are exchanged, in heavy water, by deuterium, belong to carboxylic groups -COOH, amides $-NH_2$, the =NH of the backbone of the protein chain. Some of the hydrogens may never exchange, unless the protein is unfolded, e.g. by extreme pH. The exchange rates vary by orders of magnitude for different chemical groups. In our table we assumed the time to be long enough to allow exchange of **all** dissociating hydrogens of the polar groups mentioned above.

Scattering density of the solvent H_2O: there are $N_A/M = 6.02 \times 10^{23} / 18 = 3.34 \times 10^{22}$ water molecules per cm^3. Each of these molecules has a total scattering length of $[2 \times (-0.374) + 0.58] \times 10^{-12}$ cm $= -0.168 \times 10^{-12}$ cm (see Table VIII. 1). The scattering density ρ is the product of $\rho = -0.168 \times 10^{-12}\text{ cm} \times 3.34 \times 10^{22}\text{ cm}^{-3} = -0.56 \times 10^{10}\text{ cm}^{-2}$. For heavy water, ^2H_2O, $\rho = 6.40 \times 10^{10}\text{ cm}^{-2}$ is obtained.

The volume of the lysozyme molecule, including labile hydrogen atoms, is the specific volume x relative molecular mass / Avogadro's number. For lysozyme we obtain $0.73\text{ cm}^3/\text{g} \times 14289\text{ g} / 6.02 \times 10^{23} = 1.73 \times 10^{-20}\text{ cm}^3$. Hence

$$\rho_{solvent}V = -0.56 \times 10^{10}\text{ cm}^{-2} \times 1.73 \times 10^{-20}\text{ cm}^3 = -0.97 \times 10^{-10}\text{ cm}.$$

The *excess scattering length of one lysozyme molecule in light water* is the difference

$$\Sigma b_n - \rho_{solvent} V = (3.43 - (-0.96)) \times 10^{-10}\text{ cm} = 4.39 \times 10^{-10}\text{ cm}.$$

Dividing the excess scattering length by the particle volume yields the *external contrast*

$$\rho = 4.39 \times 10^{-10} / 1.73 \times 10^{-20} = 2.54 \times 10^{10}\ cm^{-2}.$$

The number N of *lysozyme molecules per cm³* is given by

N = weight concentration of lysozyme (here chosen to be 10 mg cm^{-3})
x Avogadro's number / relative molecular mass

$$= 0.01\ g\ cm^{-3} \times 6.02 \times 10^{23} / 14289\ g = 4.22 \times 10^{17}\ cm^{-3}.$$

Forward scattering then is described by a scattering coefficient

$$\mu = N\ |\Sigma b_n - \rho_{solvent} V|^2 = 4.22 \times 10^{17} \times (4.39 \times 10^{-10})^2\ cm^{-1} = 0.081\ cm^{-1}$$

The **relative molecular mass** of the dissolved particles can be determined from zero-angle scattering (see e.g. Glatter & Kratky, 1982).

At this point it may be of interest to estimate the probability of photons (or particles) of undergoing small-angle scattering. μ makes a significant contribution only within the half-width of small-angle scattering. An estimate of the half-width is obtained from the Guinier law:

$$I(Q) = N\sigma(1 - R^2Q^2/3 + \ldots)$$

or, - as often done in practice - by the Guinier approximation

$$I(Q) = N\sigma \exp(-R^2Q^2/3) \qquad \text{(VIII.17)}$$

R is the radius of gyration.

$$R^2 = \frac{1}{2} \frac{\int\int \rho(\mathbf{r})\rho(\mathbf{r}')\ |\mathbf{r} - \mathbf{r}'|^2\ d\mathbf{r}\ d\mathbf{r}'}{\int\int \rho(\mathbf{r})\rho(\mathbf{r}')\ d\mathbf{r}\ d\mathbf{r}'} \qquad \text{(VIII.18)}$$

where $\rho(\mathbf{r})$ is the spatial contrast distribution of the intraparticle structure described by (VIII.2).

A rough estimate of R is obtained from the half-width Q_H:

$$R = 1.4 / Q_H \qquad \text{(VIII.19)}$$

For lysozyme Q_H is 0.10 $Å^{-1}$. Hence R = 14 Å.

We assume that small angle scattering with μ at Q = 0 covers a solid angle of $(2\theta)^2$. Using a narrow wavelength spectrum of thermal neutrons centred at a wavelength λ= 8 Å we obtain

$$T_H = \lambda Q_H / 2\pi = 1.4\lambda / 2\pi R \qquad \text{(VIII.20)}$$

$$= 1.4 \times 8 / 6.28 \times 15.6 = 0.125 = 7.2°$$

The probability W of a neutron being scattered at small Q can approximately be described by

$$W = \mu\ d\ (2\theta_H)^2 \qquad \text{(VIII.21)}$$

d is the thickness of the sample. With d = 0.2 cm and μ = 0.081 cm^{-1} (see previous page) we obtain

$$W = 0.081\ \text{cm}^{-1} \times 0.2\ \text{cm} \times (0.25)^2$$
$$= 1.01 \times 10^{-3} \approx 10^{-3}$$

There is only very weak absorption of the neutron beam by small-angle scattering in this case, far less than by incoherent scattering (see appendix A). This is frequently the case with biological macromolecules. You may choose conditions where W is of the order of 1. Then multiple scattering has to be considered. This will happen when neutrons of still longer wavelength are scattered by very large particles. An example is the study of red blood cells by neutrons with an average wavelength of 18 Å (Kleeberger et al., 1981). There is practically no multiple scattering when the same suspension is studied by 1.54 Å X-ray photons (Stasiecki and Stuhrmann, 1978). Note that W in VIII.20 increases with the square of the wavelength, e.g. by $(18/1.54)^2 = 136$ in the above case. For a more general discussion of total small-angle scattering see e.g. Guinier & Fournet (1955).

As mentioned above, it is the incoherent scattering process that in general will dominate. It gives rise to a scattering intensity distribution that hardly varies with the scattering angle. As incoherent scattering does not bear any structural information it is called 'background scattering' by the people who are interested in structure determination (like us) and it is subtracted from the total scattering intensity. The remaining difference in the intensities is the coherent scattering, which will be discussed in section VIII.4.

There is an important technical aspect of incoherent scattering: since it dominates it will limit the thickness d of the sample. In VIII.20, W increases with d, but certainly it will not do so to infinitely large d. Incoherent scattering attenuates small-angle neutron scattering, i.e. W in VIII.20 has to be multiplied by $\exp(-\mu_{total} d)$. μ_{total} is much more efficient than the μ of small-angle scattering as it extends over the whole solid angle 4π. The attenuation of the neutron beam in water (H_2O) is characterised by $\mu_{total} = N\sigma_{total} = 3.34 \times 10^{22}\ \text{cm}^{-3} \times 225 \times 10^{-24}\ \text{cm}^2 = 7.5\ \text{cm}^{-1}$. When $d = (1/7.5)$ cm the intensity will have dropped to $1/e = 0.368$ of its initial value. As the scattering intensity varies with the sample thickness d according to $d \times \exp(-\mu_{total} d)$, maximum intensity of small-angle scattering is obtained when $d = 1/\mu_{total}$. The optimum thickness of a water sample then amounts to (1/7.5) cm = 1.3 mm.

In heavy water the attenuation of a neutron beam is smaller by an order of magnitude. Samples of 1.3 cm thickness yield maximum scattering intensity of small-angle scattering. In practice thinner samples are preferred. A sample of 3 mm thickness yields 50% of the maximum scattering intensity at 23% of the optimum sample thickness (and cost of sample material).

VIII.4. Structure factors of contrast variation

Let V(**Q**) be the structure factor of the label v(**r**) and U(**Q**) that of the unlabelled particle u(**r**). Then the intensity of coherent scattering will consist of three terms

$$S(\mathbf{Q}) = |U(\mathbf{Q})|^2 + 2X\,\mathrm{Re}[U(\mathbf{Q})V^*(\mathbf{Q})] + 2Y\,\mathrm{Im}[U(\mathbf{Q})V^*(\mathbf{Q})] + Z\,|V(\mathbf{Q})|^2 \qquad \text{(VIII.22)}$$
$$\equiv S_U(\mathbf{Q}) + X\,S_{UV}(\mathbf{Q}) + Z\,S_V(\mathbf{Q})$$

The factors X,Y and Z activate the structure factor V(Q) of the label in various ways as summarised below:

Table VIII. 2. Methods of contrast variation

method	X =	Y=	Z =
chemical or isotopic replacement (fraction x)			
e.g. H/D exchange	$(b_H-b_D)x$	-	$(b_H-b_D)^2x^2$
solvent exchange	ρ	-	ρ^2
nuclear spin dependent scattering (P = nuclear, p= neutron polarisation)	p P IB	-	$P^2 I^2 B^2$
anomalous dispersion	b'(E)	b"(E)	$b'^2(E) + b''^2(E)$

Note that $Z = X^2$ holds for substitution methods only.

VIII.4.1. Solvent exchange

This method was developed first in X-ray scattering (Stuhrmann & Kirste 1964) and it came into full bloom in neutron scattering. In the latter case the scattering density of a hydrogenous solvent is easily changed through its H/D ratio (Fig. VIII.1). External contrast variation reveals the architecture of composite particles, like copolymers, nucleoproteins, lipoproteins and many others. As the range of scattering densities of H_2O/D_2O mixtures is so large (from - 0.56 to + 6.4 x 10^{10} cm^{-2}) the scattering densities of many solutes fall into this interval (Fig. VIII.1). The square root of zero-angle scattering is linear in the solvent composition. This law also holds at any scattering angle for particles of uniform density.

The scattering density of proteins and ribonucleic/deoxyribonucleic acids increases with the D_2O content of the solvent (Fig. VIII.1). This is due to the dissociating hydrogens of the solute which assume an H/D ratio very close to that of the solvent. The hydrogens of the aliphatic chains of lipids are not exchanged.

Inhomogeneities of the scattering density due to the intraparticle structure will give rise to contrast distributions that look quite different at various levels of the solvent density. Assume a particle made up of proteins and nucleic acids. The components of this nucleoprotein differ in their scattering density and hence also in their contrast. From Fig. VIII.1 it is seen that proteins have no contrast in a H_2O/D_2O mixture containing 40% D_2O. Only the contrast of nucleic acid would contribute to forward scattering. Similarly the contrast of nucleic acids can be matched by a different solvent mixture (Fig. VIII.1) and the proteins of the nucleoprotein would contribute to small-angle scattering.

The scattering curves of the partial structures of the nucleoprotein in general will be different and so will be their radii of gyration. Thus, the radius of gyration will vary with the solvent composition, i.e. with the average contrast ρ

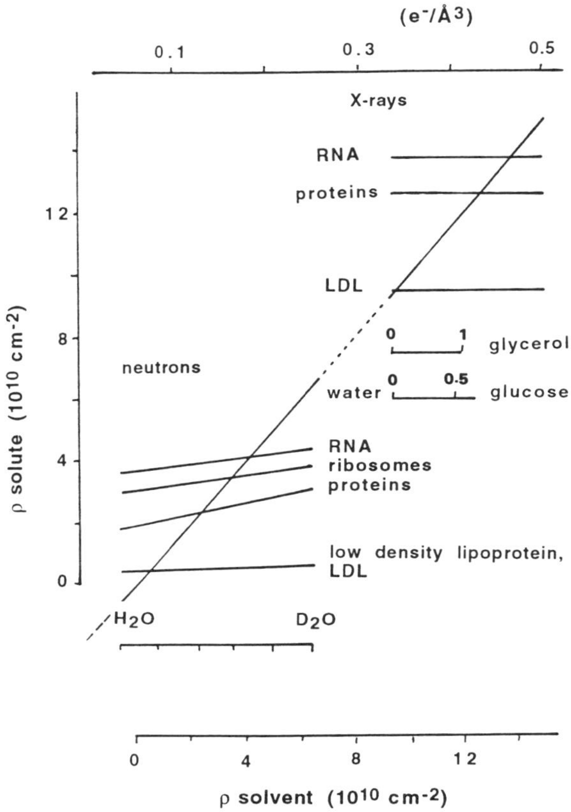

Figure VIII.1. — Scattering length density of various biomolecules.

of the dissolved particle. A general description of this observation, which is not limited to a particle consisting of two components, has been given by Ibel and Stuhrmann (1975).

$$R^2 = R_V^2 + \alpha/\rho - \beta/\rho^2 \qquad \text{(VIII.23)}$$

where

$$\alpha = 1/v \int u(\mathbf{r})\, r^2\, d\mathbf{r} \qquad \beta = 1/v^2 \int\int u(\mathbf{r})\, u(\mathbf{r'})\, \mathbf{r}\, \mathbf{r'}\, d\mathbf{r}\, d\mathbf{r'}$$

$$R_V^2 = 1/v \int v(\mathbf{r})\, r^2\, d\mathbf{r} \qquad v = \int v(\mathbf{r})\, d\mathbf{r} \qquad \int u(\mathbf{r})\, d\mathbf{r} = 0$$

$u(\mathbf{r})$ and $v(\mathbf{r})$ are defined by (VIII.3).

In a plot of R^2 versus $1/\rho$, α is obtained from the initial slope (Fig. VIII.2). Positive values of α are encountered with proteins and more strongly with

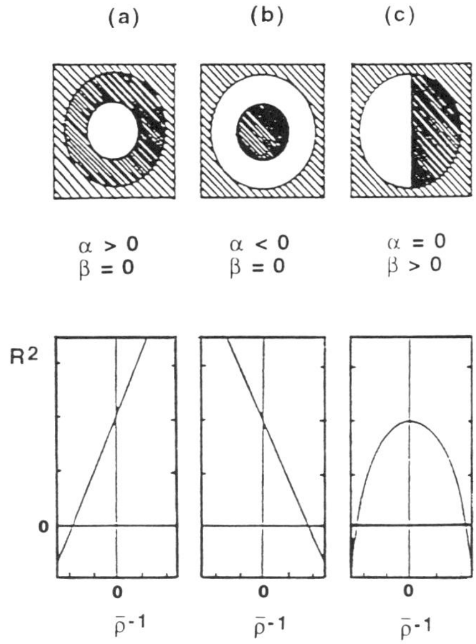

Fig. VIII.2. — The radius of gyration R of composite particles depends on the contrast ρ as described by VIII.22. The spherical structures a) and b) are simplified presentations of spherical averages of the scattering particle (monopole). c) is the schematic drawing of a dipolar structure, containing no contribution from a monopole. Real particles can be described by a mixture from many multipoles (3).

lipoproteins (Stuhrmann et al. 1975) and with chromatin (Hjelm et al. 1977). These particles are characterised by a core of relatively low scattering density. Particles with a high density core, like ferritin and ribosomes, give rise to negative α. β is a measure of the shift of the apparent centre of mass of the scattering particle. Particles giving rise to β are asymmetric (Heumann et al. 1988. Lederer et al. 1992). β is always positive. For a review see Perkins (1988) and Table VIII.3.

When the label is not congruent with the shape of the dissolved particle the dependence of R on the contrast is described more conveniently by the parallel-axes theorem (Serdyuk & Grenader, 1975):

$$R^2 = x_U R_U{}^2 + (1-x_U) R_V{}^2 + x_U (1-x_U) L^2 \qquad \text{(VIII.24)}$$

R_U and R_V are the radii of gyration of u(**r**) and v(**r**) relative to the centre of mass of each. L is the distance between the centres of mass of the two components. $x_U = n_U v_U / (n_U v_U + n_V v_V)$ is the relative scattering fraction from u(**r**). v_U, v_V and ρ_U, ρ_V are the volumes and the excess scattering densities of u(**r**) and v(**r**) respectively.

Table VIII.3. Comparison of R_V, α, β for macromolecules (from Perkins, 1988)

	10^{-3}M	non-protein	R_V	$10^5 \times \alpha$	$10^{12} \times \beta$
Myoglobin	18	0%	14.8	3.5	-
a_1-Acid glycoprotein	37	44% CHO	22.4	7±2	-
C3 of complement	187	4% CHO	51	30±10	-
Clq of complement	460	7% CHO	128	-100±100	-
Low density lipoprotein	1900	79% lipid	80	280	-
Cytochrome reductase					
Triton X100	670	10% detergent	57.3	48.6	80±240
Deuterated Cemulsol	660	10% detergent	56.7	45±7	280±40
Chromatin nucleosomes	235	53% DNA	40.4	45±2	-
30S ribosomal sub-unit	850	59% RNA	69-74	-100	-
with deuterated RNA	850	59% RNA	69	0	290±20
70S ribsomes	2300	64% RNA	81	-190	-
with deuterated 30S	2300	64% RNA	85	50	420
Southern bean mosaic virus	6700	21% RNA	123	-280	-
Tomato bushy stunt virus	8700	17% RNA	133	-240	-
Influenza B virus	176000	5% RNA 27% lipid	448	500	-
Frog virus	520000	20% DNA 10% lipid	720	-	-

VIII.4.2. Nuclear spin polarisation

The way how neutrons colliding with nuclei will interact depends on their initial spin states. Denoting the neutron spin operator by **s** and that of the nucleus j by $\mathbf{I}_j$ we may write the Fermi scattering length operator (scattering amplitude operator) for a free nucleus j (Abragam and Goldman, 1982):

$$\mathbf{A} = b_j + 2\, B_j\, \mathbf{I} \cdot \mathbf{s} \tag{VIII.25}$$

and define the two structure factors

$$U(Q) = \sum^{M} b_j \exp(i\mathbf{Q} \cdot \mathbf{r}_j)$$
$$V(Q) = \sum^{N} P_j I_j B_j \exp(iQ \sum r_j) \tag{VIII.26}$$

where U(**Q**) is the invariant and V(**Q**) is the polarisation (P) dependent structure factor; $\mathbf{r}_j$ is the position of the nucleus j. b and B are scattering lengths.

For the hydrogen isotopes and nitrogen we have (in units of 10^{-12} cm)

^{1}H: (I = 1/2)	b = -0.372	B = 2.92
^{2}H: (I = 1)	b = +0.66	B = 0.28
^{14}N: (I = 1)	b = 0.94	B = 0.14 (usually very weak)

Other nuclei occurring in biological matter do not exhibit any significant contribution to B in (VIII.26). The intensity of coherent scattering then depends on the polarisation of proton spins (P_H), deuteron spins (P_D) and of neutron spins p. Omitting **Q** and denoting Re[UV*] by [UV] we have

$$S = |U|^2 + pP_H [UV_H] + pP_D [UV_D] + |P_H V_H + P_D V_D|^2 \qquad \text{(VIII.27)}$$

In practice: $p = +1$ or -0.96; $-0.95 \geq P_H \geq +0.95$; $-0.4 \geq P_D \geq +0.4$.

For incoherent neutron scattering see appendix A.

The determination of the basic scattering functions starts from a series of measurements of S(**Q**) at different P and p. S_U(**Q**) is observed at P = 0. At a given nuclear polarisation P, two directions of the neutron spin polarisation, p and -p, can be used most easily in an alternating way (Knop et al. 1986) :

$$S(\uparrow\uparrow) = S_U + p\,P\,S_{UV} + P^2 S_V$$

$$S(\uparrow\downarrow) = S_U - p\,P\,S_{UV} + P^2 S_V \qquad \text{(VIII.28)}$$

From the difference $S(\uparrow\uparrow) - S(\uparrow\downarrow)$ one obtains

$$S_{UV} = [\,S(\uparrow\uparrow) - S(\uparrow\downarrow)]\,/\,(2\,p\,P) \qquad \text{(VIII.29)}$$

The sum of $S(\uparrow\uparrow)$ and $S(\uparrow\downarrow)$ yields

$$S(\uparrow\uparrow) + S(\uparrow\downarrow) = 2\,S_U + 2\,P^2 S_V \qquad \text{(VIII.30)}$$

S_V can also be obtained from experiments using unpolarised neutrons (Koghi et al. 1987, Glättli et al. 1989).

The square root of zero-angle scattering is a linear function of proton or deuteron spin polarisation. The contrast of the unpolarised target is changed by nuclear polarisation. The polarisation added contrast is called **nuclear spin contrast** (Fig. VIII.3).

Nuclear spins are polarised at low temperatures in a strong magnetic field. At T = 10 mK and B=10 T a proton spin polarisation of 77% is expected. In most hydrogenous substances the nuclear spin system will get reasonably close to the thermal equilibrium (e.g. P_H=0.77) after a very long time (in years).

At very low temperatures the nuclear spin system is thermally disconnected from the lattice. This is the necessary prerequisite of another method called dynamic nuclear polarisation (DNP) which is widely used for nuclear spin polarisation in dielectrics.

The polymer or protein is dissolved in a solvent which is known to be a glass former. Antifreezes, like glycerol/water (1:1), are convenient. A small amount of paramagnetic centres (less than 1 wt% of an organic radical) is added. This substance is essential for DNP. DNP works very efficiently with some organic compounds of Cr(V), like sodium bis(2-ethyl-2-hydroxybutyrato)-oxochromate(V) monohydrate, $Na[Cr(C_6H_{10}O_3)_2O].H_2O$, which is easily synthesised from dry sodium dichromate and 2-ethyl-2-hydroxybutyric acid in acetone.

This radical is quite stable at low pH. At pH = 7, however, it decays with a half life of 10 minutes. Solutions to which Cr(V) has been added therefore should be frozen quickly and they should be kept at liquid nitrogen temperature. The organic radical has no detectable influence on the small-angle scattering of the

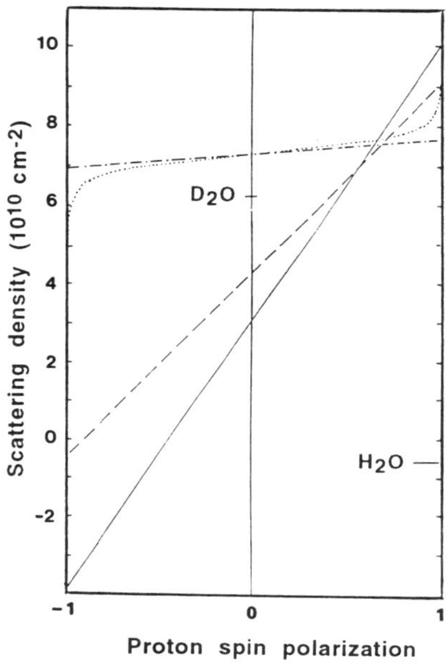

Fig. VIII.3. — The scattering length density of proteins and RNA and of the deuterated solvent (glycerol/water 1:1) as a function of nuclear spin polarization.

dissolved particles studied so far (bovine serum albumin (Knop et al. 1989), apoferritin (Knop et al. 1992), large sub-unit of E.coli ribosomes (Knop et al. 1991), myoglobin, tRNA) if the solutions are frozen very soon (in less than 1 minute) after the addition of Cr(V). For polymer solutions ortho terphenyl is a convenient solvent (van der Grinten et al. 1993).

At temperatures below 1 K and in a standard magnetic field of 2.5 T, the electronic spins will align almost completely. Irradiation of the sample by 4 mm microwaves will induce simultaneous spin flips of electrons and nuclei. Nuclear spins will be oriented parallel (anti parallel) to the external magnetic field at a microwave frequency of 69.1 (69.4) GHz. As the oriented nuclei relax much more slowly than electron spins, considerable nuclear polarisation can be achieved by a kind of pumping process (DNP). The spins of protons are oriented to a higher degree than those of the deuterons, as the nuclear magnetic moment of the latter is about six times smaller than that of a proton spin.

Moreover, nuclear spins can be selectively depolarised by saturation of NMR (106.4 MHz for proton spins, 16.4 MHz for deuteron spins). By this method targets (samples with polarised nuclei) can be prepared with only one kind of nuclear spin polarised, i.e. the polarised protons coexist with the unpolarised deuteron spin system and vice versa. At temperatures below 130 mK these targets

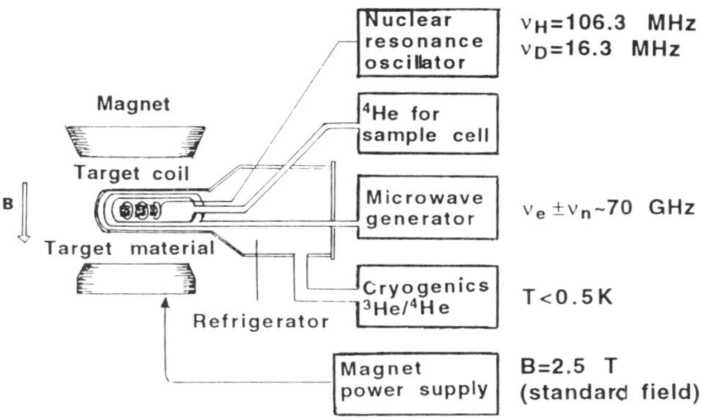

Fig. VIII.4. — A schematic presentation of the polarised target facility

are stable for days. Thus, there is time enough for detailed studies by polarised neutrons. Fig. VIII.4 shows a set up for DNP.

VIII.4.3. Anomalous dispersion

Contrast variation by anomalous dispersion is different from the other methods as there is an imaginary part of the scattering length, ib''. The anti-centro-symmetric term Im[UV*] is important in crystallography (for a review see Hendrickson, 1992). It is absent in scattering from powders and solutions.

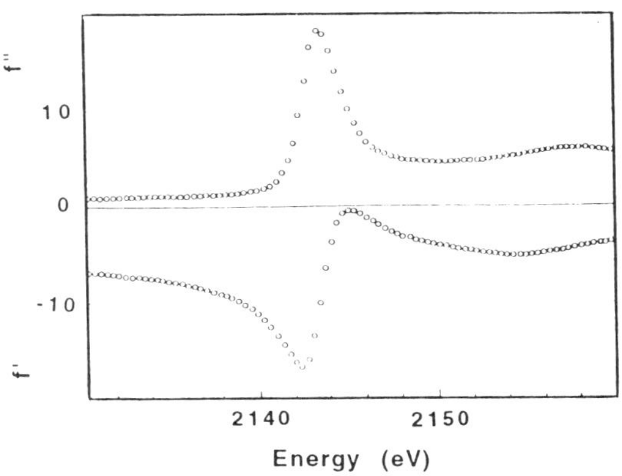

Fig. VIII.5. — The spectrum of X-ray absorption of the large sub-unit of E.coli ribosomes near the K-absorption edge of phosphorus. From the dispersion of X-ray small-angle scattering (Fig. VIII.2) it appears that b' of phosphorus varies by 15 electrons, i.e. by nearly the total number of its atomic electrons.

The imaginary part b" is related to the total cross section s by the optical theorem:

$$\sigma(E) = 2\lambda\, b''(E). \qquad \text{(VIII.31)}$$

The dispersion of b' is obtained from b"(E) by Kramers-Kronig relation

$$b'(E) = \frac{2}{\pi}\int_{e=0}^{\infty} e\, b''(e)\left[e^2-E^2\right]^{-1} de \qquad \text{(VIII.32)}$$

The dispersion of b' resembles the first derivative of b"(E) (Fig. VIII.5). Near an X-ray absorption edge b'(E) reaches a deep, narrow minimum of negative value. Its width dE/E is of the order of 0.001. There is hardly any overlap of anomalous dispersion between different chemical elements. Limitations in the use of anomalous dispersion techniques in X-ray diffraction are mainly due to absorption restricting the use to elements with $Z \geq 14$. An instrument capable of handling a very broad X-ray spectrum from $\lambda = 1$ to 7 Å is shown in Fig.VIII.6.

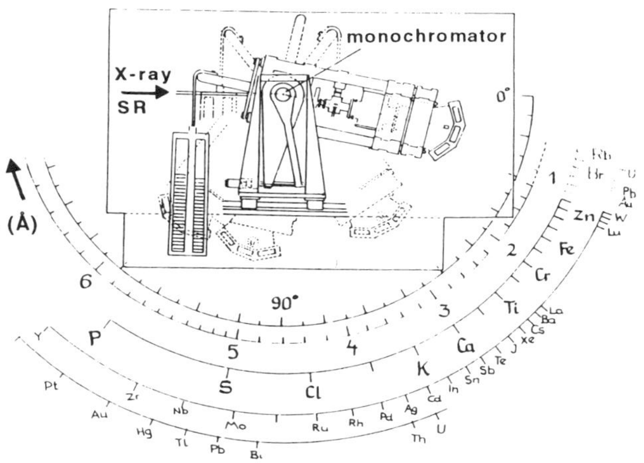

Fig. VIII.6. — The wavelength range of the diffractometer at beam line A1 of HASYLAB. Using the 111 plane of a germanium single crystal monochromator wavelengths up to 6 Å can be reached. Thus the K-absorption edges down to phosphorus (Z=15), the L_3-absorption edges down to Yttrium (Z=39) and the M_5 absorption edges down to iridium (Z=78) can be used. Replacing Ge-111 by InSb-111 the K-absorption edge of silicon can be included.

Nuclear resonant neutron scattering is described by the Breit-Wigner formula. There are only few nuclei that show a resonance in the thermal neutron energy range, e.g. ^{133}Cd, ^{149}Sm, ^{157}Gd,..

The change of b' and b" (in units of 10^{-12} cm) due to anomalous dispersion is

1 to 1.5 at most of the X-ray K absorption edges
2 to 3 at some X-ray K absorption edges (e.g. P in RNA)
at most of the X-ray L_3 absorption edges
up to 6 at L_3 edges of elements with $54 < Z < 79$ and M_5 edges,
and with resonant nuclear neutron scattering.
100 Mössbauer scattering from ^{57}Fe (Mössbauer, 1973)

The X-ray scattering length of one electron is 0.28×10^{-12} cm.

According to (VIII.31) anomalous dispersion of the atomic scattering length is a consequence of the photoelectric effect, leading to absorption and fluorescence. Usually the dispersion of absorption will be stronger than or comparable to that due to coherent scattering. This may have some influence on the choice of the wavelengths for the measurement of anomalous scattering. Corrections for absorption can be reduced by using thinner samples. Fluorescence may be eliminated by electronic energy discrimination of the detector system.

Anomalous X-ray small-angle scattering has been measured from iron in ferritin (Stuhrmann, 1980), terbium in parvalbumin (Miake-Lye, Doniach, Hodgson 1983) and more recently from sulphur in crystalline purple membrane (Munk, 1988), bacteriophages (Kühnholz, 1989) and phosphorus in ribosomes (Hütsch, 1993) (Fig. VIII.7).

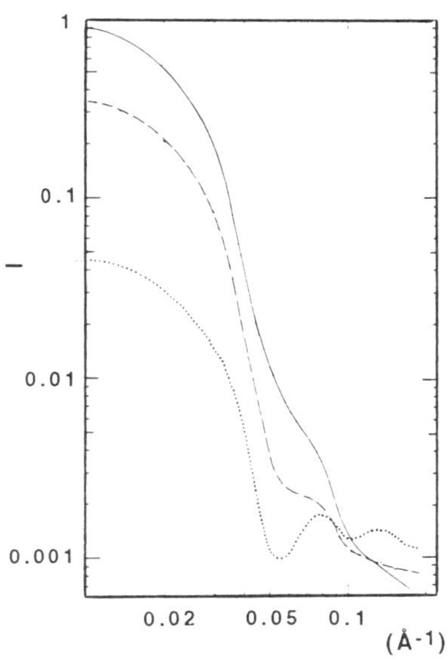

Fig. VIII.7.. — The basic scattering functions of contrast variation from anomalous dispersion X-ray small-angle scattering of the large sub-unit of E.coli ribosomes near the K-absorption edge of phosphorus (Hütsch 1993).

Since b" increases as λ^2, the penetration depth of X-rays decreases as $1/\lambda^3$ (see (22)). For water, it amounts to 1mm at λ= 1.5 Å and it decreases to 30 μm at wavelengths near the K-absorption edge of sulphur (λ_K = 5 Å). Liquid samples are held between two elastically stretched Mylar foils of 6 μm thickness. The sample cell has to be vacuum tight, as the camera is evacuated (Stuhrmann, Goerigk, Munk 1991).

VIII.4.4. The atomic density

The atomic density is still another contrast factor.

Hydrogen is by far the most abundant atom in biomolecules and polymers (Table VIII. 4). It is therefore a most convenient contrasting agent, provided its scattering cross section is reasonably large. Neutrons are strongly scattered by hydrogen and much less by its heavier isotope deuterium. Moreover, there is a big difference in the spin dependence of the isotopic scattering lengths. This provides powerful labels that can be tailored to any size.

Carbon, nitrogen and oxygen are not used as labels in scattering experiments. There is a possibility to use the nuclear spin contrast of nitrogen.

Table VIII. 4. **Abundance of atoms in biomolecules and use as labels:**

atom	source	concentration atoms/nm^3	mol/l	method
H	water	66	110	isotopic
	proteins	56	93	substitution
	nucleic acids	35	58	& spin contrast
C	proteins	36	60	(λ_K = 43.68 Å)
	nucleic acids	30	50	
N	proteins	11	18	spin contrast
	nucleic acids	12	21	(λ_K = 31.0 Å)
O	proteins	10	17	(λ_K = 23.32 Å)
	nucleic acids	22	36	
P	RNA	3	5	λ_K = 5.76 Å
S	proteins	0.5	0.8	λ_K = 5.02 Å
Ca/Tb	parvalbumin	0.2	0.3	λ_K = 3.07 Å (Ca)
				λ_{L3}= 1.48 Å (Tb)
Cs	gramicidin			λ_{L3}= 2.47 Å
Fe	haemoglobin	0.05	0.08	λ_K = 1.74 Å
	catalase	0.01	0.02	

VIII.5. The *in situ* structure

The basic scattering functions are very well suited for the analysis of contrast variation experiments.

The basic scattering functions reflect correlations in real space.

structure	u(**r**)		v(**r**)
	/ \		/ \
correlation:	u * u	u * v	v * v
intensity:	$\vert U \vert^2$	$2\,\mathrm{Re}[UV^*]$	$\vert V \vert^2$

u(**r**) is a structure which may be known to some extent, e.g. a low resolution model. v(**r**) is the structure of the label, which has been introduced, usually by some chemical means.

There are two cases:

1) u(**r**) is considered as unknown. Then only $|V|^2$ is of interest

2) u(**r**) is known at lower resolution, e.g. from electron microscopy. Then all basic scattering functions are of interest. This is similar to the difference Fourier method

In reality other considerations intervene: feasibility, costs, beam time,... Label contrast variation by isotopic substitution demands three different levels of contrast. The preparative effort still might be jeopardised by the lack of accurate data at intermediate contrasts. In practice, labels with high contrast are used, e.g. a completely deuterated label in a protonated matrix (= unlabelled part of the solute + solvent). This kind of work essentially follows case 1. Examples are:

- the label triangulation method proposed by Hoppe(1972) and Engelman & Moore (1972).
- the 'transparent ribosome' approach by Nierhaus (May et al. 1992)
- the 'triple isotopic substitution method' (Serdyuk & Pavlov,1988)

The authors of the last method emphasise the interest of $|V|^2$.

Case 2 does not require three levels of contrast in each case. An example is the difference Fourier method (see textbooks of protein crystallography). The essential basic scattering functions are $|U|^2$ and $2\mathrm{Re}[UV^*]$. The native particle and one derivative (i.e. 2 levels of label contrast) are sufficient. To analyse the data one needs to know the structure u(**r**) to some extent and one needs to know how handle the cross-term $2\mathrm{Re}[UV^*]$. The analysis of coherent scattering from solutions differs in some points from the difference Fourier method used in crystallography (see below).

Label contrast variation is most efficiently performed by means of nuclear spin contrast variation. This technique is presently available at CEN Saclay and at GKSS Geesthacht. The polarised target station at Geesthacht is run in collaboration with CERN, Geneva.

VIII.5.1. Labels on a transparent background

This is a most direct approach for the structural characterisation of the labelled part of a larger particle. The basic assumption is that u(**r**) does not give rise to small-angle scattering. Hence $|U|^2$ and $2Re[UV^*]$ will disappear. Ideally, the scattering intensity observed, $|V|^2$, is then due entirely to the label. This technique was proposed by Nierhaus and it is presently used for the *in situ* structure determination of ribosomal proteins of the large ribosomal sub-unit (May et al. 1992).

The success of this method relies on the skill of biochemists who deuterate the components of the unlabelled ribosome, rRNA and total protein (TP) to such an extent (different for rRNA and TP), that both will have the same scattering density as that of an H_2O/D_2O mixture containing 92% D_2O. Neutron scattering of the labelled and unlabelled particle is measured. There should be no scattering from the unlabelled ribosome.

VIII.5.2. Label triangulation

As can be deduced from its name, the sites of labels are deduced from inter-label distances. For the spatial arrangement of N (N > 3) labels 4N-10 distances between labels need to be determined, the construction being unique except for its handedness. This method has been used for the determination of the sites of the 21 ribosomal proteins of the small ribosomal sub-unit by P. Moore and his collaborators at Yale and Brookhaven (Capel et al. 1988).

The method follows case 1. The labels 1 and 2 are described by $v_1(\mathbf{r})$ and $v_2(\mathbf{r})$. They give rise to $|V_1 + V_2|^2$.

In order to determine the distance from the centre of mass of label 1 to that of label 2 the following samples have to be prepared:

(1) label 1 deuterated,	label 2 deuterated	yields	$I_1(Q)$
(2) label 1 native,	label 2 native	yields	$I_2(Q)$
(3) label 1 deuterated,	label 2 native	yields	$I_3(Q)$
(4) label 1 native,	label 2 deuterated	yields	$I_4(Q)$

The inter-label distance is determined from

$$I_D(Q) = [I_1(Q) + I_2(Q)] - [I_3(Q) + I_4(Q)] \qquad \text{(VIII.33)}$$

Explanation:

$$\begin{aligned}
(1):\ (U + V_1+V_2)^2 &= U^2 + 2\,U(V_1+V_2) + V_1^2 + 2\,V_1V_2 + V_2^2 \\
(2):\ U^2 &= U^2 \\
(3):\ -(U + V_1)^2 &= -U^2 - 2\,UV_1 \qquad - V_1^2 \\
(4):\ -(U + V_2)^2 &= -U^2 \qquad - 2\,UV_2 \qquad -V_2^2 \\
\hline
\Sigma:\quad I_D &= 2\,V_1V_2
\end{aligned}$$

The difference between the scattering intensities [(1)+(2)] and that of [(3)+(4)] not only yields the distinct correlation function giving rise to $V_1(Q)V_2(Q)$ but it also eliminates interparticle scattering (Hoppe, 1972). The distinct correlation function is the set of vectors connecting the scatterers of the label 1 with those of the label 2. In a first approximation, it is the average length of these vectors that can be measured.

VIII.5.3. Triple isotopic substitution

The method as proposed by Serdyuk and Pavlov (1988) follows the case 1. In a simplified way it works as follows.

(1) unlabelled $U^2 = U^2$ (VIII.34)

(2) label at maximum contrast $(U + V)^2 = U^2 + 2\,UV + V^2$

(3) label at half contrast $-2\,(U + V/2)^2 = -2\,U^2 - 2\,UV - V^2/2$

---

scattering from label: $= V^2/2$

As the scattering of (3) is subtracted from that of [(1)+(2)] the scattering intensity of the label V^2 is obtained cleanly, i.e. without interparticle scattering (Serdyuk & Pavlov 1988). Needless to say, UV can be deduced as well.

VIII.5.4. Nuclear spin contrast variation.

This method follows case 2. All basic scattering functions are available. Moreover these are determined from different target types of the same sample.

As mentioned in VIII.4.2, we use the word target in the sense of nuclear and high energy physics. A proton target, for instance, is a sample with its proton spins polarised.

Example: the deuterated large sub-unit of E.coli ribosomes in a deuterated solvent and its derivatives protonated in the ribosomal proteins L3, L4 and (L3+L4):

The following basic scattering functions have been determined from each of the 4 samples

$$|U|^2,\ 2\mathrm{Re}[UV_H^*],\ 2\mathrm{Re}[UV_D^*],\ 2\mathrm{Re}[U(V_H+V_D)^*],\ |V_H|^2,\ |V_H+V_D|^2$$

Those of the polarised proton spin target will be discussed here briefly. A complete discussion will be given elsewhere (Zhao et al., 1993).

In order to determine the site of a single labelled ribosomal protein, we need to know to some extent the structure of the host particle. If the latter is unknown, then it is only its centre of mass that can be taken as a reference, and it is the distance between the centre of mass of the label and that of its host particle that can be determined. As the structure of the large sub-unit of E.coli ribosomes is known at low resolution from electron microscopy and from neutron small angle scattering we shall take it into account by writing the structure factor in the following form:

$$U(Q) = c_{TP}\,A_{TP}(Q) + c_{RNA}\,A_{RNA}(Q) + c_{label}\,A_{label}(Q) \qquad \text{(VIII.35)}$$
$$V(Q) = k_{TP}\,A_{TP}(Q) + k_{RNA}\,A_{RNA}(Q) + k_{label}\,A_{Label}(Q)$$

where "TP" denotes the total protein (34 different proteins named L1 to L34) of the large sub-unit of E.coli ribosomes. The structure factors $A_{TP}(Q)$ and $A_{RNA}(Q)$ have been calculated from the model of the large ribosomal sub-unit of E.coli published by Wagenknecht, Grassucci & Frank (1989). The contrast c_{TP} of the deuterated ribosomal proteins and c_{RNA} of the ribosomal RNA of the unpolarised target, and the polarisation added contrasts k_{TP} and k_{RNA} could be derived from the comparison of the data from nuclear spin contrast variation with the scattering curves calculated from the model. We prefer to use the deuteration of TP and rRNA as variables since these determine the contrasts. The root mean square deviation of the calculated scattering functions from the measured data is minimised when the deuteration of TP is 0.97 and that of rRNA is 0.87 (Fig. VIII.8). The apparently different values of deuteration for TP and rRNA are derived from nuclear spin contrast variation. Its average of 0.92 is in agreement with data obtained by contrast variation in H_2O/D_2O mixtures (zero

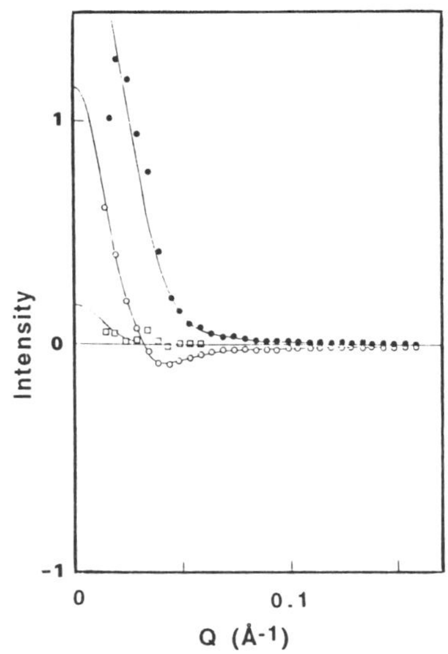

Fig. VIII.8. — The basic scattering functions of the deuterated large ribosomal sub-unit dissolved in a mixture of heavy water and deuterated glycerol (9/11).

□ $|U|^2$ (un-polarised target)

o $2\,\mathrm{Re}[UV^*]$

• $|V|^2$

The lines are calculated scattering curves.

The units of the intensity are n/s received by a pixel of the detector at 0.7 m distance from the sample. Pixel size: 9x9 mm^2. Collimation length: 2m. Cross section of the neutron guide: 30x40 mm^2. Irradiated sample volume: 15x15x3 mm^3. Wavelength: 8.5 Å (10% wavelength resolution). Only proton spins were polarised. The intensity of incoherent scattering of the unpolarised target is 1 n/s.

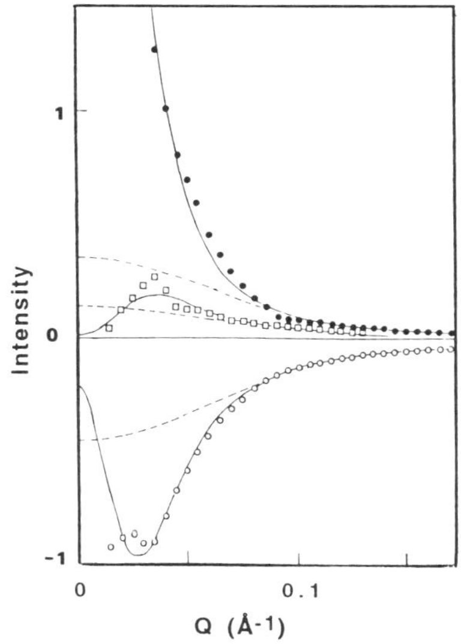

Fig. VIII.9. — The basic scattering functions of the L3 derivative. The notations are the same as in Fig. VIII.8. The broken lines are the calculated scattering curves of the L3 protein in the absence of any ribosomal structure. At larger Q they follow closely the basic scattering functions.

angle scattering of the deuterated ribosome vanishes at the extrapolated D_2O content of '124%', corresponding to an average deuteration of 0.95). As the proton density of TP equals to the proton density of the solvent, only the protons of the rRNA will contribute to V_H. This result from the polarised proton spin target enters into the analysis of polarised neutron scattering from the derivatives.

The basic scattering functions of neutron scattering from the proton spin polarised L3-derivative (deuterated ribosome with protonated L3 protein) are shown in Fig. VIII.9. They are quite different from those of the unlabelled large ribosomal sub-unit. The large contrast c_{label} and proton spin contrast k_{label} of the ribosome bound L3 protein induces significant qualitative changes in all three basic scattering functions of proton spin contrast variation. The only unknown in (VIII.35) is the structure factor $A_{label}(Q)$ reflecting the *in situ* structure of the L3 protein. A method for positioning and mutually orienting sub-particles has been proposed by Svergun (1989). We use it here in a simplified way. In a first step the L3 protein is assumed to have a spherical structure described by its centre of mass and its radius of gyration. The deviation of the calculated scattering functions from the measured data are minimised by variation of the site and the size of the label. In the course of the iteration the co-ordinates of the L3 protein prefer to lie reasonably close to the site suggested by Stöffler-Meilicke and Stöffler (1990) using cross-links (Fig.VIII.10).

The radius of gyration of L3 as determined from the spherical model

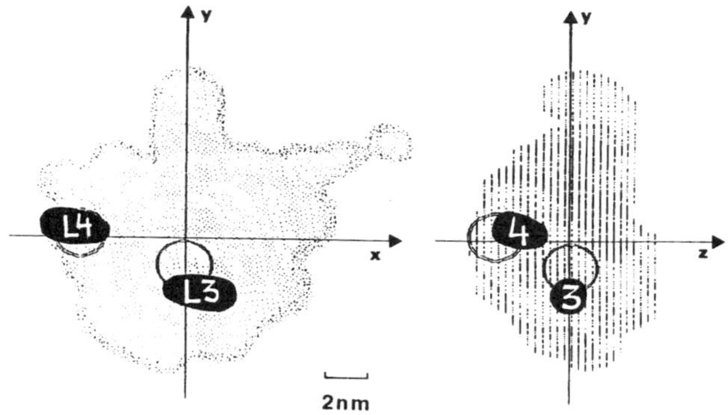

Fig. VIII.10. — The binding sites of ribosomal proteins L3 and L4 in the large sub-unit of E.coli ribosomes as determined by nuclear spin contrast variation (blackened areas) and by methods of electron microscopy (circled areas) from Stöffler-Meilicke and Stöffler (1990). The shape of the large sub-unit is that described by Wagenknecht, Grassucci and Frank (1988).

approximation is 21 Å. If L3 were a sphere then its radius of gyration would be 14.5 Å. Thus, the shape of L3 is far from being spherical. We therefore describe L3 by two spherical structures. There is a further decrease of the deviation of the calculated intensities from the measured data when the distance between these two sub-particles of L3 is 25 Å. The present state of analysis allows the determination of the direction of the elongated ellipsoid-like L3-particle with respect to the large ribosomal sub-unit (Fig. VIII.10). The radius of gyration of this L3 model is not larger than 19 Å. The site and shape of L3 are determined from neutron scattering of the L3 derivative only.

In a similar way the co-ordinates of the ribosomal protein L4 have been determined (Fig. VIII.10). As the radius of gyration of this particle is the same as that of L3, a similar deviation of its structure from a spherical one may be expected. L4 is aligned along a radius vector. The analysis of the data is in progress.

The analysis of the (L3+L4) derivative confirms the sites of L3 and L4 as determined from the single protein derivatives of the large ribosomal sub-unit. The distance between the centres of mass of the ribosomal proteins L3 and L4 is (125±10) Å.

Although the analysis of the neutron scattering data leads to a surprisingly detailed model of the proteins L3 and L4 that is not in conflict with the model derived from electron microscopy, the reasons for this are not yet fully understood. At low contrast of TP and rRNA other structural features, such as scattering density fluctuations inside the rRNA phase of the 50S sub-unit and at the particle surface due to the expected glycerol-free hydration layer (Lehmann & Zaccai 1982), might be comparable with those of the boundaries between rRNA and TP. The influence of these structural features of the deuterated ribosomal sub-unit on the results of the *in situ* structure determination of labelled components is still being investigated.

It is of interest to compare the intensity of polarised neutron scattering from the polarised proton spin target of the L3-derivative with that of the L3 protein in a 'transparent' ribosome as used by Nierhaus and his co-workers for the determination of the sites of the ribosomal proteins of the large sub-unit (May et al. 1992). For this purpose the scattering of the L3 protein of the polarised proton target in the absence of any ribosome structure is shown in Fig. VIII.9 (broken lines). The scattering $|U|^2$ from the unpolarised target compares with that of L3 in the transparent ribosome. Considerably larger scattering intensities of the L3 protein are observed in the other basic scattering functions of proton spin contrast variation. These are obtained with high accuracy as systematic errors due to background subtraction are much smaller than with the unpolarised target, the absolute statistical errors being equal for all three basic scattering functions. This is one of the reasons why the *in situ* structure of the label can be determined by nuclear spin contrast variation to a higher resolution than has been achieved by any other method.

The volume fraction of the ribosomal proteins L3 or L4 is about 0.02 of the large ribosomal sub-unit and it is typically 10^{-4} of the sample volume. Functionally important components, like tRNA or fragments of mRNA bound to the ribosome occupy a smaller fraction of the particle than one of the proteins mentioned above and they are less contrasted than protonated ribosomal proteins (Knop et al. 1991). The present data of polarised neutron scattering from polarised targets suggest that more challenging structural problems of molecular biology may be solved by nuclear spin contrast variation.

VIII.6. Appendices

VIII.6.1. Appendix A

INCOHERENT NEUTRON SCATTERING

The variation of coherent scattering with the nuclear spin polarisation as described in (VIII.27) is accompanied by a change of the incoherent cross section of neutron scattering:

$$(d\sigma/d\Omega)_{inc} = 4\pi \sum B_j^2 [I_j(I_j+1) - \mathbf{p} \bullet \mathbf{P}_j I_j - P_j^2 I_j^2] \qquad \text{(VIII.36)}$$

The spin quantum numbers I and the coefficients b and B of some elements are given in section VIII.4.2.

Protons show the most pronounced changes of scattering cross section with the polarisation P_H. According to (VIII.36) their incoherent cross section is in units of 10^{-24} cm^2

$$\sigma_{inc} = 106.5\,(0.75 - 0.5\,\mathbf{p}\bullet\mathbf{P}_H - 0.25\,P_H^2) \qquad \text{(VIII.37)}$$

For deuterons, incoherent scattering is nearly two orders of magnitude smaller:

$$\sigma_{inc} = 1.02\,(2 - \mathbf{p}\bullet\mathbf{P}_D - P_D^2) \qquad \text{(VIII.38)}$$

Incoherent cross section is zero for $\mathbf{p}\bullet\mathbf{P}_H = 1$ (or $\mathbf{p}\bullet\mathbf{P}_D = 1$). In this case, only coherent scattering is observed and no spin flip occurs. For opposite polarisation directions of neutron spins and nuclear spins, i.e. $\mathbf{p}\bullet\mathbf{P} = -1$, a maximum of incoherent scattering is observed, but coherent scattering may be strong as well.

The incoherent cross section largely determines the absorption of the neutrons by proton-rich samples. Polarised targets of hydrogenous material are therefore used as neutron spin filters or neutron polarisers (Khogi et al. 1987). The variation of the transmission of thermal neutrons by highly deuterated samples is due to comparable amounts of both incoherent and coherent scattering. Only a minor part of coherent scattering is observed as small-angle scattering.

VIII.6.2. Appendix B

POLARISATION OF DEUTERONS AND PROTONS AT THERMAL EQUILIBRIUM.

The polarisation P is a function of the ratio of the magnetic field B and the temperature T:

$$P_I(\mu B/kT) = (2I+1)/2I \coth[(2I+1)\mu B/2IkT] - 1/2I \coth(\mu B/2IkT) \quad \text{(VIII.39)}$$

At thermal equilibrium, the static polarisation of proton spins ($I=1/2$, $\mu = 2.793$ nuclear magnetons) differs from that of the deuteron spins ($I=1$, $\mu=0.857$ nuclear magnetons). How does the ratio $P_{1/2}(\mu_H B/kT)$ / $P_1(\mu_D B/kT)$ vary with the proton spin polarisation P_H? An empirical formula relating the proton spin polarisation P_H to that of the deuteron is

$$P_D^* = P_H / [4.88 - 1.8 P_H^2 - P_H^{10}] \quad \text{(VIII.40)}$$

The relative deviation from the exact value of P_D, $(P_D-P_D^*)/P_D$ will not exceed 2% at $P_H \leq 0.96$.

At $P_H \leq 0.75$ a shorter formula yields the same precision:

$$P_D^* = P_H / (4.88 - 1.8 P_H^2) \quad \text{(VIII.41)}$$

REFERENCES

ABRAGAM A., GOLDMAN M. (1982) Nuclear Magnetism: Order and Disorder, Clarendon Press, Oxford

BLUNDELL T.L., L.N. JOHNSON (1976) Protein Crystallography, pp 404, Academic Press, London

BRAGG W.L., M.F. PERUTZ (1952) Proc. Royal Soc. **A213** 425

BROWN J. (1980) in "Electron and Magnetization Densities in Molecules and Crystals", P. Becker Ed. Plenum Press

CAPEL M.S., D.M. ENGELMAN, B.R. FREEBORN, M. KJELDGAARD, J.A. LANGER, V. RAMAKRISHNAN, D.G. SCHINDLER, D.K. SCHNEIDER, B.P. SCHOENBORN, I.-Y. SILLERS, S. YABUKI, P.B. MOORE (1988) Makromol. Chem., Macromol. Symp., **15** 123-130

DEBYE P. (1915) Ann. Phys. **28**, 809-823

EDMONDS, A.R. (1957) Angular Momentum in Quantum Mechanics, Princeton University Press, Princeton

ENGELMAN D.M., P.B. MOORE (1972) Proc. Nat. Acad. Sci. USA **69** 1997

ENGELMAN D.M., G. ZACCAI (1980) Proc. Nat. Acad. Sci. USA **77** 5894

FEIGIN L.A., D.I. SVERGUN (1987) Structure Analysis by Small-Angle X-ray and Neutron Scattering , edited by G.W. Taylor, Plenum Press, New York

GLATTER O., O. KRATKY (1982) Small-Angle X-ray Scattering, Academic Press, London

GLÄTTLI H., C. FERMON, M. EISENKREMER (1989) J. Physique **50**, 2375-2388

GUINIER A. (1938) Compt. Rend. **206** 1374

HARRISON S.C. (1969) J. Mol. Biol. **42**, 457-483

HEUMANN H., H. LEDERER, G. BAER, R.P. MAY, J.K. KJEMS, H.L. CRESPI (1988) J. Mol. Biol. **201** 115-125

HJELM R.P., G. KNEALE, G.G. SUAU, J.P. BALDWIN, E.M. BRADBURY AND K. IBEL (1977) Cell **10** 139

HOPPE W. (1972) Israel J. Chem. **10** 321

HÜTSCH M. (1993) Thesis, University of Hamburg

JACROT B. (1976) Rep. Prog. Phys. **39** 911

SEIFF F., I. WALLAT, P. ERMANN, M.P. HEYN (1985) P.N.A.S. **82** 3227-3231

HENDRICKSON W.A. (1986) in "Structural Biological Applications of X-ray Absorption, Scattering and Diffraction" pp 285, Academic Press, London

HENDRICKSON W.A. (1992) Science **254**, 51-58

HEYN M.P. J. WESTERHAUSEN, I. WALLAT, F. SEYFF (1988) **85**

IBEL K. , H.B. STUHRMANN (1975) J. Mol. Biol. **93** 255

KAHN R., R. FOURME, R. BOSSARD, M. CHIADMI, J.-L. RISLER, S. BRUNIE, J.-P. WERY, O. DIDEBERG, J. JANIN (1986) in "Structural Biological Applications of X-Ray Absorption, Scattering and Diffraction", pp 297, Academic Press, London

KLEEBERGER L., H.U. WOLF & H.B. STUHRMANN (1981) Colloid & Polymer Sci. **259** 227-237

KNOP W., K.H. NIERHAUS, V. NOWOTNY, T.O. NIINIKOSKI, M. KRUMPOLC, M. RIEUBLAND, A. RIJLLART, O. SCHÄRPF, H.-J. SCHINK, H.B. STUHRMANN, R. WAGNER (1986) Helvetica Physica Acta **50** 741

KNOP,W., M. HIRAI, G. OLAH, W. MEERWINCK, H.-J. SCHINK, H.B. STUHRMANN, R. WAGNER, M. WENKOW-ESSOUNI, J. ZHAO, O. SCHÄRPF, R.R. CRICHTON, M. KRUMPOLC, K.H. NIERHAUS, T.O. NIINIKOSKI, A.RIJLLART (1991) Physica B **174**, 275-290

KOGHI M., M. ISHIDA, Y. ISHIKAWA, S. ISHIMOTO, Y. KANNO, A. MASAIKE, Y. MASUDA, K. MOROMOTO (1987) J. Phys. Soc. Japan **56** 2681-2688

KÜHNHOLZ O. (1991) J. Appl. Cryst. **24**, 811-814

LEDERER H., O. SCHATZ, MAY R., CRESPI H., DARLIX J.-L., LE GRICE S.F.J., HEUMANN H. (1992) EMBO J. **3**

LEHMANN M.S., G. ZACCAI (1982) Biophysics of water, Franks F. &Mathias S. Eds., pp 134-136, Wiley, New York

MAY R.P., V. NOVOTNY, P. NOVOTNY, H. VOSS, K.H. NIERHAUS (1992) EMBO Journal **11**

MIAKE-LYE R.C., S. DONIACH, K.O. HODGSON (1983) Biophysical J. **41** 287

MÖSSBAUER R.L. (1973) Naturwissenschaften **60** 493

MOORE P.B., J.A.LANGER, D.M. ENGELMAN (1978) J. Appl. Cryst. **11** 479-482

MUNK B. (1988) Thesis, University of Hamburg

PERKINS S. (1988) Biochem J. **254** 313-327

SCHÄRPF O. (1991) Physica B **174**, 514.527

SCHOENBORN B.P. (1969) Nature **226** 143

SCHOENBORN B.P., A.C. NUNES, R. NATHANS (1972) Ber. Bunsenges. Physik. Chem. **74** 1202

SERDYUK I.N., A.K. GRENADER (1975) FEBS Letters **59** 133-136

SERDYUK I.N., M. YU. PAVLOV (1988) Makromol. Chem., Macromol. Symp. **15** 167

STASIECKI P. & H.B. STUHRMANN (1978) J. Appl. Cryst. **11** 1-5

STÖFFLER-MEILICKE M. AND G. STÖFFLER (1990) THE RIBOSOME, STRUCTURE, FUNCTION AND EVOLUTION, W.HILL ed., ASM, Washington, DC, pp123-133

STUHRMANN H.B., R. KIRSTE (1964) Z. f. Physik. Chem. Frankfurt **46** 247

STUHRMANN H.B. (1970) Acta Cryst. A **26**, 367-375

STUHRMANN H.B., A. TARDIEU, L. MATEU, C. SARDET, V. AGGERBECK, A.M. SCANU (1975) Proc. Nat. Acad. Sci. USA **72** 2270

STUHRMANN H.B. (1980) J. Appl. Cryst. **A36** 996 - 1001

STUHRMANN H.B., G. GOERIGK & B. MUNK (1991) Handbook on Synchrotron Radiation, Vol. **4**, edited by S. Ebashi, M. Koch & E. Rubenstein, pp. 555-580, Elsevier

Science Publishers B.V.

SVERGUN D.I. (1989) J. Appl. Cryst. **24**, 485-492

VAN DER GRINTEN, M., GLÄTTLI, H., FERMON,C., PINOT,M. (1993) Phys.Rev.Lett. (in press)

WAGENKNECHT,T., R. GRASSUCCI AND J. FRANK (1988) J. Mol. Biol. **199** 137-147

ZHAO, J., N. BURKHARDT, W. MEERWINCK, K.H. NIERHAUS, M. RÜHL, M. SCHMITT, H.

VOSS, R. WILLUMEIT and H.B. STUHRMANN (1993) manuscript in preparation

CHAPTER IX

THERMODYNAMICS AND STRUCTURE - CONCENTRATED SOLUTIONS - STRUCTURED DISORDER IN VISION

A. TARDIEU

IX.1. Introduction

The functional properties of biological macromolecules, essential for life, are first determined by a special sequence of the basic units: amino acids for proteins, purines and pyrimidines (i.e. basic aromatic heterocycles) for nucleic acids. The linear information contained in a nucleic acid sequence already accounts for many aspects of the genetic program. The folding of a peptide chain creates the protein enzymatic properties. The distribution of hydrophilic and hydrophobic surfaces gives rise to specific intermacromolecular interactions. Such features are extensively studied. Recently, however, our studies on the eye lens proteins led us to analyse the **nonspecific interactions** between macromolecules, that occur in all solutions but may significantly change the **macroscopic properties** (solubility, transparency, viscosity etc..), when the solutions are concentrated like in cell cytoplasm (i.e. inside the cell, with the exception of the cell nucleus). The protein interactions may be described to a large extent by using the formalism that was developed to study either polymer or colloid interactions. These physical objects are, however, in many respects different from proteins. The study of protein interactions therefore presents some characteristic features that we shall attempt to discuss.

To relate the macroscopic properties of protein solutions to protein structural parameters, the coupling of approaches sensitive either to macroscopic or to structural properties is essential. On the structural side, small angle scattering, either of X-rays or of neutrons, is particularly convenient to study proteins in solution. On the thermodynamic side, as will be shown below, dilute or concentrated protein solutions have to be distinguished. Sedimentation equilibrium and densimetry experiments can be coupled to structural approaches in the dilute case, whereas light scattering and osmotic pressure can be used in the concentrated case. Those aspects will be developed in part IX.2. The results that can be obtained on concentrated solutions will be illustrated in part IX.3 with the lens proteins. As an example of structure / function relationship, part IX.4 will describe how lens protein properties are responsible for the optical properties of the eye lens.

IX.2. Theoretical relationships between thermodynamic and structural approaches

IX.2.1. Structural approaches : general presentation

On the structural side, small angle scattering, either of X-rays or of neutrons, is particularly convenient to study proteins in solution

As illustrated in figure IX.1, the intensity scattered by one particle as a function of s (where $s = 2 \sin(\theta/\lambda)$ and 2θ is the scattering angle), usually called the particle form factor, is the Fourier transform of the spherically averaged auto-correlation function of the contrast associated with the particle. When the solution is ideal, i.e. in the absence of interactions, the total scattering, I(0,s), is the sum of the scattering of the individual particles.

With concentrated solutions of monodisperse, spherical particles, departure from ideality may be accounted for simply by a multiplying factor or interference term, S(c,s), which is a function of the scattering angle :

$$I(c,s) = I(0,s)\ S(c,s) \qquad \text{(IX.1)}$$

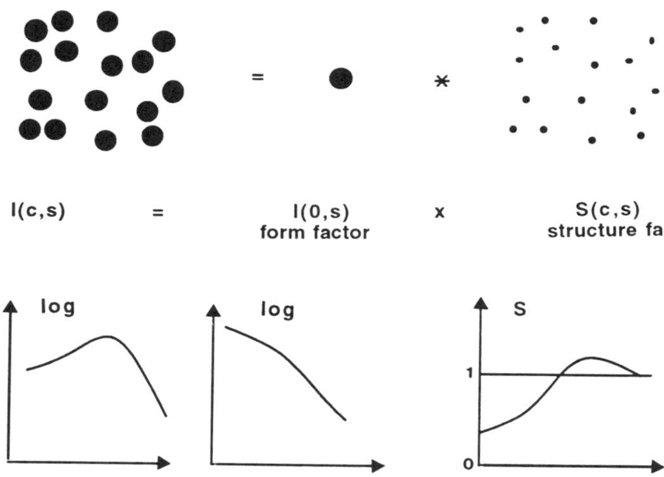

Figure IX.1. — Top : schematic representation of a concentrated solution; bottom corresponding scattered intensity plotted as a function of the scattering angle.

S(c,s) is usually called the solution structure factor. Since the solution can be described mathematically as the convolution product of a particle shape and a particle distribution (i.e. a set of delta functions placed at the particle centres), S(c,s) is the Fourier transform of the spherically averaged auto-correlation function g(r) of the particle distribution . The equation can still be considered valid, yet in a smaller s range, with quasi-spherical particles and/or polydisperse particles.

$$S(c,s) = 1 + (N/V \int 4\pi r\, 2\, (g(r) - 1)\, (\sin 2\pi rs/2\pi rs)\, dr \qquad \text{(IX.2)}$$

where (N/V) is the number of particles per unit volume.

From that simple presentation, it should be clear that two different types of studies can be performed with X-ray or neutron small angle scattering, either in dilute solutions to measure and analyse the form factor, or in more concentrated solutions to analyze the structure factor.

Protein molar masses are in the range 10 000 - 1 000 000 daltons (g/mol). Assuming globular shapes, the external diameters of the molecules are of the order of 25 - 200 Å. This range of scattering vectors is easily measured. From certain standpoints, and especially with concentrated protein solutions, X-rays are more suited than neutrons. A protein is usually seen with X-rays as a hydrated object, with a hydration of the order of 20-40%. In an H_2O buffer, convenient for biological studies and for easy change of the physico-chemical conditions, the contrast is therefore quite good, of the order of 0.07 e/Å3 (0.41 e/Å3 - 0.34 e/Å3) while the background remains small, as there is no incoherent scattering with X-rays. Since synchrotron radiation facilities are now available, it is possible to make measurements in a short time, typically a few minutes, with point collimation and monochromatic radiation. The only data treatment therefore is background subtraction. It is also possible to record a large angular domain at the same time and on an absolute scale. From now, the formula will be written for the X-ray case. It is the reason why the formula are written as a function of s instead of q. Please note that as a consequence, symmetrical expressions are obtained for direct and for reciprocal space.

IX.2.2. Relationships between thermodynamic and structure

IX..2.2.1. Dilute solutions

Dilute solutions are usually assumed to behave like ideal solutions. The structural part is extensively reviewed by H. Stuhrmann (chapter VIII this volume). Let me just recall that, with biological macromolecules, the concern was initially with the form factor, i.e. particle shape and density distribution (the reader has to remember that biological macromolecules usually are not spheres or rods of uniform electron density). To analyze this density distribution the emphasis was, in the seventies for both X-ray and neutron scattering, on the contrast variation approach. Sugars or salts may be used to vary the electron density contrast. Hydrogen-deuterium exchange is, however, particularly convenient to vary the neutron scattering length density. The success of the H/D contrast variation to study biological macromolecules will also be reviewed by H. Stuhrmann (chapter VIII).

The contrast variation approach, with either X-rays or neutrons, may also be used in connection with two thermodynamics approaches: sedimentation equilibrium and densimetry. X-rays are convenient in this case since avoiding H/D exchange is an advantage for such a coupling. The X-ray relevant parameter is the extrapolated intensity to the origin (X-ray forward scattering) that measures the **electron density increment** associated to the particle with respect to the solvent. Sedimentation equilibrium measures the **density increment** associated to the particle with respect to the solvent and densimetry the **density increment of the solution** with respect to the solvent.

To make a long story short, a set of three equations are obtained where the left part is the usual writing of what is measured in, respectively, densimetry, sedimentation equilibrium and X-ray experiments (Eisenberg 1976; Tardieu et al., 1981; no textbook available for such a coupling !) :

$$(\rho - \rho_o) / c_{v2} = \partial\rho / \partial c_{v2} \qquad \text{(IX.3)}$$

$$(2RT / \omega^2)(d \ln c_2 / dr^2) = M_2 \partial\rho / \partial c_{v2} \qquad \text{(IX.4)}$$

$$i_n(0,\rho_{e0}) \;/\; c_{e2} = m_2(\partial\rho_e/\partial c_{ev2})^2 \qquad \text{(IX.5)}$$

where M and m are the molecular mass (daltons) and number of electrons of the particles, c, c_e, c_v, c_{ve} are the particle concentrations in g/g, e/e, g/cm^3, e/cm^3, and ρ, ρ_e, ρ_0, ρ_{e0}, are the solution density and electron density, and the solvent density and electron density. Suffix 2 refers to the macromolecule.

These equations (either alone or combined) are the basis of the determination of the particle molecular mass, of the particle partial specific volume and, in solvents of variable density, of the preferential associations between the particle and the diffusible components of the solvent.

IX.2.2.2. Concentrated solutions

We are now concerned with departure from ideality i.e. the structure factor S(c,s) is different from 1 (Guinier et Fournet 1955). Such studies, well known in liquid state physics, were only recently applied to biological macromolecules.

On the thermodynamics side, the relevant appoaches are light scattering and osmotic pressure (Tanford 1961; Kerker 1969; Eisenberg 1976). With the usual macromolecular sizes, light scattering (usually introduced in text books as $r^2\ I_s/I_0$) is independent of the scattering angle. A static measurement is equivalent to recording one point of the X-ray experiment : the X-ray scattering at the origin, I(c,0):

$$r^2\ I_s/I_0 \sim I(c,0) \sim McS(c,0) \qquad \text{(IX.6)}$$

In some cases, however, light scattering can be conveniently replaced by osmotic pressure since :

$$r^2\ I_s/I_0 = (2\pi^2/\lambda^4)\, n_0^2(\partial n/\partial c)^2\ (cRT/Na)\, (\partial\Pi/\partial c)^{-1} \qquad \text{(IX.7)}$$

where R (8.31 10^7 erg °K^{-1}mol^{-1}) is the gas constant, T (°K) the absolute temperature, Na the Avogadro's number, n_0 the refractive index of the solvent and $(\partial n/\partial c)$ the refractive index increment due to the macromolecular solute.

Either in structural or in thermodynamic approaches, departure from ideality may also be described as a function of β, the compressibility of the solution. The compressibility is related to fluctuations in the number N of particles around the average value. It may be expressed in terms of osmotic pressure Π, and structure factor at the origin by (Guinier et Fournet, 1955; Eisenberg, 1976):

$$\beta = (-1/V)(\partial\Pi/\partial V)^{-1} = (1/c)\,(\partial\Pi/\partial c)^{-1} \qquad \text{(IX.8)}$$

Departure from ideality therefore corresponds to departure from 1 of an interference term introduced as a multiplying factor, $[cRT\beta/M] = S(c,0)$. From equations (IX.7-8) we find convenient relationships between thermodynamics and structural approaches :

$$I(c,0) \sim Mc\,S(c,0) = cRT\,(\partial\Pi/\partial c)^{-1} \qquad \text{(IX.9)}$$

or

$$\Pi = (RT/M)\int_0^c (1/S(c,0)\,dc \qquad \text{(IX.10)}$$

Note that we considered here only monodisperse particles.

In the thermodynamics approach, the osmotic pressure is often expanded as a series of Virial coefficients

$$\Pi = cRT\,((1/M) + A_2c + A_3c^2 + ...) \qquad \text{(IX.11)}$$

The interference term (osmotic pressure derivative) in equation (IX.7) is then expressed as a function of these coefficients. The interpretation of the Virial coefficients in terms of the molecular parameters of the protein is usually that the first Virial coefficient depends on the excluded volume, the second on the various interactions within the solution (Eisenberg, 1976). Such an approach, however, usually fails at high protein concentrations. We therefore find it better to start from the scattering properties and equation (IX.10). Another possibility is to relate the osmotic pressure to the partition fonction Z(c) of the system :

$$\Pi = (RT/M)\,Z(c) \qquad \text{(IX.12)}$$

In the case of hard spheres, i.e. impenetrable spheres without any other interaction (neither repulsive nor attractive), a closed analytical expression of the partition function has been developed as a function of one parameter ϕ, which is the excluded volume fraction associated with the particles (Carnahan & Starling 1969):

$$Z(c) = c(1 + \phi + \phi^2 - \phi^3)/(1 - \phi)^3 \qquad \text{(IX.13)}$$

In the case of hard spheres, equations (IX.9 and IX.13) provide a simple expression for S(c,0) :

$$S(c,0) = (1 - \phi)^4 \,/\, (1 + 4\phi + 4\phi^2 - 4\phi^3 + \phi^4) \qquad \text{(IX.14)}$$

On the experimental side, it happens that the procedures to measure light scattering and osmotic pressure on biological macromolecules are totally different. At low protein concentrations, light scattering is particularly convenient. At high protein concentrations, osmotic pressure may be measured in a rather simple and inexpensive way using dialysis bags and a polymer solution, like dextran, of known osmotic pressure as the stressing agent and the reference standard (Prouty et al. 1985; Parsegian et al. 1986). Dialysis may be done with a small amount of material and all biological laboratories are already set up to do such measurements. (Be cautious, however, that the total osmotic pressure of a buffered solution of macromolecules is in fact made of two parts: $\Pi + \Pi_s$ where Π_s is the osmotic pressure due to the ions and small particles that are free to circulate across a dialysis membrane. Π, the osmotic pressure of the macromolecular solute which is impermeant to the membrane is often referred to as the colloidal osmotic pressure).

IX.2.3. Concentrated solutions : model analysis

The analysis of concentration effects, departure from ideality etc.. has been extensively developed in liquid state physics and resulted in the building up of various models. Building such models usually involves a number of assumptions and approximations. Essentially, a pair particle potential is written, from which a particle distribution at equilibrium is inferred. The scattering and/or macroscopic properties and thermodynamic variable are then calculated from the distribution (Hayter & Penfold 1981; Hansen & Hayter 1982; Belloni 1991).

Proteins are usually studied in aqueous buffered solutions to mimic *in vivo* conditions. Besides possible specific interactions between the various components, non specific interactions are present between proteins, between proteins and solvent (i.e. water molecules, ions and possible small solute particles), and within the solvent.

One result of these interactions or interaction potentials is to determine the protein distribution in the solution, as illustrated in figure IX.2.

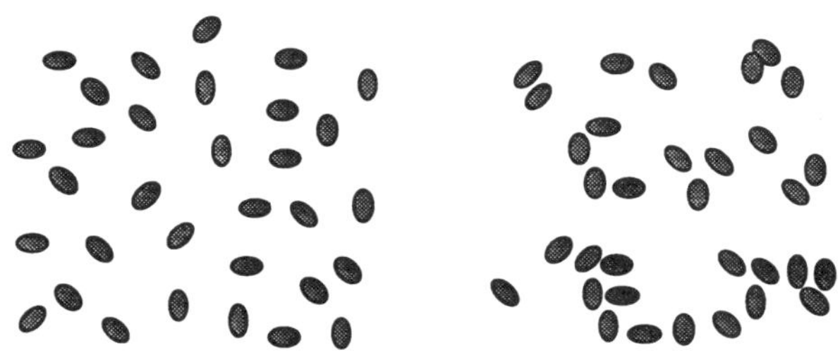

Fig IX.2. is a schematic representation of the macromolecular distribution. Left : with repulsive interactions; right : with attractive interactions.

Of course, one has to keep in mind that the 2D static illustration in figure IX.2 is a poor picture of the real situation where the proteins keep moving all the time. On one side, in the case of overall repulsive interactions, the proteins in a concentrated solution tend to remain as far as possible from each other, and they stay evenly distributed, although without any long range or periodic order, in the solution. Such a distribution is sometimes referred to as "liquid-like". At the other extreme, when attractive forces are predominant, the proteins may tend to form clusters or domains and the density fluctuations within the solution may therefore be important. The distribution is "gas-like". Eventually, attractive interactions may lead to phase separation. An exact calculation of the macroscopic properties of such solutions from the structural, molecular properties, is a difficult (not to say impossible) mathematical problem. Fortunately, to adequately describe some properties, a number of approximations were shown to be valid. It is often sufficient to assume that proteins are spheres, that ions are points and that the water solvent is a continuous medium described by its dielectric constant.

It is often sufficient to assume that the pair potential between two colloidal particles (e.g. proteins) is a DLVO (Derjaguin, Landau, Verwey, Overbeek) potential. According to the DLVO theory, two colloidal particles interact through hard sphere (i.e. the spheres are impenetrable and undeformable) and electrostatic repulsive potentials and through Van der Waals attraction. In practice, the various contributions to the pair potential are evaluated separately and added together. From such potentials, particle distribution functions are inferred, from which the solution properties are calculated.

For instance, with monodisperse solutions of spherical particles and with repulsive interactions, models are a function of 3 parameters : the excluded volume fraction ϕ, the charge Ze, and the particle diameter σ. The screened Coulomb potential between spherical particles in the MKS system is described by:

$$\upsilon(r) = \left(\pi\varepsilon_0\varepsilon\sigma^2\psi_0^2/r\right)\exp\left(\left(-k(r-\sigma)\right)\right) \quad \text{for} \quad r > \sigma \tag{IX.15}$$

where r is the interparticle centre to centre distance, ε is the dielectric constant of the solvent medium, ε_0 is the permittivity of the free space and k is the Debye-Hückel screening length :

$$k = 1/\lambda_d = (2e^2\ 10^3\ N_a I\ /\ \varepsilon_0\varepsilon k_B T)^{1/2} \tag{IX.16}$$

where e is the electronic charge, k_B is the Boltzmann constant and I is the ionic strength of the solvent (in mol). Typically, in an aqueous solution at 25°C one finds:

$$\lambda_d(\text{Å}) = 3\ /\ \sqrt{I} \tag{IX.17}$$

ψ_0 is the surface potential related in the Debye-Hückel approximation, to an "effective" particle charge Ze through:

$$\psi_0 = Ze\ /\ \pi\varepsilon_0\varepsilon_\sigma\ (2 + k\ \sigma) \tag{IX.18}$$

IX.3. Lens protein solutions

The eye lens is made of regularly stacked fiber cells, the cytoplasm of which is essentially a highly concentrated solution (20-40% by weight) of specific proteins, the cystallins (figure IX.3). The high concentration ensures the high refractive index essential for lens function.

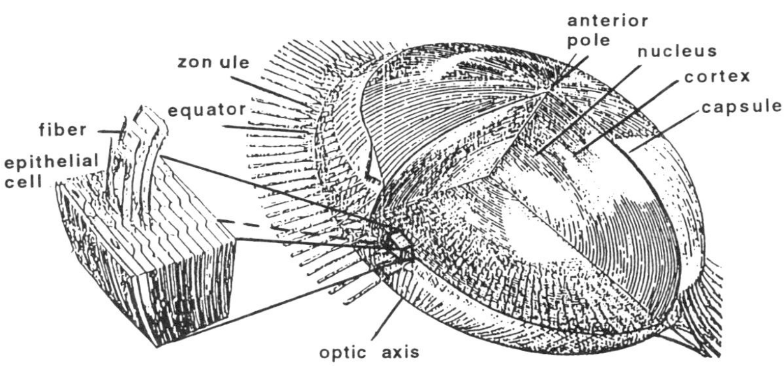

Fig IX.3. — Representation of an eye lens showing the stacking of the fiber cells and their quasi hexagonal packing viewed in a section perpendicular to the fiber long axis. From Koretz, 1988.

α-Crystallins are a monomodal distribution of polydisperse oligomers made of ‘5 ± 10 (at half width) subunits, β-crystallins are heterooligomers with a smaller ›er of subunits, and a wider subunit distribution, from 2 to 16, and γ-crystallins .. ..ionomers. All subunits are close to 20 000 daltons molecular mass each. Two types of subunits are identified in α-crystallins, but much more in β- and γ-crystallin classes. In ‘physiological conditions’, i.e. at pH 7 and temperatures between 0 and 37°C, each protein class was found to have a specific behaviour as a function of protein concentration up to about 350 mg/ml, as could be observed by both X-ray scattering and osmotic stress experiments. X-ray scattering, osmotic pressure and model analysis were coupled to analyse the non specific interactions in concentrated lens protein solutions. We operated in reciprocal space so that no Fourier transform is involved. The experimental data, therefore as rough as possible, are directly compared to a calculated model. Essentially, for one X-ray experiment one writes :

$$\mathrm{Iexp}\,(c,s) = \mathrm{Iexp}(0,s)\,.\,\mathrm{Sexp}(c,s) \qquad \text{(IX.19)}$$

Iexp(0,s), taken as the form factor, is measured at low particle concentration for which S(c,s) = 1. With the normation used, I(0,0) is proportional to the molecular mass M. Osmotic pressure was measured by dialysis versus dextran solutions.

α-Crystallins present what is usually referred to as a liquid like behaviour: the normalized X-ray scattering decreases near the origin as a function of increasing protein concentration, while a maximum becomes more and more pronounced at medium angle. Since the protein structure itself does not change with concentration, all the curves coincide at high angles (see figure IX.4). The effect is enhanced with decreasing the ionic strength of the solution, indicating the role of charges, and is barely sensitive to temperature. Accordingly, the osmotic pressure increases much more rapidly with protein concentration than in the ideal case, and still much more with decreasing ionic strength.

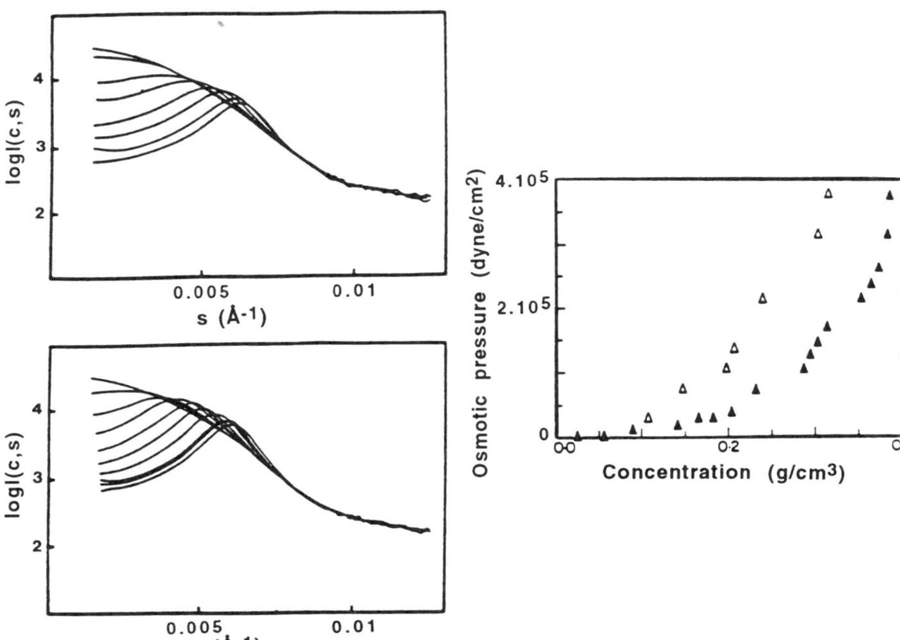

Fig IX.4 a. — Left : Xray scattering curves recorded with : (top) α-crystallins at 150 mM ionic strength, protein concentrations from top to bottom 10, 25, 84, 147, 205, 273, 308, 343 mg/ml; (bottom) α-crystallins at 17 mM ionic strength and 10, 36, 66, 107, 147, 184, 215, 272, 284, 308 mg/ml; Right : experimental osmotic pressure data recorded with α-crystallins (empty symbols 17 mM ionic strength, full symbols 150 mM), β-crystallins have the same X-ray qualitative behaviour at low angles, but because of polydispersity no maximum can be observed at higher angles (since the position of the maximum is roughly correlated to the particle diameter, the maximum disappears with large size polydispersity). The charge effect is also smaller. In the same way, the osmotic pressure variation is qualitatively similar to that of α-crystallins but quantitatively smaller. γ-Crystallins present quite different features, which are reminiscent from a gas-like behaviour. The normalized scattered intensity increases near the origin as a function of increasing concentration, while the osmotic pressure increases more slowly than in the ideal case (see figure IX.4). No clear effect of charge is detectable. Interestingly, however, the intensity near the origin increases as a function of decreasing temperature, in agreement with the well known observation that the γ-crystallins undergo a phase transition when decreasing the temperature. The phase transition is analogous to a liquid-gas phase transition : one observes the formation of domains with high protein concentration in a matrix with lower protein concentration.

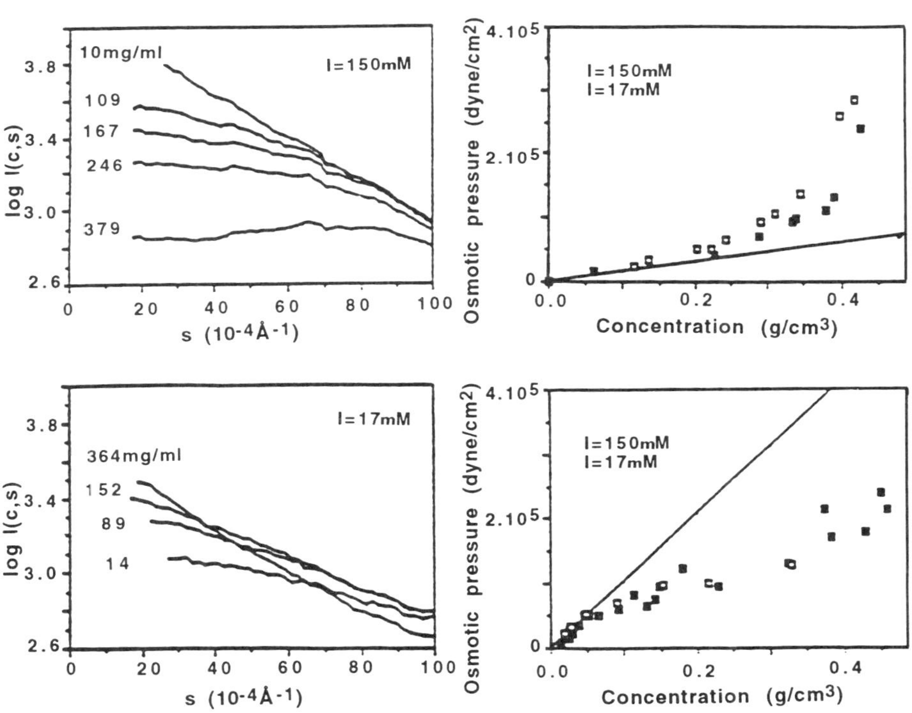

Fig IX.4 b. — Left: Xray scattering curves recorded with, top, β-crystallins; bottom, γ-crystallins. Right experimental osmotic pressure data recorded with , top, β-crystallins and, bottom, γ-crystallins.

These results therefore demonstrate that different proteins may behave rather differently (Vérétout et al. 1989, Tardieu et al. 1992). The model analysis was done on the structure factors. The result was that in different cases, it has been possible to obtain good fits of the angular dependence of the scattering, in a quite large range of scattering vectors, and with only a few parameters. Without entering into detail, the α-crystallin behavior is well described with a model of coulombic screened repulsive interactions. The relevant parameters are therefore the excluded volume fraction, the diameter and the charge which define an equivalent particle. When compared with α-crystallin quaternary structure, the excluded volume of the equivalent particle is found to be about twice the dry volume, and is close to the spherical envelope of the α-crystallin. We have here an original way to change the macroscopic solution properties by changing the protein structure which does not seem to exist in the colloid field.

The β-crystallin behaviour can also be accounted for by electrostatic repulsive screened coulombic interactions, with, however, smaller charge and excluded volume and by introducing polydispersity. It thus appears that, keeping the same type of protein interactions, quantitative variation of the interactions may be achieved by suitable combinations of charge, polydispersity, and compactness of the particles.

These results demonstrate that proteins may be designed with quite different thermodynamic properties in a given range of physico-chemical parameters. At present, it is still unclear whether an adequate change of the physico-chemical parameters would lead to reverse the protein behaviour or if each protein can only explore a limited part of a more general phase diagram. Work is also in progress to understand what features in protein sequence and folding are directly responsible for the observed thermodynamic behaviour (no textbook available).

IX.4. Structured disorder in vision

The image of a transparent and of an opaque lens is familiar to many of us. For most of us, however, opacity means cataract and therefore, a pathological event that we would prefer to avoid. From a physical standpoint, as explained by Trokel as early as 1962, lens transparency is limited by two phenomena, absorption and scattering of visible light, which are a function of lens structure and composition. Both of them reduce the light intensity transmitted through the lens. This transmitted intensity I_t may be simply written as a function of the incident intensity I_0 and of an extinction coefficient τ :

$$I_t = I_0 \exp(-\tau l) \qquad \text{(IX.20)}$$

When $\tau = 0$, one hundred per cent of the incident light is transmitted and the lens is transparent. When τ increases, because of absorption and/or scattering, I_t / I_0 decreases until the lens becomes opaque. As also explained by Trokel (1962), given the lens composition, in the normal state light absorption is negligible and lens transparency is determined by light scattering which arises from fluctuations of the refractive index. The extinction coefficient corresponds to the integral of the scattered intensity I_S:

$$\tau = \int I_S / I_0 \qquad \text{(IX.21)}$$

The normal lens is composed of two independent scattering systems, membranes and proteins. Membranes are quasi-periodically organized with periodicities much larger than light wavelengths, of the order of microns. The cytoplasm is a concentrated protein solution with scatterers much smaller than light wavelengths; γ-crystallin diameter is about 40 Å and α-crystallin diameter is 180 Å. In a transparent lens the membrane scattering was estimated to remain negligible and the light scattering is essentially due to the lens specific proteins, the crystallins. At the high protein concentration prevailing in the lens, the level of scattering is determined by the spatial organization, or local order, of the crystallin proteins. Trokel was, however, unlucky in his appreciation of the local order. He suggested 'a local order approaching a paracrystalline state'. Some years later, Benedek (1971) correctly pointed out that at the high protein concentrations present in vivo, a liquid like organization was sufficient to account theoretically for transparency.

We prefer here to present the problem in the following way. The lens cytoplasm is essentially a concentrated protein solution. The light scattered by one protein is proportional to its molecular weight. The light scattered by a protein solution is, in addition, proportional to the protein concentration. Furthermore, with concentrated solutions, the scattering is modulated because of the protein distribution.

Light scattering, I_s/I_0, of lens was first measured as a function of protein concentration by Mireille Delaye around 1981. The measurements clearly showed that light scattering does not vary linearly with protein concentration but first increases and then decreases as can be seen in figure IX.5. The corresponding percentage of light transmission was calculated (Tardieu & Delaye, 1988). The result is really illustrative. It can be inferred that the transparency first decreases and then increases with the protein concentration. In the physiological concentration range, i.e. around 350 mg/ml, for a lens 1 cm thick and green light, about 95% of the incident intensity is transmitted. In the same conditions, with the same particles but with a 'random' protein distribution, only 50% of the incident light would be transmitted and the lens would be turbid (figure IX.5b).

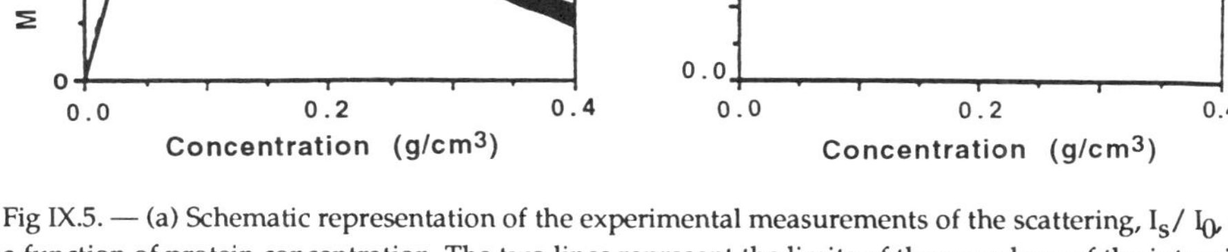

Fig IX.5. — (a) Schematic representation of the experimental measurements of the scattering, I_s/I_0, as a function of protein concentration. The two lines represent the limits of the error bars of the intensity at the origin, extrapolated from the data presented in figure IX.5. (b) Percentage of transmitted light, I_t/I_0, as a function of protein concentration, for the scattering experiments shown in figure IX.6a. The calculation was done using equations IX.18-19 for a 1cm thick lens and a wavelength of 500 nm.

Light scattering, however, does not allow us to analyze the protein distribution responsible for such results. X-ray measurements (figure IX.6), that are able to probe the local organization of the scatterers, demonstrated that the variation of the scattering as a function of c was indeed due to a short range, liquid-like order (Delaye and Tardieu, 1983).

To find out the molecular basis of eye lens transparency, it is necessary to go some steps further, and the next step may be expressed by : the protein distribution is itself determined by the protein-protein interactions that were studied according to the approaches described above.

The structural parameters are the particle 'specific excluded volume', φ/c, i.e. the ratio of the excluded volume fraction associated to the particles divided by the particle concentration, the particle diameter, ϕ, and the particle charge Z. Furthermore, the 3 parameters φ/c, ϕ and Z are almost independently determined from respectively amplitude, position and amplitude difference at both ionic strengths of the maximum of the scattering curves.The osmotic pressure and the structure factor of a solution may now be calculated from these 3 parameters and therefore, so can the extinction coefficient and the transparency (figure IX.7).

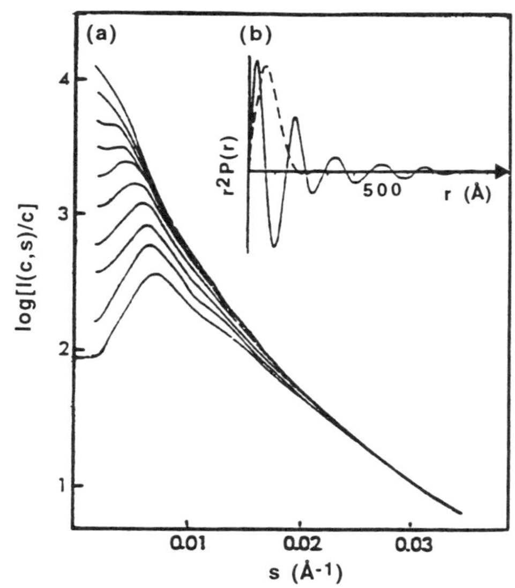

Fig IX.6. — (a) X-ray scattering curves recorded with cytoplasmic preparations as a function of protein concentration, from top to bottom : 3, 70, 150, 208, 247, 303, 355, 404, 464, 510 mg/ml. Note that, since the cytoplasmic preparations are a mixture of α-, β- and γ-crystallins, the curves do not cross each other. (b) Patterson function at low and high concentrations.

The analysis therefore shows that the parameter relevant for transparency in vivo, in addition to the concentration and the molecular weight, is essentially the φ/c ratio. We can now calculate from the scattering shown in figure IX.7 that, given their molecular weight, compact α-crystallins would be turbid while α-crystallins as a pile of subunits are perfectly compatible with transparency (figure IX.8).

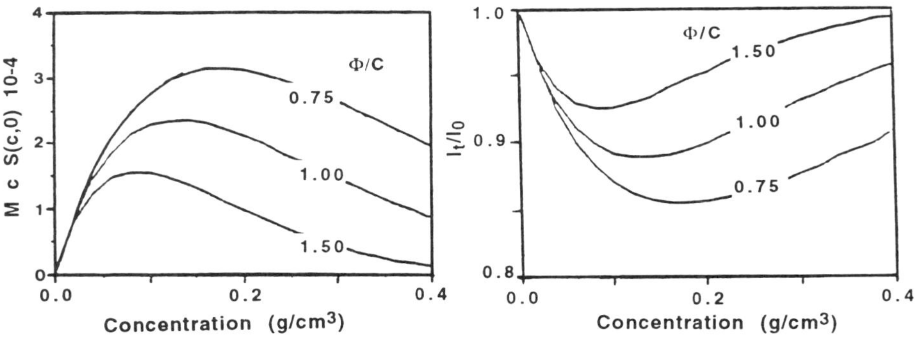

Fig IX.7. — (a) Calculation of the scattering, I_s/I_0, as a function of the protein concentration and as a function of the compactness. The scattering is represented by the product M c S(c,0). The proteins are supposed to interact through screened electrostatic repulsive interactions. **a)** The scattering was calculated for a protein molecular weight of 500 000 daltons. The scattering was calculated for different φ/c values as indicated in the figure. (b) Corresponding percentage of transmitted light, I_t/I_0, as a function of protein concentration, for the scattering experiments shown in figure IX.7a. The calculation was done for a 1cm thick lens and a wavelength of 500 nm.

Fig IX.8. — Illustration of the curves shown in figure IX.7. Proteins with a quaternary structure like 'a pile of subunits', i.e. with excluded specific volume φ/c of the order of 1.50 are compatible with transparency while proteins with compact structures, i.e. with excluded volume close to the dry volume and φ/c close to partial specific volume i.e. 0.75, would be slightly turbid.

One of the major results of our study is therefore to show that, by adequate combination of molecular weight and compactness, transparency may be achieved in many ways. Obviously, small proteins with the same type of repulsive interactions as those observed with α-crystallins, would have been better for transparency. This result certainly means that there must be other types of constraints which account for the protein mixture and protein distribution observed in vivo. One advantage of a mixture of different size proteins might be to increase the maximum concentration attainable and to decrease the viscosity. Another one is, again, that size heterogeneity is an efficient way to prevent crystallization.

In our opinion, however, the reasons for a protein mixture are to be found elsewhere. The eye lens possess other interesting characteristics. For instance, it is known for a long time that a lens acts as an almost perfect osmometer in a large range of applied osmotic pressure and changes its intracellular water content accordingly. In vivo, the lens osmotic pressure is imposed by or in equilibrium with the surrounding medium, i.e. the circulating aqueous humor and the vitreous humor.

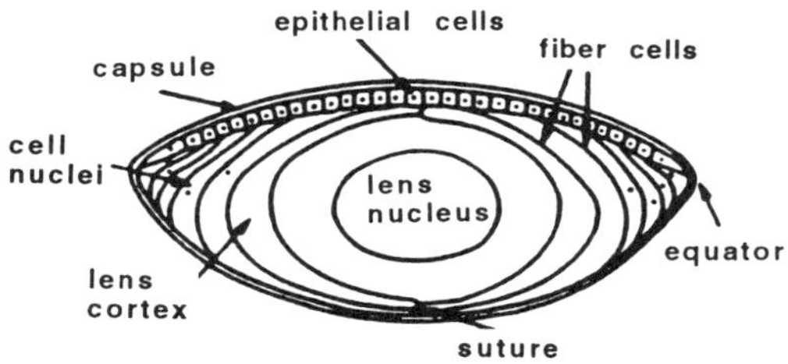

%	α	β	γ
nucleus	21	48	31
cortex	41	46	13

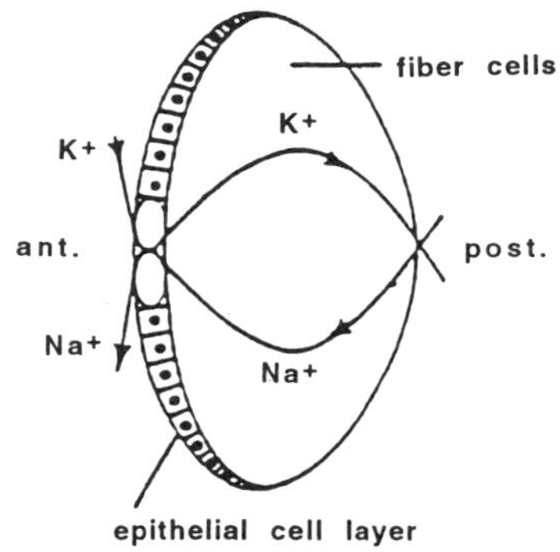

Fig IX.9. — Right : for osmotic pressure experiments, the lens behaves as a unique cell since water and small solutes are free to circulate from cell to cell through intercellular juntions (from Kinsey & Reddy, 1965). As far as protein are concerned, they cannot pass through the junctions. Therefore, the protein composition within one cell does not vary. whereas the protein composition varies from the center to the periphery (left and table). The latter variation is genetically determined.

It is also known that a protein concentration gradient exists from the center to the periphery of most lenses, and therefore a refractive index gradient, which contributes to the lens optical quality. On the biochemical side, the gradient seems to be accompanied by an uneven distribution of the crystallin classes, with the center enriched in γ-crystallins and the periphery in α-crystallins. Since we had shown that the different crystallin interactions resulted in different osmotic behaviour, the whole of the facts quoted above led us to hypothesize that the protein concentration gradient within eye lens might originate from constant osmotic pressure coupled to differential interactive properties of crystallins (Vérétout and Tardieu 1989).

One way to test the hypothesis is to measure osmotic pressure as a function of protein concentration for protein mixtures extracted either from the periphery or from the center of the lens. The results are shown in figure IX.10.

Indeed, in the physiological concentration range, for a given osmotic stress, the protein concentration at equilibrium is found to be higher in the mixture extracted from the lens center than in the mixture extracted from the lens periphery. Varying the composition of a protein mixture , where the proteins have different interactive properties, is therefore a good device to create or maintain a protein concentration gradient with constant applied osmotic pressure.

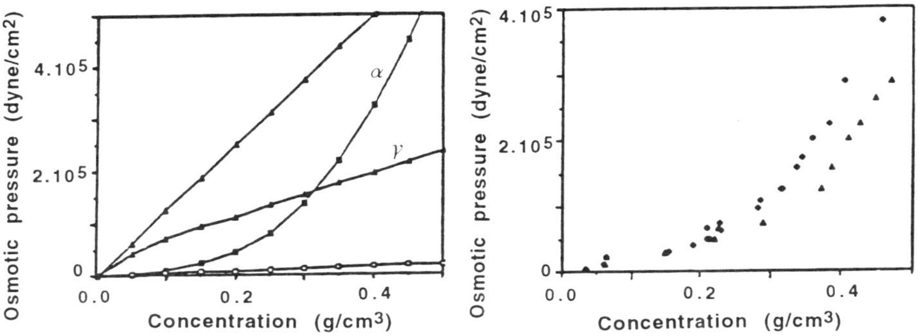

Fig IX.10. — Left : osmotic pressure versus protein concentration measured with α-crystallins (square symbols) and γ-crystallins (triangles) and corresponding osmotic pressure that would be observed with "ideal" α-crystallins and γ-crystallins, i.e. in the absence of interactions. Right : osmotic pressure versus protein concentration measured with a protein mixture extracted from the center of the lens (triangles) and from the periphery (diamonds).

IX.5. Conclusion

From a biological standpoint, the study of the lens crystallins is a good example of a functional role of non specific protein interactions in vivo. The mixture of proteins selected through evolution appears to be a compromise between different evolutionary constraints or different desired properties.

From a physicochemical standpoint, researchers should consider the advantage of proteins as model systems since they can form monodisperse solutions, they present different shapes, sizes, degree of sphericity... In addition studies can be made as a function of a variety of parameters, e.g. concentration, salt, pH, while keeping a constant form factor.

GENERAL READING

BELLONI L. in "Neutrons, X-ray and light scattering : introduction to an investigative tool for colloidal and polymeric systems"; 1991 - LINDNER P. & ZEMB T. eds, North Holland.

EISENBERG H., 1976 - Biological macromolecules and polyelectrolytes in solution - Clarendon Press, Oxford.

GUINIER A., & FOURNET, 1955 - Small angle scattering of X-rays - Wiley, New York.

KERKER M., 1969 - The scattering of light and other electromagnetic radiation - Academix Press Inc. New York

NINHAM B.W., 1985 -The background to hydration forces - Chemica Scripta 25, 3-6

TANFORD C., 1961 - Physical chemistry of macromolecules - Wiley, New York

TARDIEU A, DELAYE M., 1988 - Eye lens proteins and transparency: from light transmission theory to solution X-ray structural analysis - Ann Rev Biophys Biophys Chem 17, 47-70

TARDIEU A., VACHETTE P., GULIK A., le MAIRE M., 1981 - Biological macromolecules in solvents of variable density : characterization by sedimentation equilibrium, densimetry and X-ray forward scattering - Biochemistry 20, 4399-4406

REFERENCES

BENEDEK, G. B., 1971 - Theory of transparency of the eye - Appl. Opt. 10:459-473.

CARNAHAN N.F. & STARLING K.E., 1969 - Equation of state for non attracting rigid spheres - J. Chem. Phys. 51, 635-636

DELAYE M, TARDIEU A, 1983 - Short range order of crystallin protein accounts for eye lens transparency - Nature (London) 302, 415-417

HANSEN J.P. & HAYTER J.B., 1982 - A rescaled MSA structure factor for dilute charged colloidal dispersions - Mol. Phys. 42, 651-656

HAYTER, J. B. & PENFOLD, J., 1981 - An analytic structure factor for macroion solutions - Molec. Phys., 42, 109-118

KINSEY V.E. & REDDY D.V.N., 1965 - Studies of the crystallin lens. XI The relative role of the epithelium and the capsule in transport - Invest. Ophthalm. 4, 104-121

KORETZ J.F., 1988 - How the human lens focuses - Scientific American, 259: 64-71

PARSEGIAN V.A., RAND R.P. FULLER N.L. & RAU D.C., 1986 - Osmotic stress for the direct measurement of intramolecular forces- Methods Enzymol. 127, 400-416

PIATIGORSKY J., 1984 - Lens crystallin and their gene families - Cell 38, 620-621

PROUTY M. S., SCHECHTER, A. N., and PARSEGIAN, V. A., 1985 - Chemical potential measurements of deoxyhemoglobin S polymerization: Determination of the phase diagram of an assembling protein- J. Mol. Biol. 184, 517-528.

TROKEL, S., L., 1962 - The physical basis for transparency of the crystalline lens - Invest. Ophthalmol., 1, 493-501.

VERETOUT F., DELAYE M. & TARDIEU A., 1989 - The molecular basis of eye lens transparency: osmotic pressure and x-ray analysis of alpha-crystallin solutions - J Mol Biol 205: 713-728

VERETOUT F. & TARDIEU A., 1989 - The protein concentration gradient within eye lens might originate from constant osmotic pressure coupled to differential interactive properties of crystallins - Eur. Biophys. J. 17, 61-68

CHAPTER X

DIFFRACTION EXPERIMENTS ON LIPID BILAYERS AND BIOLOGICAL MEMBRANES

G. BÜLDT

Biological membranes are the barriers in cells and organelles against the outside medium. The two major components of membranes are lipids and proteins. Lipids are amphiphilic molecules, which in aqueous media spontaneously form double layered structures. The amphiphilic character of the lipids with polar headgroups and apolar hydrocarbon chains is responsible for the two-dimensional geometry of biological membranes. Integral membrane proteins, which span the lipid bilayer, mediate the transport of matter and signals. Soluble proteins can be associated with one side of a membrane through a lipid anchor covalently bound to the protein. A membrane can also contain protein anchors for filamentous structures, which mechanically stabilize the cell.

The first part of this lecture describes X-ray and neutron diffraction experiments on pure lipid bilayer systems, whereas the second part presents experiments concerning the structure and dynamics of membrane proteins. Scattering experiments on biological membranes or on pure lipid bilayer systems are normally carried out on aggregates like single shell vesicles or multibilayer systems. Diffraction experiments on membranes have two levels in the degree of difficulty. Recording diffraction patterns and simply measuring positions of diffraction peaks provide geometrical information on bilayer systems like the thickness of membranes, packing of lipid chains, secondary structure of membrane proteins, and membrane protein packing. Any nonspecialist can collect this information without spending too much effort. Measuring integrated intensities of diffraction peaks, solving the phase problem, and calculating density maps in order to solve structures is much more complicated.

X.1. Pure lipid bilayers

Although, in a biological membrane, proteins modify the interactions between membranes as well as between membrane components, it is the amphiphilic architecture of a lipid molecule, which is responsible for its geometric arrangement in the form of bilayers. In addition, the inherent ability of most lipids to take up a large number of conformations and their segmental mobility leads them to form an excellent two-dimensional solvent for membrane proteins. Therefore, to have a clear understanding of the forces which are responsible for the various lipid and bilayer properties, an investigation of pure lipid bilayer systems is necessary. X-ray and neutron diffraction are the most powerful tools in characterizing lipid and bilayer morphology and to follow up their changes due to alterations in the number of components and their concentrations. It should be noted that ions and other molecules in the water phase adjacent to the bilayer surface influence the physical state of membranes and should therefore be considered as components of the bilayer.

As a nicely written introductory article for the diffraction methods applied to bilayer systems, a paper of Franks and Lieb (1981) is recommended.

X.1.1. Diffraction experiments on oriented films of multilamellar stacks of bilayers

With respect to the basic principles of a diffraction experiment, there is no large difference between neutrons and X-rays. As the flux of neutron sources is several orders of magnitude smaller than for X-rays, the sample volume in a neutron diffraction experiment is typically of the order of 10 x 20 x 1 mm^3 and therefore much larger than in the X-ray case. In the paragraphs (X.1.1 and 1.2), which are introductory, the experiments described are carried out with X-rays. The same type of experiments can be done with neutrons, after modifying the instrumentation with respect to beam size, scattering volume and detectors.

In a first experiment, a sonicated dispersion of 4 mg of the lipid 1,3-DPPC (1,3-dipalmitoyl-glycero-2-phosphocholine) in 0.2 ml H_2O is spread on a glass slide and allowed to evaporate slowly. After evaporation of the bulk water a multibilayer film is formed with predominent orientation parallel to the bilayer surface. The slide with the lipid film is aligned in an X-ray beam which hits the film tangentially (Figure X.1.A.). In quantitative experiments the sample should be controlled with respect to temperature and relative humidity which is achieved by a sample chamber with windows transparent for the X-ray beam.

In Figure X.1.B. three diffraction patterns of the same 1,3-DPPC lipid film are shown. The equidistant intensity spots on the meridian of the X-ray film prove that we have a one-dimensional periodicity along the direction perpendicular to the lipid film.

From Bragg's law $|\mathbf{G}| = |\mathbf{s}|$, with $|\mathbf{G}| = \frac{h}{d}$ and $|\mathbf{s}| = \frac{2}{\lambda}\sin(\theta)$

we obtain $\frac{h}{d} = \frac{2}{\lambda}\sin(\theta)$ with $h = 1, 2, 3, \ldots$

where d is the repeat distance of the one-dimensional lattice, which is the thickness of the lipid bilayer plus the water layer. Bragg angle θ is calculated (see Figure X.1.A.) by $\mathrm{tg}\, 2\theta = \ell/L$.

The diffraction spots in Figure X.1.B.a. are extended arcs. The intensity distribution over an arc is directly proportional to the number of domains in the lipid film with different orientations relative to the glass slide. The angular distribution of all domains in the sample is called the **mosaic spread,** which in (a) is of the order of 25° FWHM (full width at half maximum). The mosaic spread for this lipid film can be reduced by an annealing procedure, in heating up the sample to 110°C for 4h and then cooling it down slowly (Büldt et al., 1979). The procedure results in an enhanced intensity of diffraction orders in this case by a factor of 20 as seen in (b). The diffraction patterns (a) and (b) were collected at 22°C, where DPPC is in the so called gel phase, which is characterized by hydrocarbon chains in all-*trans* conformation. When the temperature is raised to

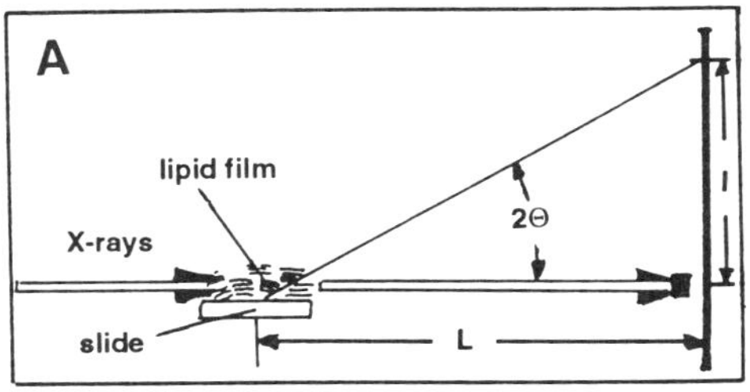

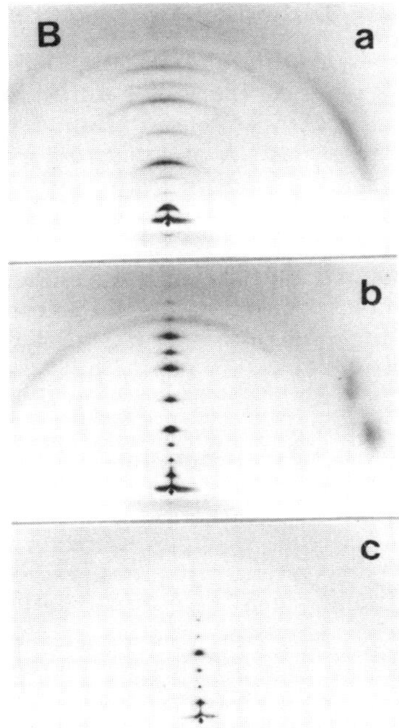

Figure X.1. — (A) Experimental set-up for an X-ray diffraction experiment on bilayer stacks oriented on a glass slide. The incident beam is absorbed by a small lead beam stop in front of the X-ray film. (L: sample to film distance, ℓ: distance of a diffraction spot on the film from the origin of the diffraction pattern. (B) Diffraction pattern of oriented bilayer stacks of 1,3-DPPC (a) in the gel phase, not annealed (broad mosaic spread), at 22°C, about 6% H_2O content; (b) same conditions as in (a), but after the annealing procedure (narrow mosaic spread); (c) in the fluid phase, 70°C, about 8% H_2O content. (Photographs from Büldt and de Haas, 1982).

70°C, i.e., above the first order phase transition temperature of DPPC, the sample is in the fluid phase with hydrocarbon chains forming *gauche+* and *gauche-* conformations which result from rotations around C-C bonds by + or - 120° degrees. The diffraction pattern of the fluid phase is shown in (c). The Bragg peaks are further apart, which means that the bilayer plus water layer becomes smaller. This is the result of a reduction in the thickness of the hydrocarbon chains due to segmental disorder. We also observe a decrease in intensity which is quite dramatic for diffraction spots greater than h = 6. This is caused by the Debye-Waller factor, which is a Gaussian function centered on s = 0 $Å^{-1}$ with a width depending on the disorder and motion of lipid segments in the bilayer.

Off-axis to the equator at 1/s = 4.2 Å one single reflection spot is visible in (a, b). This reflection originates from the packing of hydrocarbon chains. If we assume an hexagonal arrangement, the observed reflections can be indexed by h,k=1,0. Using Bragg's law

$$\frac{2}{\sqrt{3}a}\sqrt{h^2+hk+k^2} = \frac{2}{\lambda}\sin(\theta)$$

the lattice constant a, which is the distance between two neighboring chains, is determined to be a = 4.85 Å. The off-axis position of the chain-packing reflection is caused by a tilt of hydrocarbon chains with respect to the bilayer normal (Tardieu et al., 1973). This reflection becomes more pronounced at smaller mosaic spread (b) and is very weak in the fluid phase. The increase of disorder in this phase causes a broadening and a shift of the chain reflection from 4.2 Å to 4.5 Å.

X.1.2. Diffraction experiments on powders of multilamellar stacks of bilayers

For sample preparation 7.5 mg of DPPC is mixed with 2.5 mg H_2O and filled in a glass capillary of 1 mm diameter with 0.01 mm wall thickness. The capillary is closed by wax and aligned perpendicular to an X-ray beam. The X-ray film shows a pattern of concentric equidistant rings centered around the origin of the diffraction pattern. The ring system originates from the diffraction of multilamellar bilayer stacks in the powder. The chain packing gives a ring at 4.2 Å. The lamellar reflections in the diffraction pattern of the fluid phase, which can be obtained by heating the capillary with hot air, are shifted away from the center which means that the thickness of the bilayer plus water became smaller. The chain reflection is broadend and shifted to 4.5 Å, as observed for the oriented sample.

X.1.3. The conformation of a phospholipid molecule in a bilayer as studied by neutron diffraction

In the previous sections, we have derived geometrical information on bilayer systems. In this paragraph it will be shown how coordinates of certain segments in a lipid molecule in the bilayer are obtained, applying methods of one-dimensional crystallography to the one-dimensional periodicity of multilamellar stacks of bilayers. Structure factors, derived from lamellar diffraction spots, give only profile information of the bilayer which is the projection of density on the bilayer normal. With X-ray diffraction it is not possible to see positions of certain segments in the bilayer profile in the gel or fluid phase. The fluctuations of individual segments along the bilayer normal are quite large and overlapping with neighbouring segments occurs, so that the determination of the average position of a segment is impossible. This problem was solved by neutron diffraction in combination with selective deuteration of a molecular group, thus enhancing the local contrast. In this way the average position and the extent of fluctuations of a segment were determined from a difference of two density maps with and without the deuterated group. In general, one great advantage for the application of neutrons over X-rays to biological structures is the possibility of contrast enhancement by isotope exchange which normally is a very small disturbance to the structure. For biological material in most cases the difference in coherent scattering length between hydrogen ($b = -0.374 \cdot 10^{-12}$ cm) and deuterium ($b = +6.67 \cdot 10^{-12}$ cm) is used. A neutron diffraction experiment on an oriented lipid film is performed in a similar way to an X-ray experiment. However, the area of the lipid film in a neutron diffraction experiment has to be quite large (about 25x50 mm^2). The diffraction pattern of such a film is shown in Figure X.2..

Experimental details are given in the figure legend. The differences in relative height of the diffraction orders demonstrate the sensitivity of the method with respect to label positions, e.g., the change in label position to a neighbouring segment leads already to a remarkable change in relative intensities (Figure X.2.). From integrated intensities of diffraction spots the absolute values of structure factors are derived. Since pure lipid bilayers are centrosymmetric, the phase problem reduces to a distinction between phase angle 0 or π. This problem can be solved for the lower order reflections by H_2O/D_2O exchange experiments using the water layer between membranes as an isomorphous replacement site.

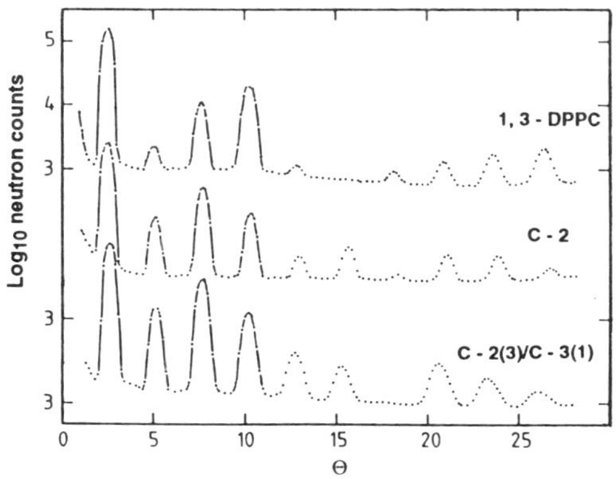

Figure X.2. — Log plots of neutron counts versus Bragg angle (θ) for 1,3-DPPC in H_2O measured on the diffractometer D16 at the ILL Grenoble (λ = 4.5 Å): The upper plot: non-deuterated 1,3-DPPC; the pattern in the middle: 1,3-DPPC deuterated at the first methylene segment or at the second carbon atom (C-2) in both chains; the third plot: the same lipid, deuterated at the second carbon atom in chain 3 (C-2(3)) and at the third carbon atom in chain 1 (C-3(1)). All three patterns are taken in the gel phase at 20°C and 6% H_2O content, lamellar spacing 52.2 Å. (From Büldt and deHaas, 1982).

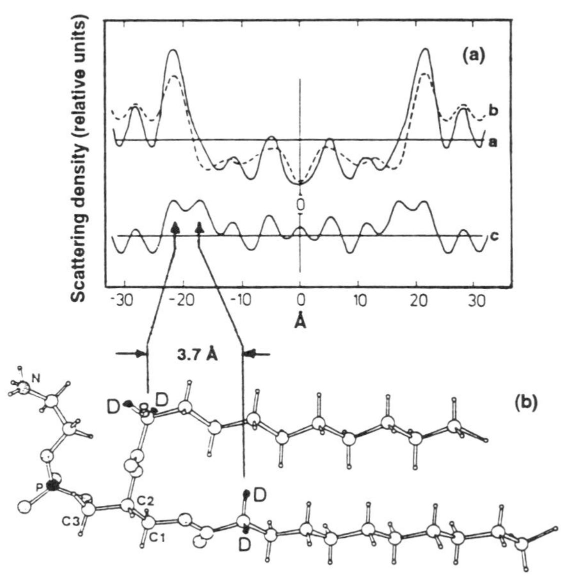

Figure X.3. — (A) Neutron scattering length density profiles calculated from 10 structure factors of (a) DPPE deuterated at the C-2 position, (b) of the non-deuterated sample, and (c) the difference of these profiles. (Plot from Büldt and Seelig, 1980, with permission).

(B) Molecular model of a phospholipid in a bilayer according to the distance of the first methylene segments in both hydrocarbon chains.

Figure X.3.A. shows the profile of DPPE (1,2-dipalmitoyl-*sn*-glycero-3-phosphoethanolamine) selectively deuterated in the first methylene segments of both chains together with the profile of the non-deuterated DPPE. The resulting difference profile (lower trace) shows two peaks at position 17.4 Å and 21.1 Å.

Although the profile is quite noisy, the numbers can be determined very accurately from a refinement procedure on the structure factors. The main result of this profile is that the first methylene segments in both hydrocarbon chains are shifted with respect to each other by 3.7 Å along the bilayer normal (Büldt and Seelig, 1980).

The consequence of this experimental finding on the molecular model of a phospholipid is illustrated in Figure X.3.B.. The glycerol backbone is oriented perpendicular to the membrane plane with the sn-1 chain following this direction whereas the sn-2 chain becomes parallel to sn-1 after a sharp bent. Using Deuterium Magnetic Resonance this conformation was shown to be general for all phospholipids in all phases and membranes (Seelig and Seelig, 1980), and even in its crystal structure where it was first discovered (Hitchcock et al., 1974).

X.2. Membrane proteins

There exists a wide range of methodical approaches for the investigation of membrane proteins with neutrons or X-ray synchrotron radiation. A number of membrane proteins have now been crystallised but only two fundamentally different structures have been solved to high resolution by X-ray crystallography (Deisenhofer et al., 1985; Weiss et al., 1991). In this paragraph we restrict ourselves to a few examples of diffraction studies on stacks of so called "Purple Membranes" (PM) which are part of the cell membrane of *Halobacterium halobium*.

The integral membrane protein bacteriorhodopsin (BR) is the sole protein in the purple membrane and acts as a light-driven proton pump. Although many attempts have been made (Michel, 1982), up to now the formation of well-diffracting 3-dimensional crystals of BR has been unsuccessful. Numerous structural and spectroscopic methods have been applied to this protein, in order to obtain various pieces of information, which together form a now nearly complete mosaic of the structure-function relationship. Diffraction methods as well as electron microscopy studies on this protein benefitted from its aggregation into two-dimensional hexagonal crystals (purple membranes) in the cell membrane of these bacteria. The unit cell contains a trimer of BR (26500 Da) and 21 to 27 lipids. The purple colour reflects the interaction of the chromophore retinal in the protein pocket of BR. It is the absorption of a photon by the chromophore followed by a sequence of isomerisation steps which are coupled to the translocation of one proton across one BR molecule, per functional cycle (Grzesiek and Dencher, 1986).

The characterization of the aggregation state of BR in the membrane, as well as the amount and orientation of its secondary structure was given by X-ray diffraction methods (Blaurock, 1975; Henderson, 1975). The basic picture of the tertiary structure, describing the three-dimensional arrangement of α-helices spanning the membrane, was given in several papers on low dose electron microscopy (e.g., Unwin and Henderson, 1975; Henderson and Unwin, 1975).

The knowledge of the amino-acid sequence (Ovchinnikov et al., 1979) allowed the prediction of a folding model by determining seven α-helical stretches in the polypeptide chain which were assigned to the seven α-helices seen in the electron microscopical structure (Engelman et al., 1980). In the following years neutron diffraction was the method of choice to prove this model using BR molecules containing certain perdeuterated amino acids (Engelman and Zaccai, 1980; Trewhella et al., 1983) and partially deuterated α-helices (Trewhella et al.,1986; Popot et al., 1989). This technique was also successful in the localization of perdeuterated or partially deuterated retinal within the protein (Jubb et al., 1984; Heyn et al.,1988). Heavy-atom labelled retinal analogues incorporated into the protein confirmed these results by X-ray diffraction (Büldt et al., 1991). In early neutron diffraction experiments the hydration of purple membranes was studied (Zaccai and Gilmore, 1979) and more recently the location of the proton channel in BR was determined by neutron diffraction in combination with H_2O/D_2O exchange experiments (Papadopoulos et al., 1990).

In 1990, cryo-electron microscopy provided an additional breakthrough providing a structure of BR at 3.5 Å resolution in the X-Y plane and 7 Å in the Z-direction. These pictures allowed for the first time, including all the information from other methods mentioned above, to construct a molecular model of BR (Henderson et al., 1990).

The examples given in paragraph X.2.2. and X.2.3. will demonstrate that diffraction by neutrons and X-rays give complementary information which is not obtained from cryo-electron microscopy.

X.2.1. Diffraction experiments on oriented films of two-dimensional protein crystals

This section gives an introduction to diffraction experiments on the two-dimensional hexagonal lattices of BR. Figure X.4. illustrates in (A) a two-dimensional protein latice in real space and its counterpart in reciprocal space. The h,k-plane, which corresponds to the plane of two-dimensional periodicity in real space, has a distinct intensity distribution at reciprocal lattice points. Perpendicular to these points in the l-direction continuous intensity distributions, the so called Bragg rods, are obtained which correspond to the direction with no periodicity in real space. This is the situation for electron diffraction in the electron microscope where a diffraction pattern from a single protein lattice can be recorded (Henderson and Unwin, 1975), since the interaction of an electron with matter is much larger than for X-rays or neutrons. The latter case is illustrated in (B), where several millions of lattices form a film of purple membranes. In this film, lattices have small but different angles to the supporting surface which determines the mosaic spread as in the case of oriented pure lipid bilayers. Within each stack, lattices are randomly oriented to each other with respect to an axis perpendicular to the membrane plane. The corresponding reciprocal lattices are also randomly oriented with respect to the l-axis and form a system of concentric cylinders, which have a continuous intensity distribution in the l-direction. Only the Bragg rod at h,k = 0,0, which is on the l-axis, is the Fourier transform of the projected density of the membranes (the profile) and is therefore periodic showing distinct intensity spots. In scattering geometry (i), where the incident beam runs parallel to the purple membranes, "Ewald's sphere" or "the sphere of reflection" intersects reciprocal space tangentially to the l-axis at the origin. The intersection picture is shown on the X-ray film with the lamellar spots on the meridian and the intersection lines of Ewald´s sphere with the concentric cylinders on the equator. In configuration (ii), with the incident beam perpendicular to purple membranes, Ewald's sphere intersects the reciprocal space tangentially to the h,k-plane at the origin, resulting in concentric rings on the X-ray film.

X.2.2. The location of the chromophore retinal in bacteriorhodopsin by neutron diffraction

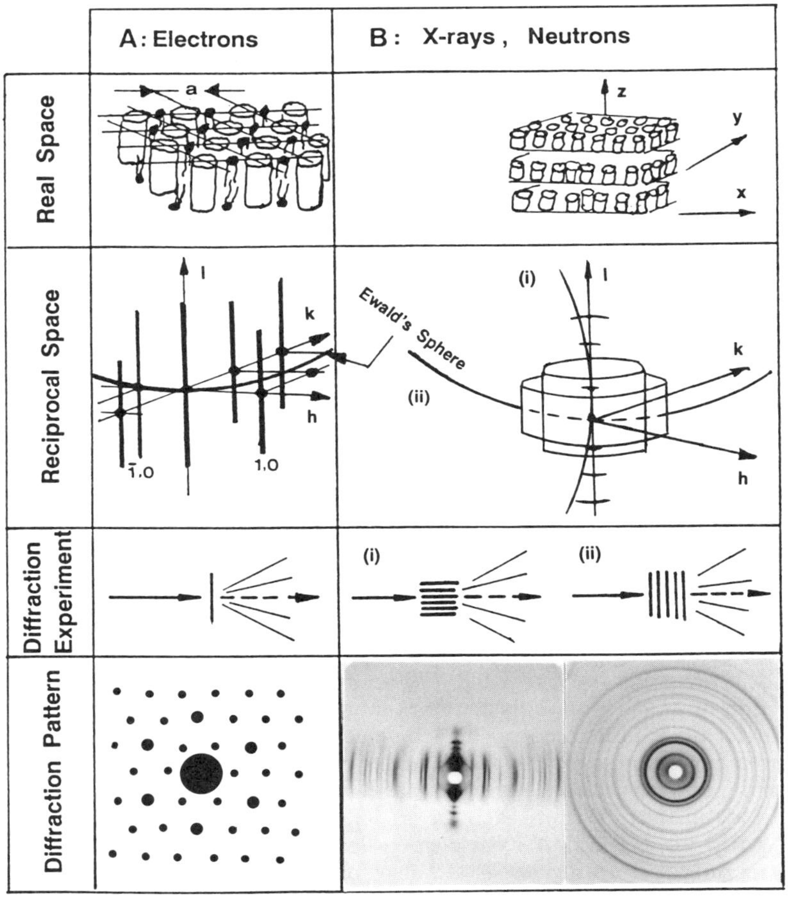

Figure X.4. — (A) Density in real space and intensity distribution in reciprocal space of a single protein lattice and an electron diffraction experiment.

(B) Intensity distribution of a stack of lattices randomly oriented with respect to Z-axis. X-ray films show diffraction pattern for stacks of purple membranes in two different scattering geometries (i) and (ii).

A nice example of the enhancement of local contrast by selective deuteration is the localization of the chromophore retinal in bacteriorhodopsin by neutron diffraction. The first successful solution of this problem was done by *in vivo* incorporation of perdeuterated retinal into BR by Jubb et al. (1984). Figure X.5.A. shows diffraction peaks in the radial direction of rings as illustrated in Figure X.4. (B, ii). The neutron wavelength used was 4.5 Å and therefore much higher than the copper K_α-line of 1.5 Å in X-ray experiments. Therefore, to insure that the scattering vector **s** is always parallel to the membrane plane a so called θ-2θ scan was performed, where the sample aligned on the ω-axis of the diffractometer is rotated by θ and the detector by 2θ. At detector angle 2θ = 0 purple membranes were oriented perpendicular to the incident neutron beam.

Figure X.5.A. shows two diffraction patterns for purple membrane films with deuterated and non-deuterated retinal. The integrated intensity was obtained for each reflection after substraction of background and multiplication of a Lorentz factor. As seen in the previous section rotation of Bragg rods around the l-axis leads to a superposition of reflections with the same value in h^2+hk+k^2, e.g. (h,k) and (k,h). Therefore the intensity $I(h^2+hk+k^2)$ = const. was split between the two (or more) intensities I(h,k), I(k,h) etc., according to their ratio in electron diffraction (Henderson and Unwin, 1975). Phase angles determined by electron microscopy were used. Figure X.5.B. shows the resulting difference Fourier map. The highest contour line depicts the location of the center of mass of deuterons in the retinal.

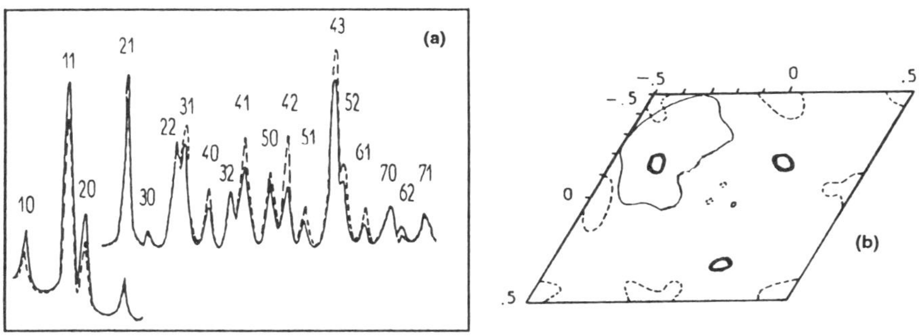

Figure X.5. — (a). Neutron diffraction pattern of an oriented purple membrane film. Dashed line: sample with non-deuterated retinal, full line: sample with perdeuterated retinal.
(b). Difference Fourier map: highest positive bold contour lines show the center of mass of the deuterated retinal, the thin line marks the protein area. (Photographs from Jubb et al., 1984, with permission).

In the following years, these experiments were extended and refined using retinals which were partially deuterated at three different positions within the chromophore (Heyn et al., 1988). With these retinals the orientation of the chromophore within the unit cell was determined in projection on to the plane of the membrane as well as in the profile from lamellar diffraction (Hauß et al., 1990).

Heavy atom labelled retinals in PM used in X-ray diffraction experiments confirmed these results (Büldt et al., 1991). The heavy atom labelled retinals are now used in EXAFS-experiments to explore the fine structure of the retinal binding pocket in BR for different intermediate states of the photocycle.

X.2.3. Conformational changes in bacteriorhodopsin during the photocycle as seen by neutron and time-resolved X-ray diffraction

Bacteriorhodopsin translocates protons over the membrane in a cyclic way using the energy of light. The ground state BR_{568} of this cycle is characterized by an optical absorption band centered at 568 nm. In this state the retinal is in an all-*trans* conformation. Absorption of a photon by the retinal starts the photocycle, which after several short living intermediates comes to the long living M_{412}-state which under normal conditions decays within several milliseconds. In this state, the chromophore is in a 13-*cis* conformation with an absorption maximum at 412 nm. The photocycle, which is easily followed by changes in the color of the chromophore, is linked to the proton translocation process. In the BR_{568}, the Schiff's base linkage of the retinal to lysine 216 is protonated and becomes deprotonated during M-formation. In this way, the movement of the Schiff's base proton starts the pumping cycle resulting in the ejection of a proton through the external channel of BR to the outside, which is followed by the uptake of another proton from the cytoplasmic side of BR, which leads to a reprotonation of the Schiff's base.

The mechanistic details of this cycle are not yet understood. Central for the unterstanding of this mechanism is the question, whether conformational changes of the protein occur and are essential for the cycle, or whether the protein stays in its ground state conformation with a rigid channel, where the chromophore, by switching from all-*trans* to 13-*cis,* controls the passage of protons. Time resolved X-ray experiments using an intense X-ray plasma source gave indications for very large conformational changes during the photocycle in BR (Frankel and Forsyth, 1985), whereas electron diffraction experiments by Glaeser et al. (1986) detected no changes in diffraction orders in the resolution regime from 60-7 Å. Neutron diffraction experiments on oriented films of purple membranes trapped at liquid nitrogen temperatures in the M_{412} and in the ground state showed moderate intensity changes between both states of the order of 9% in $\Sigma|\Delta I|/\Sigma I$. (Dencher et al., 1989). Model calculation revealed that small changes of 1 or 2 degrees in the tilt angle of several α-helices could account for these differences.

This indicated for the first time that small conformational changes in the tertiary structure are involved in proton translocation. In order to show that these structural changes correlate with relaxation processes in the photo- and pumping cycle of BR, the transition from the M-state to the ground state was followed by a time-resolved X-ray diffraction analysis using intense synchrotron radiation (Koch et al., 1991). The M_{412} decay rate was slowed down by employing mutated BR Asp96Asn in purple membranes.

The time course of flash-induced changes for three reflections at neutral pH is illustrated in Figure X.6.(A). The diffraction pattern was recorded with a linear detector on the X33 double focusing mirror-monochromator camera of the EMBL at HASYLAB (Hamburg). The diffraction experiments were performed as

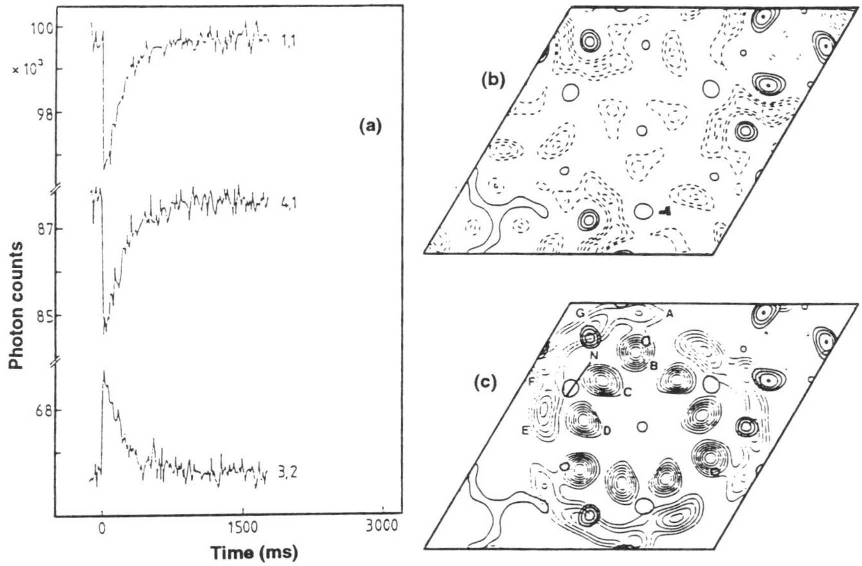

Figure X.6. — (a) Time course, after a short light flash (0.75 ms FWHM) at t = 0 s, of the integrated intensity including back ground under the (1,1), (4,1) and (3,2) reflections for mutated BR Asp96Asn at neutral pH.

(b) Two-dimensional difference Fourier map between the electron densities of the M- and the ground state calculated from amplitudes derived from intensity changes in (A). Continuous contour lines represent positive densities, dashed lines negative densities.

(c) The highest positive difference density levels (bold contours) are plotted on the projected structure of BR in the M-state (thin lines). N marks the position of the Schiff's base nitrogen at the end of the retinal (bold bar). (Photographs from Koch et al., 1991, with permission).

illustrated in Figure X.4.(B.ii). The changes in individual reflections before and immediately after the light flash are consistent both in amplitude and in direction with the earlier neutron results.

Figure X.6.B. shows the difference electron density map obtained from these intensity alterations. The most pronounced changes occur in the vicinity of helix G and between helices E and F (Figure X.6.C.), as also found by neutron diffraction (Dencher et al., 1989). At present it cannot be decided, if the additional large density peak emerging close to helix F, which was not observed by neutron diffraction, reflects a difference in the structural changes of mutated BR. The relaxation times of individual reflections in Figure X.6.A. are very similar having half-times $\tau_{1/2}$ = 150 ms, which are in aggreement with a large relaxation component of absorbance changes in the transition from the M- to the ground state with half-time $\tau_{1/2}$ = 208 ms. This indicates that structural changes correlate in their time behaviour to processes seen by the chromophore in the binding pocket of BR.

Recent cryo-electron diffraction experiments (Subramaniam et al., 1993) fully agree with the above neutron and X-ray results. In these experiments, the M-state was obtained by illumination of a wet purple membrane on a microscopy grid just before injecting it into liquid ethane cooled by a bath of liquid nitrogen. In this way, in difference to earlier electron diffraction experiments, BR is trapped in the M-state embedded in amorphous ice which prevents the drying of the sample.

REFERENCES: Pure lipid bilayers

BLAUROCK, A.E., 1982 - Biochim. Biophys. Acta 650, 167

BÜLDT, G., GALLY, H.U., SEELIG, J., ZACCAI, G., 1979 - Nature 271, 182

BÜLDT, G., SEELIG, J., 1980 - Biochemistry. 19, 6170

BÜLDT, G., WOHLGEMUTH, R.J., 1981 - J. Membrane Biol. 58, 81

BÜLDT, G., DE HAAS, G.H.J., 1982 - J. Mol. Biol. 158, 55

FRANKS, N.P., LIEB, W.R., 1981.- in Liposomes, page 243-270, ed. by C.G. Knight, Cambridge, Elsevier

HITCHCOCK, P.B., MASON, R., SHIPLEY, G.G., 1975 - J. Mol. Biol. 94, 297

SEELIG, J., SEELIG, A., 1980 - Quarterly Rewiews of Biophysics 13, 19

TARDIEU, A., LUZZATI, V., REMAN, F.C., 1973 - J. Mol. Biol. 75, 711

REFERENCES: Membrane proteins

BLAUROCK, A.E., 1975 - J. Mol. Biol. 93, 139

BÜLDT, G., KONNO, K., NAKANISHI, K., PLÖHN, H.-J., RAO, B.N., DENCHER, N.A., 1991 - Photochem. Photobiol. 54, 873

DEISENHOFER, J., EPP, O., MIKI, K., HUBER, R. and MICHEL, H., 1985 - Nature 318, 618

DENCHER, N.A., DRESSELHAUS, D., ZACCAI, G., BÜLDT, G., 1989 - Proc. Natl. Acad. Sci. 86, 7876

ENGELMAN, D.M., ZACCAI, G., 1980 - Proc. Natl. Acad. Sci. 77, 5894

ENGELMAN, D.M., HENDERSON, R., MCLACHLAN, A.D., WALLACE, B.A., Proc. Natl. Acad. Sci. 77, 2023-2027 (1980).

FRANKEL,R.D., FORSYTH; J.M., 1985 - Biophys. J. 47, 387

GLAESER, R.M., BALDWIN, J., CESKA, T.A., HENDERSON, R., 1986 - Biophys. J. 50, 913

GRZESIEK, S., DENCHER, N.A., 1986 - FEBS Lett. 208, 337

HENDERSON, R., 1975 - J. Mol. Biol. 93, 123

HENDERSON, R.,UNWIN, P.N.T., 1975 - Nature 257, 28

HENDERSON, R., BALDWIN, J.M., CESKA, T.A., ZEMLIN, F., BECKMANN, E., DOWNING, K.H., 1990 - J. Mol. Biol. 213, 899

HEYN, M.P., WESTERHAUSEN, J., WALLAT, I., SEIFF, F., 1988 - Proc. Natl. Acd. Sci. 85, 2146

JUBB, J.S., WORCESTER, D.L., CRESPI, H.L., ZACCAI, G., 1984 - EMBO J. 3, 1455

KOCH, M.H.J., DENCHER, N.A., OESTERHELT, D., PLÖHN, H.-J., RAPP, G. BÜLDT, G., 1990 - EMBO J. 10, 521

Michel, H., EMBO J. 1, 1276-1271 (1982)

OVCHINNIKOV, YU., ABDULAEV, N., FERGIRA, M., KISELEV, A., LOBANOV, N., 1979 - FEBS Lett. 100, 219

PAPADOPOULOS, G., DENCHER, N.A., ZACCAI, G., BÜLDT, G., 1990 - J. Mol. Biol. 214, 15

POPOT, J.-L., ENGELMAN, D.M., GUREL, O., ZACCAI, G., 1989 - J. Mol. Biol. 210, 829

SUBRAMANIAM, S., GERSTEIN, M., OESTERHELT, D., HENDERSON, R., 1993 - EMBO J. 12, 1

TREWHELLA, J., ANDERSON, S., FOX, R., GOGOL, E., KHAN, S., ENGELMAN, D., ZACCAI, G., 1983 - Biophys. J. 42, 233

TREWHELLA, J., POPOT, J.-L., ZACCAI, G., ENGELMAN, D.M., 1986 - EMBO J. 5, 3045

UNWIN, P.N.T., HENDERSON, R.J., 1975 - Mol. Biol. 94, 425

ZACCAI, G., GILMORE, D.J., 1979 - J. Mol. Biol. 132, 181

WEISS, M.S., KREUSCH, A., SCHILTZ, E., NESTEL, U., WELTE, W., WECKESSER, J. & SCHULTZ, G.E., 1991 - FEBS Lett. 280, 379.

CHAPTER XI

MACROMOLECULAR CRYSTALLOGRAPHY WITH SYNCHROTRON RADIATION

R. FOURME

XI.1. Introduction

Synchrotron radiation is used in X-ray crystallographic studies of biological macromolecules for several applications (monograph by Helliwell, 1992).

The first and basic class of applications is measurement of diffraction data with the highest possible quality, using existing crystals. Monochromatized radiation is used as it provides the highest signal to noise ratio and allows a complete survey of the reciprocal lattice.

A sub-class is concerned by the use of anomalous dispersion. Here, the emphasis is on accurate measurements of small differences between the intensities of Bragg reflections in anomalous pairs.

The second class of applications is kinetic crystallography and fast data collection methods, using either monochromatic or polychromatic radiation.

XI.2. Accurate and high resolution data collection

The conclusion of the first steps of the analysis of every macromolecular structure by x-ray diffraction is the production of an ordered list of structure factor amplitudes $|F(\mathbf{h})|$ for the native crystal plus, generally, other lists for derivative crystals (Blundell & Johnson, 1976). The quality of final structural results, and often the possibility of solving the structure, are directly related to resolution, accuracy and completeness of these data.

Here, two pieces of equipment play a key role: the x-ray source and the area detector.

Acquisition of diffraction data with monochromatic radiation proceeds in general by recording a succession of diffraction images (frames) through adjacent rotations of amplitude $\Delta\phi$ of the crystal. Frames are partitioned into elements (pixels) and the pixel content is stored in digital form. Extraction of structure amplitudes is obtained by off-line analysis of the record of digitized frames, using for each reflection a three dimensional box (voxel) centred on the reflection at x, y and ϕ with dimensions defined by three integers m x n x p, where m and n are numbers of pixels defining the integration area on the detector (boxel) and p is the number of increments $\Delta\phi$. The total count in each voxel is obtained by summing contributions from contributing frames. The background is estimated from the count in an adjacent voxel free of diffraction signal and substracted from the total count to get the integral diffraction signal. This scheme has several variants, and is generally followed by post-refinement based on profile analysis of strong reflections. The best signal to noise ratio is obtained when the background is integrated over an angular interval no larger than the reflection width, which is usually a fraction of a degree; in this respect, an increment $\Delta\phi$ of 0.05 - 0.1° is often most appropriate.

Table XI.1 contains basic considerations on data collection with monochromatic radiation (Fourme et al., 1991).

The first and second columns show respectively facts of biocrystallographer's life and main problems to be solved. In short, we are faced to (a) a large number of diffracted beams (b) spatially very close (c) with intensities spanning a dynamical range of 5 decades and peaking over a substantial background; we would like to measure these intensities with one to a few per cent accuracy. Crystals are subject to radiation damage which is dependent on wavelength, total dose and dose rate; for a given dose, a higher dose rate is in general beneficial; shorter wavelengths are favorable in all respects, improving accuracy - less absorption, better signal to noise due to longer crystal to detector distance - and the amount of information acquired per sample.

The third column shows the technical goals which we would like to achieve:

(i) an intense, quasi-parallel and highly monochromatic x-ray beam, which requires a source with a high spectral brilliance. In this respect, synchrotron sources outcome by far any other source. The total energy in a diffracted beam **h** for a crystal rotating at a constant angular velocity ω is given by:

$$E(\mathbf{h}) = C\ \omega^{-1}\ I_0\ \lambda^3\ LP\ A\ (V_x/V_0^2)\ |F(\mathbf{h})|^2$$

where C is a constant, I_0 the intensity of the incident x-ray beam of wavelength λ, LP the Lorentz-polarization correction, A a correction for absorption, V_x the volume of the sample, V_0 the volume of the unit cell and $|F(\mathbf{h})|$ the structure factor amplitude. When the expression of L is taken in account, $E(\mathbf{h})$ is proportional to $\sim \lambda^2$. So, the need for high brilliance is comparatively more important for work at short wavelengths, below 1Å.

(ii) the reduction of various sources of parasitic scattering (crystal mounting, evacuated paths, beam catcher design....). A longer crystal-to-detector distance d is also useful in this respect. Due to the low divergence of the x-ray beam and the narrow width of Bragg reflections from most macromolecular crystals, the boxel size on the detector may be kept constant; in contrast, the solid angle subtended by the boxel decreases as $1/d^2$. Note that this argumentation is simplified, and results will depend on each instrumental configuration; during tests with an image plate detector, we found an effective gain in signal to noise with increasing values of d, but less than predicted by the inverse square law.

(iii) a good detector. This is a challenging problem for the physicist, because requirements are of a very high level and often conflicting. There are as yet no detector which can meet all requirements, so that the potential of even current synchrotron radiation sources is only partially exploited; but there are several lines of development to make improved detectors within the next few years. The data collection procedure just described is typical of electronic detectors such as multiwire proportional chambers (Kahn et al., 1986) or detectors based on a charge-coupled device (CCD) (such as developed by Westbrook et al., 1989 or by J. P. Moy and coll. in the ESRF Detector Group), in which the dead time between two frames may be a small fraction of the exposure time per frame. With current imaging plate detectors (Miyahara et al.,1986), the dead time due to readout and

features	goals	solutions
global structural method		intense X-ray beam, hence higher dose rate
several data sets (MIR)	collect vast amounts of data in short times	2D detector with high efficiency even at short λ high count rate fast readout low noise large number of pixels small point spread function
large number of reflections excited simultaneously	avoid spatial overlap	monochromatic and parallel beam small oscillations per frame large crystal-to-detector distance
static and dynamical disorder in crystal		
diffuse scattering by crystal various parisitic scattering	improve S/N = increase number of reflections per crystal	improve crystal mounting evacuated paths
radiation damage in crystal		short λ (0.3 Å) cryocrystallography

Table XI.1. Data collection using a monochromatic X-ray beam, or how to improve the signal to noise ratio (S/N), i.e. the ratio of diffraction signal to non-diffraction signal (after Fourme et al., 1991, modified)

erasing is relatively long, and $\Delta\phi$ increments of typically 1° are used in order to get a reasonable duty cycle. Large increments tend to increase background and produce overlapping spots. These drawbacks are minimized using a Weissenberg camera with an imaging plate (Sakabe, 1991). This instrument combines sharp collimation of the beam, translation of the detector in one or two directions coupled with the crystal rotation, long crystal-to-film distance, made possible by the use of a large single plate or an array of plates and helium paths.

As a conclusion of table XI.1, it is important to note that it describes essentially *how to improve the ratio (diffraction signal/non-diffraction signal)*

Figure XI.1 shows a practical example of the comparison between several data sets obtained with various X-ray sources and various detectors with crystals from the same batch.

In conclusion, the future of accurate and high resolution data collection with synchrotron radiation is especially bright. For fundamental as well as applied research, the demand in structural results - now and extrapolated to at least the end of this century - is increasing quickly (Smith and Watenpaugh, 1991). Work that can be done with rotating anode sources may be improved by using synchrotron radiation. Synchrotron radiation has contributed to set new

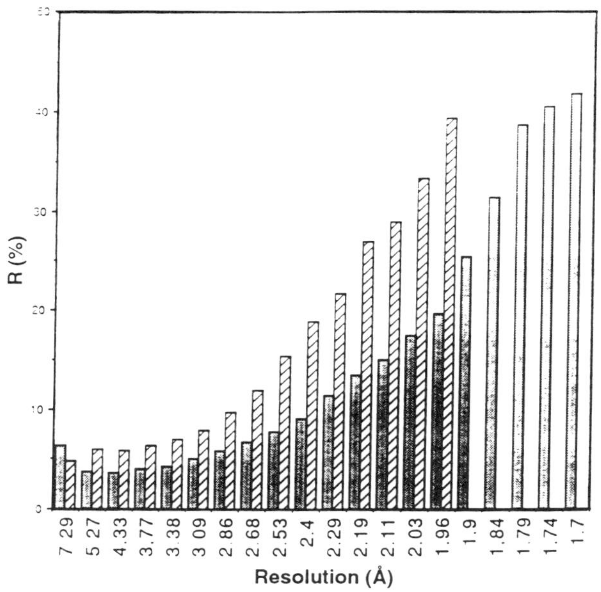

Fig. XI.1. — Crystals of the antigen-binding fragment of an immunoglobulin (Fab Hil); space group $P2_12_12_1$, a = 110.6 Å, b = 127.4 Å, c = 66.50 Å.
High resolution data collection using:
- a commercial multiwire proportional chamber and a rotating anode tube (hatched)
- a spherical drift proportional chamber (Kahn et al. 1986) and synchrotron radiation at LURE (grey).

$$R = \sum_{\mathbf{h}} \sum_{i} |I_i(\mathbf{h}) - \langle I(\mathbf{h}) \rangle| / \sum_{\mathbf{h}} \sum_{i} I_i(\mathbf{h}) \times 100$$

where $I_i(\mathbf{h})$ is the ith measurement of the intensity of reflexion **h** and $\langle I(\mathbf{h}) \rangle$ is the weighted mean of all measurements of $I(\mathbf{h})$ (courtesy of F. Saul, Immunologie Structurale, Institut Pasteur, Paris).

standards in the quality of structural work. In this respect, one could consider synchrotron radiation as a basic supply as necessary and as trivial as laboratory fluids like water or liquid nitrogen. For more challenging projects (small crystals, ultra high resolution, complex macromolecular systems, very large unit cell...), undulators on 6-8 GeV third generation storage rings, with the fundamental emission in the X-ray range, are the real breakthrough. This has been demonstrated by protein cristallography experiments performed at the ESRF in early 1993, using the unfocused third or fifth harmonics from the ID10 undulator (ESRF report "Proteins on Troika", reporter M. Lehmann, february-march 1993).

XI.3. Applications of anomalous dispersion

The spectral brilliance over a wide range of wavelengths, of synchrotron radiation sources has stimulated appplications of anomalous dispersion in many fields. Concerning macromolecular crystallography, familiar applications of anomalous dispersion to the phase problem (single isomorphous replacement with anomalous scattering or SIRAS, multiple isomorphous replacement with anomalous scattering or MIRAS) can now be made optimal by appropriate wavelength selection (see Dumas et al., 1992 for an example of optimised SIRAS). A new method (multiwavelength anomalous diffraction or MAD) exploits anomalous diffraction from an ordered constellation of atoms to help solving the phase problem. Last, multiwavelength anomalous solvent contrast (MASC) exploits the wavelength-dependent scattering from anomalous scatterers dispersed in the solvent for determination of the macromolecular envelope and phasing of low resolution reflexions. MASC measurements can be analysed by a MAD-like formalism (Fourme et al., 1993) and experiments are in preparation at LURE.

XI.3.1. Basic physics of anomalous dispersion

The x-ray scattering factor of an isolated atom is the amplitude of the wave scattered by this atom, taking as reference the amplitude of the wave scattered by a free electron located at the same point. Assuming that electrons in the atom are free, one gets the "normal" scattering factor f^0 which is the Fourier transform of the electron density distribution of the atom. f^0 is a real number which, at $s = 2\sin\theta/\lambda = 0$, is equal to the atomic number Z; it drops off rapidly with increasing values of s because the diffuse part of the electronic charge distribution becomes more and more ineffective in scattering.

In fact, the electrons are bound to the atom and, when submitted to the electromagnetic field, are not free oscillators. Using a simple classical mechanical model (see for example James, 1965), the atom scatters as a collection of damped oscillators whose natural frequencies are those of the absorption edges of the electronic shells. As a result, the atomic scattering factor f is a complex number. This effect is termed anomalous (or resonance) scattering. It is customary to write

$$^{\lambda}f = f^0 + {}^{\lambda}f' + i\,{}^{\lambda}f'' \qquad \text{(XI.1)}$$

where f' and f'' (the superscript λ is implicit) are the real and imaginary anomalous dispersion corrections, respectively, of the normal scattering factor.

Let us recall a few important properties of f' and f'':

- f' is the (in-phase) dispersive term. f'', (the out-of-phase) absorptive term, is proportional to the product of the photon energy by the atomic absorption coefficient, so that a measurement of this coefficient will give, after proper scaling, the corresponding value of f''.

- f' and f'' are related through the Kramers-Kronig dispersion relation. This relation may be used to calculate f' from f'' values.

- f' and f'' are only weakly s-dependent, in contrast to f^0. In effect, anomalous dispersion is due to innermost electronic shells which, for atoms used as anomalous scatterers, have a radius much smaller than the x-ray wavelength. f' and f'' are usually assumed as independent of s. Subsequently, the relative importance of anomalous dispersion is increasing with the scattering angle.

- f' and f'' are wavelength-dependent. They vary most rapidly and reach their maximum magnitudes in the immediate vicinity of absorption edges. In fact, the domain of wavelengths close to absorption edges (say from 100 eV on the low energy side to 800eV on the high energy side) is of particular interest. The atomic absorption spectra (and hence f' and f") displays features which are dependent on the molecular bonding environment and on the local structure around the atom.

- f' and f" are polarization-dependent

With polarized x-rays emitted by synchrotron sources, absorption spectra in the near-edge region depend on the orientation of the local chemical environment of the anomalous scatterer with respect to the polarization direction of the x-ray beam (see e.g. Templeton and Templeton,1985). This effect, which is averaged out in the case of a disordered sample, can be observed with a single crystal and may affect substantially both the edge position and the magnitude of the absorption. There is a related anisotropy in the anomalous scattering (pleiochroism). As an example, the anomalous scattering of selenium covalently bound to two carbon atoms is highly anisotropic in the vicinity of the K-edge of Se; at 12,654.9 eV, the effective f" lies between 0.8 and 7.7 electrons depending on the orientation of the crystal with respect to the polarization direction of the incident and diffracted x-ray beams (Templeton and Templeton,1988). A tensorial formalism to deal with the anisotropy of anomalous scattering has been developed by Templeton and Templeton (1982).

The f' and f" curves for isolated atoms of a few elements are shown in figure XI.2A. These examples illustrate several points:

- edge position for a given orbital occur at shorter wavelengths as the atomic number increases.

- apart from the energy of transition, all K edges are essentially alike and all L edges are alike.

- L_{III} edges, which are associated with the six 2p electrons have anomalous scattering factor magnitudes about three times greater than those for K edges, which are associated with the two 1s electrons. As shown in fig XI.2B & XI.2C, the profiles observed for molecules are typically more strongly featured than the calculated ones for isolated atoms. These resonant features ("white lines"), which correspond to unoccupied molecular orbitals, can be about threefold greater than those expected for isolated atoms; of real practical interest are the 2p-5d resonances in elements from Cs to Pt, including the lanthanides.

Nearly all biocrystallographic experiments using anomalous dispersion have been performed using wavelengths between 0.6Å and 1.8Å (see table XI.2 for MAD experiments). Most elements used commonly as heavy atoms in MIR have a K or L edge in this window, plus many which are too light for use as MIR derivatives but are suitable as anomalous scatterers; such elements either occur naturally in macromolecules or can be introduced in a variety of ways. Experiments at long wavelengths are compromised by absorption; shorter wavelengths should become more readily usable with developments in sources, optics and detectors.

The anomalous scattering factors must be known for the analysis of MAD data to proceed. Initials values of f' and f" at the various wavelengths may be obtained from a combination of theory and experiment. Away from absorption edges, calculations from quantum mechanics provide reliable results (Cromer and Liberman, 1970). At wavelengths close to absorption edges, atomic absorption coefficients can be obtained from x-ray absorption spectra measured by fluorescence from the very crystal and the very x-ray beam used in the experiment

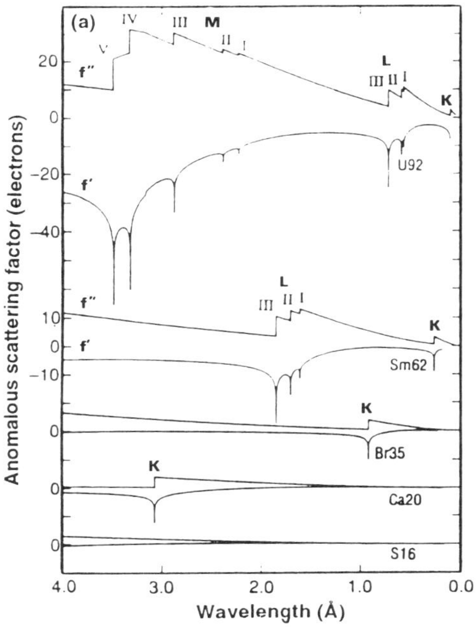

Fig. XI.2. — Anomalous scattering factor spectra for selected elements (after Hendrickson, 1991, modified).

(A) Calculated spectra for some isolated atoms. For each element, the imaginary component f" is drawn in the upper curve and the real component (f') in the lower curve. Origins for the five elements are displaced vertically as indicated.

(B) Experimental values derived from an x-ray absorption spectrum of Eu-(PhAcAc)3 (Ph, phenyl, and Ac, acetyl). The resonances from left to right are associated with the L_{III}, L_{II} and L_{I} absorption edges.

(C) Experimental values derived from an x-ray absorption spectrum of selenomethionyl thioredoxin from *E. coli*.

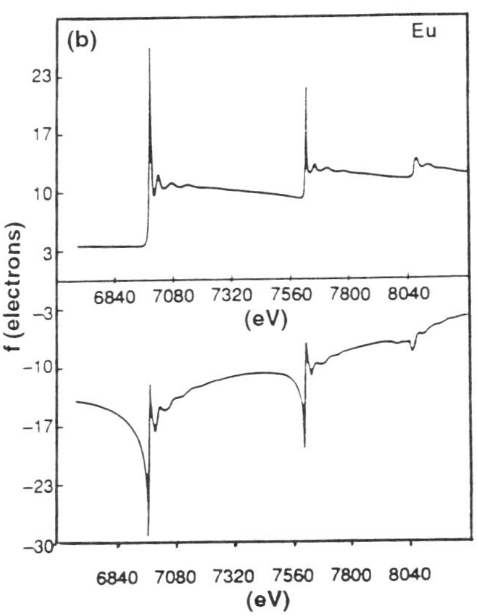

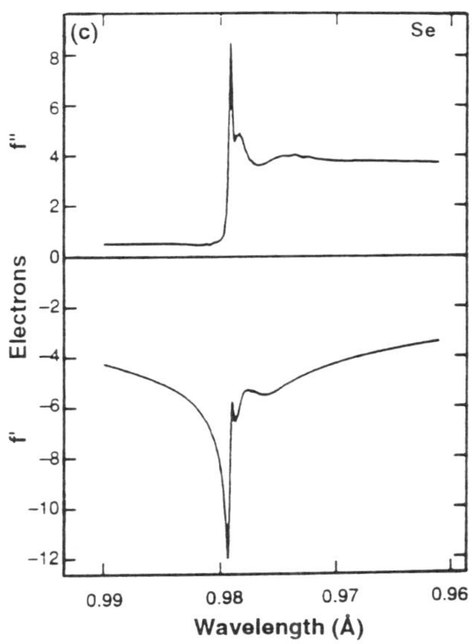

Category	Typical element
Metalloproteins	
Transition metals	Fe, Cu, Zn, Mn
Other	Ca, Mo
Metal replacements	
Lanthanides for Ca^{2+}, Mg^{2+}	Tb, Ho, Yb
Mercury for zinc	Hg
Ligands analogs	
Arnsenates, selenoethers, organohalides	As, Se, Br
Heavy-atom complexes	
Common protein derivatives	Pt, Au, Hg, U
Cluster compounds	Ta, W
Building-unit analogs	
Selenomethionyl residues	Se
Brominated nucleotides	Br

Table XI.2. — Categories of MAD applications. An accessible x-ray range from 0.3 to 3.0 Å wavelengths covers the resonant transition from atomic numbers 20 (Ca) to 58 (Ce) for K edges and from 51 (Sb) to 92 (U) for L_{III} edges. Apart from Ca, all of the likely candidates listed in the table fall within a range of wavelengths from 0.6 to 1.8 Å for which experiments are especially convenient. (after Hendrickson, 1991, modified).

(in the case of significant anisotropy effects, several spectra can be recorded for various orientations of the crystal and subsequently averaged to get an isotropic coefficient as a starting point). The reduction of these data to f" values involves background and scaling adjustments that can be tied to theoretical values calculated at points remote from the absorption edge. f' is calculated from f" by the Kramers-Kronig transformation. These starting f' and f" values may be refined against the experimental data; this is the case in MAD experiments, in which f' and f" values are often refined together with other global parameters (partial structure of anomalous scatterers) and local variables. This refinement procedure may give correct f' and f" values in case, for instance, of systematic errors due to monochromator missetting (e.g. Weis et al., 1991).

XI.3.2. Effects of anomalous dispersion on Bragg Diffraction

We assume that the structure contains many ordered atoms (P) for which only normal scattering is detectable and a few ordered atoms (A) which have detectable anomalous scattering .

Reflections **h** and **-h** are known as Friedel mates, and in the absence of anomalous scattering they have identical intensities. With significant anomalous scattering, this is no longer the case, as shown in figure XI.3.

The actual difference

$$\Delta F_{\pm \mathbf{h}} = |{}^{\lambda}F(\mathbf{h})| - |{}^{\lambda}F(-\mathbf{h})|$$

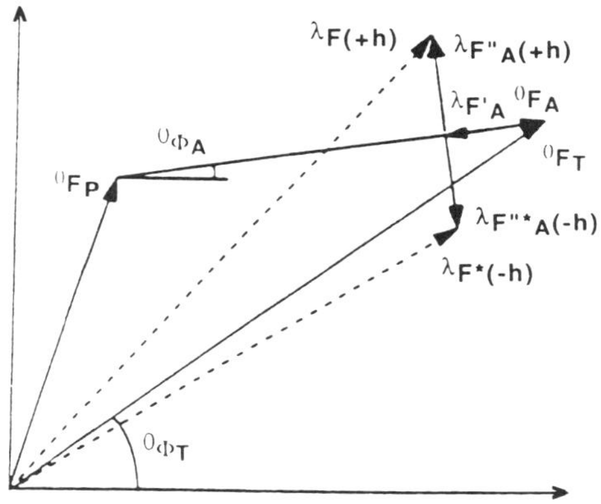

Fig. X.3. — Vector diagram in the complex plane of structure factors for one Bragg reflection **h** at one wavelength λ (see text). The structure factor for the total normal scattering, $|{}^0F_T| \exp {}^0\phi_T$, is the vector sum of normal scattering from atoms of type P and those from atoms of type A. The real and imaginary components of anomalous scattering are ${}^\lambda F'_A$ and ${}^\lambda F''_A$ respectively; when added to ${}^0F_T$, the resultant is the observed structure factor ${}^\lambda F$.

The complex conjugates of the structure factors from the **-h** reflection are also shown to illustrate the origin of the anomalous signal ($2\,{}^\lambda f''$). Note that the diagram pertains to the case for a single type of anomalous scatterer. (after Smith, 1991, modified).

between the structure amplitudes of Friedel mates or respective rotational symmetry equivalent of them, is designated as the Bijvoet or anomalous difference.

The second effect of anomalous scattering on the diffraction pattern is that the magnitudes of a given reflection differ with wavelength, because f' and f" are wavelength-dependent. We define a dispersive difference at wavelengths λ_i and λ_j as:

$$\Delta F_{\Delta\lambda} = |{}^{\lambda_i}F| - |{}^{\lambda_j}F|$$

where

$$|{}^\lambda F| = [\,|{}^\lambda F(\mathbf{h})| + |{}^\lambda F(-\mathbf{h})|\,]/2$$

Provided that the anomalous scattering is relatively weak compared with protein scattering, in the case of a single anomalous scattering type, the expected values of anomalous and dispersive ratio are (Hendrickson et al., 1985):

$$\text{rms}\,(\Delta F_{\pm h})/\text{rms}\,|F| = 2q\,{}^\lambda f'' \qquad \text{(XI.2)}$$

and

$$\text{rms}\,(\Delta F_{\Delta\lambda})/\text{rms}\,|F| = q\,|\,{}^{\lambda_i}f' - {}^{\lambda_j}f'\,| \qquad \text{(XI.3)}$$

respectively, where rms means root mean square and $q = [(2N_T/N_A)^{1/2}Z_{eff}]^{-1}$. Here, N_A is the number of anomalous scatterers per molecule, N_T is the total number of nonhydrogen atoms in the molecule; Z_{eff} is the effective normal atomic scattering for a given value of s. For increasing values of s, Z_{eff} decreases thus the expected diffraction ratios are enhanced. Note also that the scattering strength of a cluster of n anomalous scatterers is enhanced by a factor of $n^{1/2}$ if data are limited to spacings insufficient to resolve the individual atoms.

The anomalous and dispersive ratios are usually small. As an example, in the case of core streptavidin, a protein with 925 "normal" atoms and 1 Se atom as anomalous scatterer, and using three wavelengths around the Se K absorption edge, the calculated per cent ratios at s = 0 range from 0.7 to 2.5 (Hendrickson et al., 1989).

XI.3.3. The Multiple Wavelength Anomalous Diffraction method (MAD).

XI.3.3.1. The phase problem

The phase problem is a severe rate-limiting step in macromolecular structure determination. Simply stated, the diffraction pattern of a crystal gives only the amplitudes of the Fourier transforms of the crystal contents, the phase angle is lost. In principle, provided enough unique diffraction amplitudes are known, the many constraints imposed by a knowledge of general molecular structure make the inversion of the known structure amplitudes, in the absence of any phase information, into real space electron density an overdetermined problem. However, direct methods that exploit this overdeterminacy are presently only available for relatively small molecules, although there is hope to extend eventually their application to the field of macromolecular structures (for a panel of recent research in this domain, see Glusker, 1993). To date, the success of macromolecular crystallography for structures not related to ones already known is founded almost entirely on multiple isomorphous replacement (MIR). Differences in diffracted intensities from the derivative and native crystals are used to locate the heavy atoms. The calculated contributions from these centers serve as reference waves for evaluation of phases for the crystal structure. In general, the phase information from single isomorphous replacement (SIR) is ambiguous. Thus, MIR or other methods such single isomorphous replacement with anomalous scattering (SIRAS), solvent flattening or molecular averaging are used to resolve the ambiguity. But MIR phasing has its drawbacks. The preparation and interpretation of several heavy atom derivatives is a main factor limiting the speed and accuracy with which biological structures can be determined by x-ray crystallography. In this method, changes in diffraction patterns arise from chemical and structural differences, and strict isomorphism between the original structure and the derivative structure is exceptional. The ideal situation would be to rely on changes in the x-ray diffraction patterns arising from physical differences in a single sample. Anomalous scattering can indeed induce such differences. The corresponding method is called MAD (for reviews, see Fourme & Hendrickson, 1990; Hendrickson, 1991; Smith, 1991).

The principle of MAD phasing has been recognized more than 30 years ago (Okaya and Pepinsky, 1956). A landmark experiment was performed with x-ray tubes and a single-counter diffractometer (Hoppe and Jakubowsky, 1975); this work pointed out the need for a continuously tunable synchrotron radiation source and for an area detector. X-ray synchrotron sources became available in the early 1970. The development of EXAFS produced instrumentation and procedures which were directly applicable to MAD experiments. From EXAFS an' least squares refinement of crystallographic measurements, large resonance effects were demonstrated at L-edges of cesium (Phillips et al., 1978) and lanthanides (Templeton et al., 1980). Computational tests using these results showed the large potential of MAD methods (Phillips and Hodgson,1980). The realization of this potential has been fairly long to materialize, and this was due essentially to technical reasons. A MAD data collection is a complex experiment which requires extreme care; the main sources of trouble are: area detectors, crucial ingredients for which progress has been slow and uncoordinated, and which are still a main limiting factor; instabilities of synchrotron radiation sources, which became apparent when aiming at high accuracy data; the hectic pace of work and beam-time limitations at first generation synchrotron radiation facilities, which has not been optimal to perform long and careful experiments; and finally, the variety of problems associated with biological crystals such as radiation damage. The bulk of calculations performed during MAD data reduction is primarily concerned with the correction of these deficiencies. An algebraically exact theory of phase evaluation from MAD data, developed by Karle (1980) and implemented by Hendrickson (1985), has been central for the determination of most structures resolved up to now by the MAD method.

The first crystal structure solved by the MAD technique was parvalbumin from *Opsanus tau* (Kahn et al., 1985); the structures of a blue copper protein from cucumber (Guss et al.,1988) and of core streptavidin (Hendrickson et al., 1989) were more stringent tests of the MAD technique as the structures of the proteins involved were completely unknown. The incorporation of selenomethionine in place of methionine, suggested by Hendrickson (1985), can be a general vehicle for protein structure determination (Horton and Hendrickson, 1989); similarly, brominated bases can generally be used to provide suitable labels for nucleic acid structures and the structure of a DNA-drug complex has been determined by this method (Ogata et al., 1989). To date, a number of structure determinations have been reported in abstract or in extended form (table XI.3). MAD experiments are now available or in active developments at several synchrotron radiation facilities.

XI.3.3.2. Theoretical Foundation of MAD method

XI.3.3.2.1. Introduction

MAD is one of *de novo* phasing methods such as SIRAS or MIR. In these methods, the phasing procedure can be divided for clarity in two basic steps. The first step is the determination of the structural parameters of a subset of atoms which have, with respect to the rest of atoms, particular properties such as a large atomic number and/or significant anomalous scattering at the X-ray wavelength(s) used in data collection. The second step is the use of these parameters in the estimation of Fourier coefficients suitable for the calculation of an electron density map of the total structure. The MAD method relies on a subset of atoms with measurable anomalous scattering. This subset of atoms, A, is called hereafter the partial structure. The pertinent parameters of the partial structure are atomic coordinates, occupancies, temperature factors and anomalous components of atomic scattering factors at each wavelength. The first step of a MAD phasing is the determination of these parameters. The second step, for each reflexion **h**, makes a pivotal use of a reference structure factor *calculated*

Molecule	#AS / kD	source	detector	d_{min} (MAD)	reference
1/ Tests					
Erythrocruorin	1Fe / 16	tube lines	scintil	3.0	1975
Rubredoxin	1Fe / 6	SSRL	films	2.1	1977
2/ Published MAD phased electron density					
Parvalbumin	1.3 Tb / 10	LURE	MWPC	2.3	1985
Cytochrome c'	2Fe / (2x14)	PF	scintil.	6.0	1986
Azurin	1Cu / 14	CHESS	films	3.0	1987
Lamprey hemoglobin	1Fe / 18	SSRL	MWPC	3.0	1988
Ferrodoxin	8Fe / 5.9	SSRL	MWPC	5.0	1988
Cucumber basic protein*	1Cu /11	SSRL	MWPC	3.0	1988
Streptavidin *	2Se / (2x14)	PF	scintil.	3.3	1989
DNA:chromomycin*	4Br / (4x3.5)	PF	IP	2.9	1989
Ribonuclease H*	3Se / 18	PF	IP	2.2	1990
Cytochrome c-553*	1Fe / 8.7	PF	IP	2.2	1990
Urechis hemoglobin*	2Fe / 31	SSRL	MWPC	3.0	1991
Rat lectin CRD domain*	4Ho sites / 27	LURE	MWPC	2.5	1991
Tenascin FnIII domain*	2Se	CHESS	IP	2.5	1993
3/ Published MIR+MAD phased electron density					
CD4 fragment (MIR+MAD)	2Pt sites / 18	Au lines	MWPC	3.5	1990
Interleukin 1α (MIR + MAD)	3Se / 18	PF	scintil.	3.3	1990
4/ Data sets for MAD phasing under evaluation					
Cutinase	Hg	LURE	MWPC	1.7	
Hemocyanin Og unit	Cu	PF/LURE	IP/MWPC	3.0	
Thioredoxin	Se,Cu	SSRL	MWPC	3.0	

* structure solved by direct eveluation from MAD data, without any other information to develop the structure

Table X.3. — Results in MAD method

from these parameters in order to estimate the phase (and possibly modulus) of the Fourier coefficient to be used in the calculation of the initial electron density map. In all recent applications of MAD, the reference structure factor is the wavelength-dependent part of the structure factor of the partial structure, and the result is an estimation of the 'normal' part, ${}^{0}F_T$, of the structure factor of the total structure (the subscript T refers to the total structure, and the superscript 0 to a wavelength-independent quantity)(fig. XI.3). MAD has two advantages over SIRAS or MIR. First, it is a method in which the theory of phasing a wave from interference with several reference waves is strictly realised in practice, due to the fact that all measurements are made on a single sample. Second, the algebraic analysis described below gives the coefficients for the calculation of a Patterson map of the partial structure, instead of approximations of these coefficients.

XI.3.3.2.2. Determination of the partial structure of anomalous scatterers

We assume that the collection, analysis and scaling of multiple wavelength diffraction data have been performed, providing for each reflection h, n intensity measurements $|{}^{\lambda}F_T(\pm h)|^2$ (the superscript λ is the wavelength of the measurement; ±h refers to the anomalous pair h, -h).The determination of the structural parameters of the partial structure may be performed using single wavelength data, although not in optimal conditions. In effect, the Fourier map calculated with coefficients $[|{}^{\lambda}F_T(\mathbf{h})| - |{}^{\lambda}F_T(-\mathbf{h})|]^2$ (Rossmann, 1961) is a sharpened Patterson that represents the distribution of interatomic vectors for the anomalous scatterers, but with a degraded signal to noise ratio (Karle, 1980). An intrinsically more powerful method relies on the analysis of multiple wavelength data by an algebraic procedure first proposed by Karle (1980). Karle expressed all quantities in terms of two structure factors ${}^0F_A$ and ${}^0F_P$ that have contributions from normal scattering only from respectively anomalous scatterers and the rest of the atoms. Hendrickson (1985) used the formulation in terms of ${}^0F_A$ and ${}^0F_T$ in his implementation of the algebraic method. With notations as in figure XI.3, equations are the following for a single type of anomalous scatterers:

$$\begin{aligned} |{}^{\lambda}F(\pm\mathbf{h})|^2 = \; & |{}^0F_T|^2 + a(\lambda)\,|{}^0F_A|^2 \\ & + b(\lambda)\,|{}^0F_T|\,|{}^0F_A| \cos({}^0\varphi_T - {}^0\varphi_A) \\ & \pm c(\lambda)\,|{}^0F_T|\,|{}^0F_A| \sin({}^0\varphi_T - {}^0\varphi_A) \end{aligned} \qquad \text{(XI.4)}$$

with:

$$a(\lambda) = ({}^{\lambda}f'^2 + {}^{\lambda}f''^2)/\,f^{02} \qquad b(\lambda) = 2\,({}^{\lambda}f'/f^0) \qquad c(\lambda) = 2\,({}^{\lambda}f''/f^0)$$

Least-squares procedures implemented to solve the system of equations (XI.4) (Hendrickson, 1985; Chapuis et al., 1985) take advantage of their essentially linear character. Given the coefficients a, b, and c determined by the scattering factors, each equation for the case of a single kind of anomalous scatterer has the form

$$y = p_1 + a\,p_2 + b\,p_3 + c\,p_4$$

and the system can readily be solved for the parameters $\{p_j\}$. These parameters are not independent, however, as they are related by a trigonometric identity such that

$$p_1 p_2 = p_3^2 + p_4^2 \qquad \text{(XI.5)}$$

This constraint on the system can be imposed by introducing an undetermined Lagrange multiplier and continuing the non-linear constrained refinement to convergence. In this way, the three parameters, $|{}^0F_T|$, $|{}^0F_A|$ and $\Delta\varphi = {}^0\varphi_T - {}^0\varphi_A$, that characterize the MAD information are determined. Such an analysis is performed by the MADLSQ program (Hendrickson, 1985) for each reflection.

Although implementation of the linear least-squares phase of the procedure requires prior specification of scale factors and anomalous scattering factors, these values can subsequently be refined in the non-linear phase as global variables affecting the entire data set (Hendrickson, 1985). A program of this kind has proved effective in refining scattering factors from crude estimates (Hendrickson et al., 1989). It is also possible to impose the constraint of equation (XI.5) implicitly through explicit refinement of the primary variables $|{}^0F_T|$, $|{}^0F_A|$ and $\Delta\varphi$.

From the set of $|{}^0F_A|^2$, the structure of anomalous scatterers can be solved by standard methods such as Patterson or direct methods, and then refined by least squares techniques. The entire set of $|{}^0F_A|$ values, excluding outliers (i.e. abnormal values), are valid data for this refinement in contrast to the limited subset of Bijvoet differences that can be used in the single-wavelength case (Hendrickson & Teeter, 1981). An alternative to refinement of the anomalous scatterer model against these $|{}^0F_A|$ estimates would be to refine directly against the set of $|{}^\lambda F|$ observations (Hendrickson, 1985). This scheme is like the phase refinement procedure used in isomorphous replacement (Dickerson et al., 1961), but here $|{}^0F_T|$ and ${}^0\varphi_T$ are refined as local parameters for each reflection and ${}^0F_A$ is expressed through the global parameters of the atomic model for anomalous scattering centers. In this context, it is possible to return to the general expression of algebraic equations for several anomalous scattering species and allow an arbitrarily complex structure of anomalous scatterers. A program called MADDST has been written to implement this procedure and to account for anisotropy in anomalous scattering (Fanchon and Hendrickson, 1989).

XI.3.3.2.3. Estimation of phases

The second step makes use of the reference waves from the partial structure to get estimates of ${}^0F_T$.

In the context of the algebraic analysis (Hendrickson, 1985), the phase ${}^0\varphi_A$ calculated from the refined model of the partial structure is used to derive ${}^0\varphi_T$ from the phase difference $\Delta\varphi$. The phase and amplitude of the Fourier coefficient ${}^0F_T$ are thus determined. Two electron densities based on weighted ${}^0F_T$ coefficients are traced with ${}^0\varphi_A$ calculated for the two enantiomorphs of the partial structure. The correct map is distinguished by chemical reasonableness or by symmetry considerations. Weights are based on residuals in the least squares fit of the phase equations to the observations.

Another approach is based on probability methods. Once the partial structure parameters have been obtained, there remains the problem that the resulting phase indications given by MAD (and other de novo methods) are most often ambiguous because of varying degrees of bimodality in most of the acentric phase probability distributions. The standard solution is to use centroid structure factors (Blow & Crick, 1959) which minimize the root mean square error in the final map and may thus be viewed as a damage limitation measure. Following the description of Hendrickson and Lattman (1970), a Gaussian distribution is ascribed to errors of closure in $|F|^2$ values instead of $|F|$. A probability distribution function for the phase information is derived from the intensity measurements. Fourier coefficients used for the calculation of the electron density map have figure of merit-weighted moduli and centroid phases. The correct enantiomorph of the partial structure is selected as in the algebraic method. This approach has been used either with simulated MAD data (Phillips and Hodgson, 1980) or real data (Kahn et al., 1985; Harada et al., 1986; Korszun, 1987; Guss et al., 1988). Pahler et al. (1990) used it as an alternative way to derive weights for the ${}^0F_T$ Fourier synthesis in the algebraic analysis and showed that the probability distribution $P(\varphi)$ can be cast in the A, B, C, D representation of Hendrickson and Lattman (1970):

$$P(\varphi) = N\,(A \cos\varphi + B \sin\varphi + C \cos 2\varphi + D \sin 2\varphi) \qquad \text{(XI.6)}$$

The A, B, C, D formulation facilitates the combination of MAD information with information from other sources, such as isomorphous replacement, solvent flattening, non-crystallographic symmetry and partial models.

Instead of the calculation of the centroid phase, the latest version of this standard solution is the calculation of the real and imaginary components of the Fourier coefficients which will provide the best electron density map (Chiadmi et al., 1993).

Direct phase determination for macromolecular structure determination by entropy maximization and likelihood ranking is being gradually implemented. A status report and perspectives of this 'Bayesian programme' can be found in Bricogne (1993). As discussed in this article, this statistical theory, which encompasses all phasing procedures, may be important for MIR and MAD. Recourse to maximum-likelihood provides an optimal and definitive answer to the heavy atom parameter refinement in MIR (Bricogne, 1991). This approach can also be applied as well to the refinement of parameters in the MAD method where it provides a sounder statistical treatment than is currently available. In order to imlement it, a program has been written which can accomodate both MIR and MAD data, or any mixture of them (de La Fortelle et al., 1993). Maximum entropy and maximum likelihood methods can also be used for the systematic resolution of phase ambiguities, assuming that the phase information available at each stage from the MIR or MAD methods have been encoded by means of A, B, C, D coefficients. Further, because MAD exploits only one configuration of anomalous scatterers, it is unable to phase reflexions for which the transform of that constellation of scatterers is weak. A corollary to this observation is that, even in the most favourable circumstances, MAD would have to rely on some other procedure for propagating phase information, from those reflexions where it is strongly indicated, to those where it is essentially absent. Again, maximum entropy and maximum likelihood method will perform this task in an optimal fashion.

XI.3.3.3. Instrumentation

In macromolecular crystallography, even when the anomalous effects are maximized, the changes in intensities are small: the scattering from the anomalous scatterers is typically a rather small perturbation of the total scattering, but, as seen previously, it is in differences between intensity values at different wavelengths and between the Bijvoet mates that the phasing power resides. Accurate measurements of small intensity differences is thus essential; achieving this goal depends on the interplay between a careful experimental design, a strategy for efficient acquisition of data with appropriate counting statistics, and care to minimize or correct systematic errors at various steps, from sample preparation and data acquisition to reduction of raw data.

XI.3.3.3.1. X-ray source. X-ray tubes have been used for MAD measurements. But the spectral brightness of synchrotron radiation is better suited for MAD work. The bending magnet radiation from several existing sources (at CHESS, NSLS, SSRL, Photon Factory, LURE, VEPP 3, DESY, SRS and ESRF) provide adequate flux for many experiments. Wigglers give enhanced flux and shift the spectrum to higher energy; accordingly, they extend the range of accessible absorption edges and they also can make third-generation, low energy sources, suitable for MAD (at ALS, MAX-II, SRRC, LNLS). Undulators on third generation, high energy sources (available currently at ESRF and later at SPRING 8 and APS) have a ultrahigh brightness in a restricted wavelength range which may be purposely adjusted for a given anomalous scattering species. In all cases, the stability of the beam is a critical point, especially for low emittance storage rings, in order to get a stable focal point, a constant polarization of the x-ray beam at the sample location and no wavelength shifts.

XI.3.3.3.2. X-ray optics. The optical system has specifications comparable to those of many EXAFS spectrometers: i) monochromatic beam with an energy band-pass of 1-2 eV; ii) free of higher order harmonics; iii) accurately and repeatedly tunable; (iv) fixed focal point. Designs of variable costs and complexities have been proposed to meet these requirements. The emphasis is on stability (thermal stability, mechanical stability, elimination of backlash, filtering of vibrations, beam-position monitoring etc...).

XI.3.3.3.3. Goniometer. A multi-circle goniometer is required to provide maximum flexibility. A vertical equatorial plane is optimal as the synchrotron beam is polarised in the horizontal plane. Axis movements must have angular steps of 10^{-3} deg, as usual in synchrotron radiation diffractometry. Slewing speeds should be as large as possible, as MAD measurements require frequent changes in crystal orientation, for instance to measure Friedel-related data. The intensity of the beam after the collimation system of the goniometer must be monitored. This monitor is used to rescale diffraction data collected with an x-ray beam of varying intensity; it may also be used to give the reference I_0 signal in fluorescence EXAFS experiments from which the variation of f" and f' in the near-edge region will be derived.

XI.3.3.3.4. Detector. The x-ray detector is the most important piece of equipment in a MAD setup. Both a single counter detector and area detectors have been used for successful MAD studies.

Single counter diffractometers have the advantage of flexibility, which is useful for complex data collection schemes, and provide a good signal-to-noise ratio because peak and background measurements can be optimised. The low efficiency is acceptable if the sample is stable and beam time available (e.g. structure of core streptavidin by Hendrickson et al., 1989)

Area detectors measure many reflections simultaneously, so that these devices are an integral feature of MAD experiments. Electronic area detectors (EAD) acquire diffraction images directly in digital form, like a single channel diffractometer; as the dead time between two frames is very short, the data collection can be subdivided into many frames, thus optimising the signal-to-noise ratio. Multiwire proportional chambers at SSRL(Phizackerley et al., 1986) or LURE (Kahn et al., 1986) have been used for high accuracy MAD measurements but they are presently count rate limited and not suitable for very short wavelengths. These defaults could be cured by microstrip technology, fast readout systems and pressurized detectors (for a review see e.g. Walenta, 1991). A commercial television detector is used at NSLS. Charge coupled device detectors have a high dynamic range, a high quantum efficiency and a small point spread function; a prototype system is available at NSLS (Westbrook et al., 1989) and another one, using an X-ray image intensifier, is being developed at ESRF (Moy and Gibney, 1992). Imaging plates have several outstanding characteristics: unlimited count rate, large dynamic range, very uniform response, high efficiency over the whole useful wavelength range, small point spread function (Miyahara et al., 1986). The main default, which is perhaps more serious for MAD than for normal data collection, is the long processing time of each image. Various technical solutions have been proposed to overcome this problem: a carrousel of several detectors each with its built-in laser scanner, a fast plate exchanger or even a belt-type imaging plate (Sakabe, 1991).

XI.3.3.3.5. Additional experimental apparatus The degradation of crystals is a source of systematic errors in MAD measurements. The crystal lifetime is frequently increased by keeping the the crystal at a constant temperature close to 0°C. This can be obtained by commercial devices which provide a directed jet of cool gas which is coaxial with the glass capillary tube in which the sample is mounted. Another interesting possibility is shock-cooling of crystals at cryogenic temperatures, by which crystal degradation is in many cases virtually eliminated (Hope, 1988).

As discussed previously, it is important, prior to intensity measurements, to record XANES and EXAFS spectra from the sample crystal itself. A scintillation detector, mounted on one side of the diffractometer, measures fluorescence from the sample crystal as the x-ray energy is scanned across the absorption edge. The detector, positioned very close to the crystal, is mounted in the horizontal plane containing the beam and at 90° to the beam in order to decrease the intensity of the elastically scattered x-rays but not the fluorescence.

XI.3.3.4. Diffraction Data Collection and Processing

The first step of the procedure is the determination of the variation of anomalous scattering factors as described previously. Fluorescence x-ray spectra in the near-edge region are reduced to f" values and the Kramers-Kronig transformation produces f' values. Both of these calculations are made subject to constraints to values obtained from quantum calculations at points relatively remote from the edge. In case of significant pleiochroism in the near-edge region, the fluorescence spectra are recorded for several orientations of the crystal with respect to the polarized incident beam and averaged values of f" and f' are calculated.

Then the choice of wavelengths is performed. Large values of f" and large values of differences ${}^{\lambda_i}f' - {}^{\lambda_j}f'$ are both required to get the best phasing power; hence, from eq. (XI.2) and (XI.3), both Bijvoet and dispersive ratios must be large.

In fact, Bijvoet and dispersive differences provide orthogonal phase information and are thus complementary; moreover, the strength of these diffferences depends in a predictable way on the magnitudes of anomalous scattering factors. The selection of wavelengths includes at least a pair of wavelengths very close to an absorption edge. One of these wavelengths corresponds in general to the largest value of f", and the other one to the inflection point in the rise of absorption at the edge which corresponds to the largest magnitude of f'. A third wavelength at a point remote from the edge (preferably on the high energy side, where f" is larger and the absorption lower) is added, in order to contrast the extreme value of f' that occurs at the inflection point with the smaller magnitude observed away from the edge.

Data collection can then proceed. From expected Bijvoet and dispersive ratios, counting times are set so that counting statistics will not be limiting. As usual, diffraction intensities must be processed to account for a number of factors including the Lorentz factor, polarization effects (Kahn et al., 1982), detector count loss, source intensity variation, absorption and radiation damage.

Through an appropriate experimental strategy, it is often possible to arrange that remaining common systematic errors will tend to be eliminated by differencing. Thus, one obtains greater precision in phase determination than the independent accuracy of measurement would warrant. Experimental control of systematic errors has several components:

i) All data, or at least all data pertaining to a given phase, should be taken from the same crystal, eliminating the intercrystal absorption and scaling errors that limit the accuracy of the isomorphous replacement method.

ii) The data for Bijvoet mates and for all wavelengths should be collected close together in time; this procedure minimizes both the impact of radiation damage and of slow drifts in the state of polarization of the beam at the sample location, in the x-ray optics and in the detector and beam monitor response.

iii) Crystals are mounted so as to exploit their symmetry and morphology to reduce absorption errors arising from different path lengths for the mates of a Bijvoet pair

iv) Absorption differences at different wavelengths can be taken into account in experimental transmission factors.

Many of the residual experimental errors in Bijvoet and dispersive differences can be further reduced numerically by parameterized local scaling of adjacent reflections as originally developed for highly precise, single-wavelength measurements (Hendrickson and Teeter, 1981). Afterwards, data from separate wavelengths are brought to a common scale; this is first done by bringing the various data sets into least-square agreement by the local scaling algorithm, then additional scale factors that account for variation in total scattering strength must also be applied. At each step, an appropriate rejection of outliers is imposed. Finally, all data are placed on a quasi-absolute scale by Wilson's statistics, sorted, and equivalent and redundant data are merged. It is emphasized that this merging occurs at a late stage, and may be further deferred until after phase determination for separate sets.

XI.3.3.5. Applications

Although the possibility for phase determination from MAD data has been appreciated for decades, the realization of this promise has been quite slow in coming. Recently, however, the pace has quickened markedly. This reflects both the maturation of synchrotron instrumentation and also significant advances in methods of analysis as described in previous sections. The pioneering study by Hoppe and Jakubowski (1975), while determining phases for only a few hundred reflections, clearly demonstrated practical feasibility even with conventional x-ray sources. Another study with conventional x-rays, that on the small molecule selenolanthionine (Hendrickson, 1985), showed that accurate phases could be determined even using the continuum Bremsstrahlung from an x-ray tube. Another test study worth singling out was the analysis of lamprey hemoglobin, first reported at 5.5Å resolution (Hendrickson, 1984), in which many of the experimental and analytical procedures now in the use were first developed (Hendrickson et al., 1988).

Despite the usefulness of test demonstrations, it was only through complete analyses on macromolecules of unknown structure that the method could be proven. This has now been accomplished, as shown in table XI.3.

XI.3.3.6. Prospects

After a long period of gestation, the MAD method can fairly be said to have emerged with vitality. To date applications exploiting K-shell edges have predominated. Many candidate problems including nearly all metalloproteins fall into this category. The MAD approach with K edges has also been successful with elements such as selenium and bromine that are not normally heavy enough for the MIR method. Biologically engineered selenomethionine proteins might serve as a general vehicle for protein structure determination (Hendrickson, 1985; Horton & Hendrickson, 1989). Similarly, brominations of nucleic-acid bases (U or C) offer a rather general approach for this class of problems. L-shell transitions lead to much stronger anomalous scattering, particularly at the white lines such as occur with lanthanides; thus MAD studies would be appealing on nearly any heavy-atom derivative that might be used for MIR phasing -- but without a requirement for isomorphism. As instrumentation develops, experiments at very long wavelengths might be attractive (especially at S and P absorption edges, despite complexities due to the very high absorption coefficients). Using high energy storage rings and appropriate optics and detectors, very short wavelengths might also become more practical.

The MAD method is not without its complexities. The multiple data sets and the need for high precision in measurements in many MAD experiments are complications. Fortunately, there are advances in instrumentation, especially detectors, as previously discussed, and the availability and stability of synchrotron radiation for MAD experiments is also improving. Another potential pitfall concerns the effect of anisotropy, but studies indicate that, even when present, this need not be debilitating and can be accounted for (Fanchon and Hendrickson, 1989). In the case of selenomethionine proteins and certain other situations, the number of anomalous scatterer sites in a large macromolecule might present a daunting structural problem in its own right. However, with accurate $|{}^0F_A|$ values and the power of direct methods, one can also hope to address this problem. In the end, the prospect that, for macromolecules as for small molecules, one might enjoy assurance of a straightforward structural determination once a crystal is grown is a powerful incentive for continued development of MAD phasing technology. It is not easy to predict the future of MAD with respect to other competing phasing methods, such as MIR, molecular replacement and, perhaps, direct methods. NMR may also provide (currently for proteins of less than about 20 kD) a model for molecular replacement. In fact scientists in structural biology are pragmatic and use all that is available and efficient to elucidate three dimensional structures. Several new synchrotron beam lines and instruments dedicated to MAD experiments are being built, and some of them are completed. These instruments will certainly be crucial for the development of the method.

XI.4. Kinetic crystallography and fast data collection methods

The introduction of synchrotron radiation for structural studies by Rosenbaum & al. in 1971 brought the time resolution of X-ray techniques into the domain where biological action takes place. An intense beam is required with - especially polychromatic techniques - a very low divergence. In this respect, synchrotron sources are the best sources, even with respect to laser plasmas which produce a more intense but also more divergent burst of radiation.

A number of reports or reviews on the subject of time-resolved crystallography have appeared in recent years (e. g. Hajdu et al., 1988; Moffat, 1989; Moffat & Helliwell, 1989; Hajdu and Johnson, 1990; Pai, 1992). The proceedings of a meeting adressing these topics give a snapshot of the present status in the field (Cruickshank et al., 1992).

XI.4.1. Monochromatic techniques

With monochromatic X-rays, only a small fraction of lattice planes diffract at any particular orientation of the crystal and the crystal has to be rotated in the beam in order to record progressively all integrated intensities. Using conventional sources, data rates of 50 to 1000 reflections/s have been achieved, and undulators from third generation high energy storage rings are brighter by several orders of magnitude. Considering only the x-ray sources, there is a potential to bring monochromatic techniques into the subsecond range. Let us discuss the various technical problems. The centrifugal acceleration produced by fast rotation is not a serious problem: at 1000 rev./s (then, 90° are swept in 15 ms), an average size protein crystal (0.5 mm in diameter) experiences a maximal acceleration of only 0.27g (Hajdu & Andersson, 1992). The real difficulty is the detector and the data collection method. A method which looks promising is based on the Weissenberg technique (Sakabe, 1991; Andersson et al., 1991). In this technique, as described previously, it is posssible to record on each image Bragg reflections excited through an angular range of several degrees with few overlaps and while keeping a reasonable signal to noise ratio on the detector. With an undulator and an even larger imaging plate, it might be possible, especially in the case of a crystal of high symmetry, to record a large fraction of the data on a single image with an exposure time limited essentially by the source brilliance.

Monochromatic techniques tolerate crystals with high mosaicity; images are relatively easy to process and may provide quality data with a good completeness both at low and high resolution .

XI.4.2. White radiation technique. The Laue method.

Laue diffraction is discussed extensively in chapter XII by C. Wilkinson. Briefly summarized, Laue diffraction is produced when polychromatic x-rays illuminate a stationnary crystal. Reflections are integrated through a small wavelength range instead of a small angular range. A large number of lattice planes diffract simultaneously under these circumstances as the Bragg condition is satisfied for each of these planes by at least one wavelength of the spectrum. A short exposure time is the principal asset of the technique. In a quantitative analysis of Laue diffraction patterns recorded from the radiation emitted from a single bunch of electrons traversing an undulator Szebenyi et al. (1992) showed that data obtained in a 100 ps exposure are of useful quality. A few diffraction pictures taken at different orientations of the crystal can result in useful data sets. The exact number depends on the symmetry of the crystal; in the case of a cubic crystal, one single exposure can be sufficient. The advantage of Laue over monochromatic techniques is largely lost as soon as more than one exposure is needed to complete the data set.

The extraction of structure factor amplitudes from Laue images is in general more complicated than the processing of manochromatic data sets. There are are two main ways to extract this information. The first is based on a wavelength normalization curve to compensate for wavelength factors in the intensity measurements; this curve is derived from experimental data only, using intensities of equivalent reflections excited at various wavelengths (Helliwell et al., 1989). The second is based on a difference method, and can be used for the analysis of structural changes relative to a known starting structure.

The Laue technique suffers from two specific limitations, inherent to the physics of the method, which restrict its use in routine data collection (Hajdu et al, 1991) and have implications to the use of this method in time-resolved measurements.

- Sensitivity to disorder in the crystal

The size and shape of a reflection on a Laue photograph depends on the dimensions of reciprocal points (crystal mosaicity) and on the divergence of the x-ray beam. Even moderate mosaicity (0.1-0.2°), makes reflections radially elongated, which can cause extensive overlapping between neighbouring relections. Mosaicity of most crystals increases during a structural transition or reaction, and therefore useful structural data may not always be attainable from reactive crystals with the Laue method.

- Lack of low resolution data

The two limiting spheres of radius $1/d_{max}$ and $1/d_{min}$ are in contact at the origin of the reciprocal lattice, so the efficiency of the Laue method at low resolution is hardly better than the efficiency of monochromatic still photographs. In addition, most of the reflections in this region are harmonic overlaps and those which are singlets are at the short wavelength limit of the spectrum where the scattering power is the lowest, and the change in spectral intensity is the steepest. Finally, there is little or no reliable data in the range of about 4 - 5Å resolution to infinity. The consequence of the low resolution "hole" is that structural elements with a high B factor are difficult to identify from Laue data. The increase in motion in a structure causes a weakening of reflections at higher resolutions, so that the errors increase in parallel. Little information can be expected for structural motifs with high B factors from weak reflections at high resolution.

XI.4.3. Starting a reaction in the crystal and the synchronization problem

Although there are still problems associated with the crystallographic part of time-resolved experiments, the major obstacles now seem to be chemical in nature, and connected to the fact that the image obtained from a diffraction experiment is the average over 10^{14} -10^{15} molecules during the exposure time.

First, a simultaneous initiation of the reaction in the crystal is required. For most reactions in the crystalline state, temperature-jumps, pressure-jumps or Ph-jumps might be used to start the reaction (Pai, 1992). Very often, however, only photochemical activation is fast enough for efficient initial synchronization. Cage compounds are used for protecting the substrate against the enzyme, or the catalysis itself may be blocked through the covalent attachment of a reversible component to crucial active-site groups. Intense light pulses are needed to achieve fast conversion of caged compounds into substrates.

The kinetics of the catalyzed reaction in solution and in crystal must be investigated before starting time-resolved diffraction, and knowledge of the different rates of transformation of substrates into intermediates and finally into products should guide the design of the experiment. The course of the reaction during the diffraction experiment has also to be monitored, using for instance microspectrophotometer systems with time resolution in the millisecond range.

In conclusion, a combination of various experimental techniques - notably NMR and other spectroscopic techniques - with x-ray diffraction methods and computational procedures offers the obvious route for 4 dimensional (time being the 4th dimension, Andersson et al., 1992) structural studies in the future. Using cryotechniques, it should be possible to lengthen the time-intervals available for probing specific intermediates. Concerning diffraction methods, both monochromatic and polychromatic methods need consideration and further developments.

Acknowledgements: this chapter makes use of materials from various reviews and articles, in particular Fourme & Hendrickson (1990), Fourme et al. (1991), Hajdu & Andersson (1992) and other quoted references.

General Readings

BLUNDELL,T.L. & JOHNSON L. N., 1976 - Protein crystallography - Academic Press (London)

HELLIWELL, J.R., 1992 - Macromolecular Crystallography with synchrotron radiation - Cambridge University Press (Cambridge)

STRYER, L., 1988, third edition - Biochemistry - Freeman (New York)

REFERENCES

ANDERSSON, I., CLIFTON, I.J., FULOP, V. & HADJU, J.,1991 - in Crystallographic Computing 5: From Chemistry to Biology; eds. MORAS, D. PODJARNY, A.D. & THIERRY, J.C. - Oxford University Press (Oxford)

ANDERSSON, I., CLIFTON, I.J., EDWARDS, S.L., FULOP, V., HADFIELD, A.T., NORDLUND, P., PHIZACKERLEY, R.P., SOLTIS, S.M., WAKATSUKI, S. & HADJU, J., 1992 -in Time Resolved Macromolecular Crystallography; eds.CRUICKSHANK et al. - Oxford University Press for the Royal Society

BLOW, D. M. & CRICK, F. H. C., 1959 - Acta Cryst. 12, 794

BRICOGNE, G., 1991 - in Isomorphous Replacement and Anomalous Scattering, Proceedings of the CCP4 Study Week End jan. 25-26 1991, p. 60 - Daresbury Laboratory

BRICOGNE, G., 1993 - Acta Cryst. D49, 37

CHAPUIS, G., TEMPLETON, D. H. & TEMPLETON, L. K., 1985 - Acta Cryst. A41, 274

CHIADMI, M., KAHN, R., FOURME, R. & DE LA FORTELLE, E., 1993 - J. Appl. Cryst. (in the press)

CROMER, D.T., & LIBERMAN, D., 1970 - J. Chem. Phys. 53, 1891

CRUICKSHANK, D.W.J., HELLIWELL, J.R. & JOHNSON, L.N., 1992 - Phil. Trans. R. Soc. London (A) 340

DE LA FORTELLE, E., BRICOGNE, G., KAHN, R. & FOURME, R., 1993 - abstract to the 16th International Congress of Crystallography, Beijing

DICKERSON, R. E., KENDREW, J. C. & STRANDBERG, B. E., 1961 - Acta Cryst. 14, 1188

DUMAS, C., 1988 - Doctoral Thesis, Université Paris-Sud, Orsay, France.

DUMAS, C., LASCU, I., MORERA, S., GLASER, P., FOURME, R., WALLET, V., LACOMBE, M.L., VERON, M. & JANIN, J., 1992 - The EMBO J. 11(9), 3203

FANCHON, E. & HENDRICKSON, W.A., 1989 Abstracts of the American Crystallographic Association Meeting, series 2, vol. 17, p.122.

FOURME, R., BAHRI, A., KAHN, R. & BOSSHARD, R., 1991 - Proc. of the European Workshop on X-ray detectors for synchrotron radiation sources (Aussois), ed. WALENTA, A.H. - EEC (Siegen)

FOURME, R. & HENDRICKSON, W.A. - 1990 - in Synchrotron Radiation and Biophysics, eds. S.S. HASNAIN, p. 156 - Hellis Horwood Limited (Chichester)

FOURME, R., DE LA FORTELLE, E., SHEPARD, W. & KAHN, R. ,1993, (submitted for publication)

GLUSKER, J.P. ed., 1993 - Acta Cryst. D 49 (1)

GUSS, J.M., MERRITT, E.A., PHIZACKERLEY, R.P., HEDMAN, B., MURATA, M., HODGSON, K.O. & FREEMAN, H.C., 1988 - Science 241, 806

HAJDU, J., ACHARYA, K.R., STUART, D.I. & JOHNSON, L.N.,1988 - Trends Biochem. Sci. 13, 104

HAJDU, J., ALMO, S.C., FARBER, G.K., PRATER, J.K., PETSKO, G.A., WAKATSUKI, S., CLIFTON, I.J. & FULOP, V., 1991 - in Crystallographic Computing 5: From Chemistry to Biology; eds. MORAS, D., PODJARNY, A.D. & THIERRY, J.C. - Oxford University Press (Oxford)

HAJDU J. & ANDERSSON I., 1992 - in Approches Expérimentales de la Dynamique des Macromolécules Biologiques, FOURME, R. & VACHETTE, P. ed., p 89, Cahiers IMABIO - CNRS (Paris).

HAJDU, J. & JOHNSON, L.N., 1990 - Biochemistry 29, 1668

HARADA, S., YASUI, M., MURAKAWA, K., KASAI, N. & SATOW, Y., 1986 - J. Appl. Cryst. 19, 448

HELLIWELL, J.R., HABASH, J., CRUICHSHANK, D.W.J., HARDING, M.M., GREENHOUGH, T.J., CAMPBELL, J.W., CLIFTON, I.J., ELDER, M., MACHIN, P.A., PAPIZ, M.Z. & ZUREK, S., 1989 - J. Appl. Cryst. 22, 483

HENDRICKSON, W. A., 1984 - Acta Cryst. A40, C3.

HENDRICKSON, W.A., 1985 - Trans. Am. Crystallogr. Assoc. 21,11

HENDRICKSON, W.A.,1991 - Science 254, 51

HENDRICKSON, W. A., & LATTMAN, E. E., 1970 - Acta Cryst. B26, 136

HENDRICKSON, W.A., PAHLER, A., SMITH, J.L., SATOW, Y., MERRITT, E.A. & PHIZACKERLEY, R.P., 1989 - Proc. Natl. Acad. Sci. USA 86, 2190

HENDRICKSON, W. A., SMITH, J. L., PHIZACKERLEY, R. P. & MERRITT, E. A.,1988 - Proteins 4, 77

HENDRICKSON, W.A. & TEETER, M.M., 1981 - Nature 290, 107

HOPE, H., 1988 - Acta Cryst. B44, 22

HOPPE, W., JAKUBOWSKI, V., 1975 - in Anomalous Scattering, ed. RAMASESHAN, S. & ABRAHAMS, S. C., p. 437 - Munksgaard (Copenhagen)

HORTON, J.R. & HENDRICKSON, W.A., 1989 - Abstracts of the American Crystallographic Association Meeting, series 2, vol. 17, p.122

JAMES, R.J.,1965 - The Optical Principles of the Diffraction of X-rays - Cornell University Press (Ithaca)

KAHN, R., FOURME, R., BOSSHARD, R., CHIADMI, M., RISLER, J. L., DIDEBERG, O. & WERY, J. P., 1985 - FEBS Lett. 179, 133

KAHN, R., FOURME, R., BOSSHARD, R. & SAINTAGNE, V., 1986) - Nucl. Instrum. and Meth. A246, 596

KAHN, R., FOURME, R., GADET, A., JANIN, J., DUMAS, C. & ANDRÉ, D., 1982 - J. Appl. Crystallogr. 15, 330

KARLE, J., 1980 - Int. J. Quant. Chem. 7, 357

KORSZUN, Z. R., 1987 - J. Mol. Biol. 196, 413

MIYAHARA, J., TAKAHASHI K., AMEMIYA,Y., KAMIYA, N. & SATOW, Y.,1986 - Nucl. Instrum. and Meth. A246, 572

MOFFAT, K., 1989 - Annu. Rev. Biophys. Biophys. Chem. 18,309

MOFFAT, K. & HELLIWELL, J.R., 1989 - Top Curr Chem, 151, 61

MOY, J.P. & GIBNEY, S., 1992 - Low Energy X-ray Image Intensifier Development, Phase 1 report - ESRF (Grenoble)

OGATA, C.M., HENDRICKSON, W.A., GAO, X. & PATEL, D.J., 1989 - Abstracts of the American Crystallographic Association Meeting, series 2, vol. 17, p. 53

OKAYA, Y. & PEPINSKY, R., 1956 - Phys. Review 103, 1645

PAI, E.F., 1992 - Current Opinion in Structural Biology 2, 621

PÄHLER, A., SMITH, J. L. & HENDRICKSON, W. A.,1990 - Acta Cryst. A46, 537

PHILLIPS, J. C., & HODGSON, K.O.1980 - Acta Cryst. A36, 856

PHILLIPS, J.C., TEMPLETON, D.H., TEMPLETON, L.K. & HODGSON, K.O.,1978 -Science 201, 257

PHIZACKERLEY, R. P., CORK, C. W., MERRITT, E. A.,1986 - Nucl. Instrum. Meth. A246, 579

ROSENBAUM, G., HOLMES, K.C. & WITZ, J., 1971 - Nature 230, 434

ROSSMANN, M. G.,1961 - Acta Cryst. 14, 383

SAKABE, N., 1991 - Nucl. Instrum. and Meth. A303, 448

SMITH, J.L., 1991 - Current Opinion in Structural Biology 1,1002

SMITH, J.L. & WATENPAUGH, K. ed., 1991 - Structural biology and synchrotron radiation: assessment of resources and needs - Purdue Univ., USA.

SZEBENYI, D.M.E., BILDERBACK, D.H., LEGRAND, A., MOFFAT, K., SCHIELDKAMP, W., SMITH-TEMPLE, B. & TENG, T.-Y., 1992; J. Appl. Cryst. 25, 414

TEMPLETON, D.H. & TEMPLETON, L.K., 1982 - Acta Cryst. A38, 62

TEMPLETON, D.H. & TEMPLETON, L.K.,1985 - Acta Cryst. A41, 365

TEMPLETON, L. K. & TEMPLETON, D. H., 1988 - Acta Cryst. A 44, 1045

WALENTA, A.H. ed., 1991 - Proc. of the European Workshop on X-ray detectors for synchrotron radiation sources (Aussois) - EEC (Siegen, Germany)

WEIS, W.I., KAHN, R., FOURME, R., DRICKAMER, K. & HENDRICKSON, W.A. 1991 - Science 254, 1608

WESTBROOK, E.M., DEACON, M.L., STRAUSS, M.G., NADAY, I. & SWEET, R.M. (1989). Abstracts of the American Crystallographic Association Meeting, series 2, vol. 17.

CHAPTER XII

X-RAY AND NEUTRON LAUE DIFFRACTION

J.R. HELLIWELL AND C. WILKINSON

XII.1. Introduction

As indicated by McIntyre in Volume **I VII.2.**, the original X-ray Laue diffraction experiment was based on an idea of von Laue and conducted by Friedrich and Knipping (Friedrich et al. 1912) and earned von Laue the Nobel Prize for physics in 1914. The original diffraction photograph was from a crystal of copper sulphate.

The essential feature of the Laue method, as it became called, is that the incident X–ray beam is polychromatic (a 'white beam') and the crystal sample is held stationary. X–rays of all wavelengths from the emission tube are allowed to impinge onto the sample; no special filtering or monochromatisation is employed. The Bremsstrahlung continuum and the characteristic emission lines constitute this incident beam of X–rays. This beam hits the stationary crystal and each Bragg reflection making up the diffraction pattern is excited simultaneously, but by a different wavelength. A given set of 'reflecting planes' in the crystal selects from the beam the particular wavelength which allows constructive interference to occur. This is in contrast to the monochromatic rotating crystal method, in which each reflection is stimulated in turn by varying the angle of incidence on the specimen crystal of an X-ray beam of a single (normally characteristic) wavelength. The Laue method is described in the book by Amorós, Buerger and Amorós (1975).

The first analyses of simple crystal structures were made using Laue photographs by W.L. Bragg, who deduced the structure of sodium chloride and other alkali halide crystals (for a historical review see Bragg (1975)). Subsequently, Bragg primarily used the monochromatic X–ray beam to examine crystals in the spectrometer designed and built by his father W.H. Bragg. Enthusiastic proponents of the use of the Laue method were Wyckoff (1924) and Pauling. However, the Laue method at that time was subject to various real and imagined limitations (discussed below) and its use lapsed as a method for determining molecular structure as it became superseded by the monochromatic rotating crystal method. This was in large part due to the relative weakness of the Bremsstrahlung radiation with respect to the characteristic lines from an X-ray tube.

The spectral curve of synchrotron radiation, however, ideally lends itself to Laue geometry since a broad band of wavelengths of high intensity is emitted by the synchrotron electron beam (Helliwell (1992)). As a result, a very large number of reflections can be recorded in a remarkably short exposure time in a Laue experiment; this has led to a revival of the Laue method as a means for quantitative structure analysis, especially for kinetic studies and the analysis of structural perturbations. Time resolved investigations include the study of enzyme or zeolite electric fields, pressure and temperature on various materials. An example of a

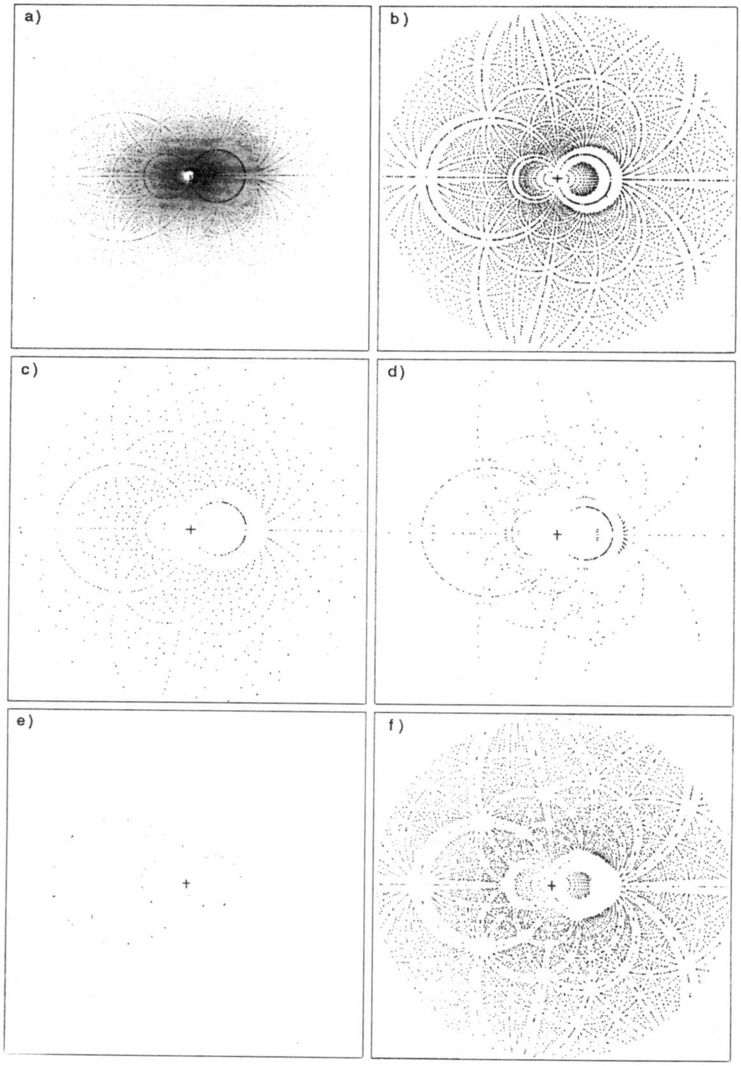

Figure XII.1. — (a) Pea lectin synchrotron X-ray Laue diffraction pattern recorded on photographic film on station 9.6 of the SRS and then in (b) the corresponding predicted Laue pattern is shown followed by those parts of the pattern which are (c) energy overlap spots, (d) spatial overlap spots, (e) spots which are both energy and spatial overlaps and, finally those that are (f) neither energy nor spatial overlaps. Crystal to film distance 95 mm λ_{min} = 0.45Å, λ_{max} = 2.6Å. Further details in Helliwell (1985).

protein crystal synchrotron Laue pattern and computer predictions is shown in figure XII.1.

XII.2. Diffraction geometry

For a stationary crystal and white radiation with $\lambda_{max}>\lambda>\lambda_{min}$ the reciprocal lattice points which are active lie between the Ewald spheres of radii $1/\lambda_{max}$ and $1/\lambda_{min}$. These spheres touch at the origin of the reciprocal lattice (figure XII.2(a)), and the wavelength at which any individual reciprocal lattice point diffracts is determined by the reciprocal radius of the Ewald sphere passing through it. Imperfect crystals are limited in the region of reciprocal space over which the reciprocal latice points are sharp and a third sphere of radius d^*_{max} about the origin limits the diffraction range to the shaded area of figure XII.2(a). A Laue diffraction spot can be composed of several reciprocal lattice points and wavelengths such that for a spot corresponding to a spacing d and a wavelength λ, Bragg reflections with d/2, d/3,.... will also be stimulated by the respective wavelengths $\lambda/2$, $\lambda/3$,.....etc (figure XII.2(b)). That is, these orders are exactly superimposed in direction. For this reason, the Laue method was formerly considered unsuitable for intensity measurement and structure determination (Bragg (1975), p. 137; Wyckoff (1924), p. 142; Amorós et al. (1975), p.13; Steinberger, Bordas and Kalman (1977), p. 1260). We consider this objection in more detail in **XII.2.1** below.

Additionally, in Laue geometry only cell ratios can be determined from the angular coordinates of spots. The absolute cell parameters therefore have to be provided from a monochromatic study. However, if the intensities of spots are also considered, then there appears to be scope to solve the problem by using knowledge of λ_{min} and the bromine K edge in the photographic film as wavelength markers. For a discussion of these and other objections see Helliwell (1989).

XII.2.1. Chromatic overlap

Figure XII.2(b) shows a ray with five orders inside the d^*_{max} sphere (the sample resolution limit), but of which only two are within the accessible region. The ray and corresponding Laue spot are therefore of actual multiplicity 2. The overall pattern, however, contains many monochromatic spots even if there is an infinite bandpass ($\lambda_{max} = \infty$, $\lambda_{min} = 0$). This problem has been considered in detail by Cruickshank et al (1987); let us examine a simple example here. Consider a zone of reciprocal lattice points which are active out to a limit d^*_{max}, with Miller indices $|h,k|=10$ (figure XII.2(c)). All reciprocal lattice points lying between $d^*_{max}/2$ and the origin of reciprocal space will be recorded as part of a multiple Laue spot and are marked with filled circles. Those lying between d^*_{max} and $d^*_{max}/2$ are part of multiples (filled circles), or are singles (open circles). The singles have indices h and k which have no common prime divisor p (=2,3,5,7 in this example). The probability that an integer h has a divisor p is 1/p, and the probability that it does not is 1-1/p. The probability that a pair of integers h,k has no common divisor p is $1-(1/p^2)$, and the probability that h,k has no common divisor of 2,3,5 or 7 is

$$Q=\left(1-\frac{1}{2^2}\right)\left(1-\frac{1}{3^2}\right)\left(1-\frac{1}{5^2}\right)\left(1-\frac{1}{7^2}\right) \qquad \text{(XII.1)}$$
$$= 0.627$$

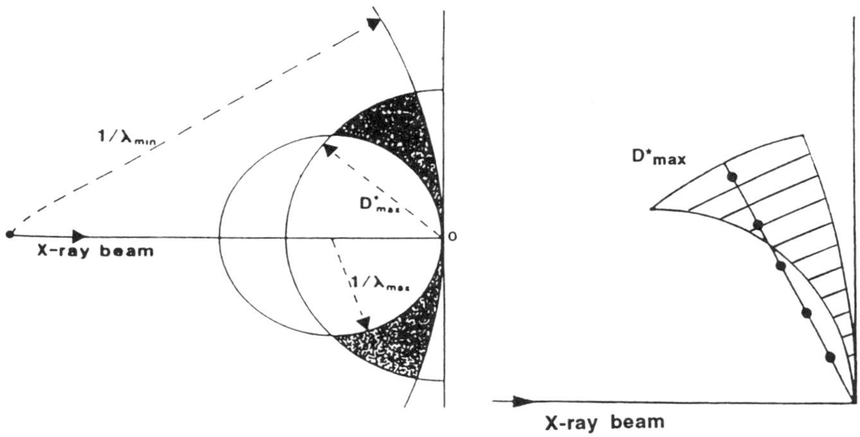

(a) (b)

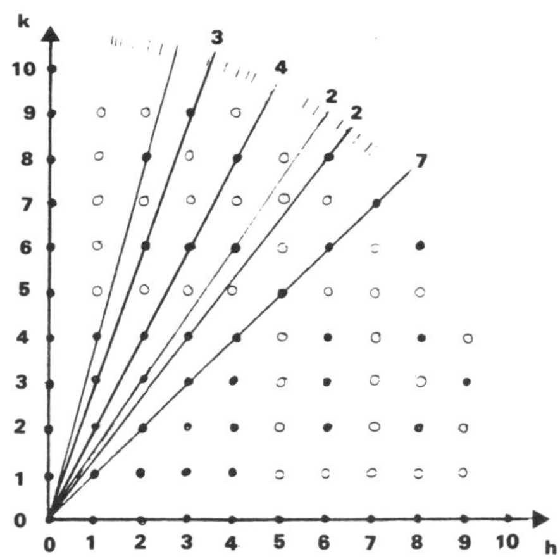

(c)

Figure XII.2(a). — Laue diffraction geometry showing the accessible region of reciprocal space between the Ewald spheres associated with λ_{min} and λ_{max} and the sample resolution limit d^*_{max}; O is the origin of reciprocal space. (b) A ray with n orders inside the d^*_{max} sphere can have a recorded multiplicity m<n where (n–m) RLPs are outside the accessible region. The diagram shows the case of n = 5 and m = 2. Only the upper section of the volume of resolution of the accessible region is shown. From Cruickshank et al. (1987). (c) Single reflections (open circles) and multiple reflections (filled circles) in a zone of reflections out to a limit d^*_{max} which corresponds to $|h| = 10$ (after Cruickshank et al. (1991)).

which gives the fraction of Laue spots which are singles. A count of the single lattice points outside $d^*_{max}/2$ in 2(c) shows that in fact they form 61% of reciprocal lattice points, satisfactory agreement considering the quantised nature of the example. Three quarters of the reciprocal lattice points out to d^*_{max} lie outside $d^*_{max}/2$ and thus the probability of any reciprocal lattice point inside d^*_{max} being single is 0.75Q, or 47.2%.

Real crystallographic problems are of course three-dimensional, and in this case the fraction of Laue spots which are singles is even greater. Cruickshank et al. (1987) showed that the probability that a randomly chosen reciprocal lattice point hkl has no common integer divisor is

$$Q = \left(1 - \frac{1}{2^3}\right)\left(1 - \frac{1}{3^3}\right)\left(1 - \frac{1}{5^3}\right)\left(1 - \frac{1}{7^3}\right)\left(1 - \frac{1}{11^3}\right) \qquad \text{(XII.2)}$$
$$= 0.832$$

Since 7/8 of all reciprocal lattice points within the sphere of radius d^*_{max} lie outside $d^*_{max}/2$, 87.5% of all Laue reflections are single rays and 7/8 x 0.832 (72.8%) of all reciprocal lattice points are singles. They also showed that 8.8% of all rays are double rays (figure XII.3).

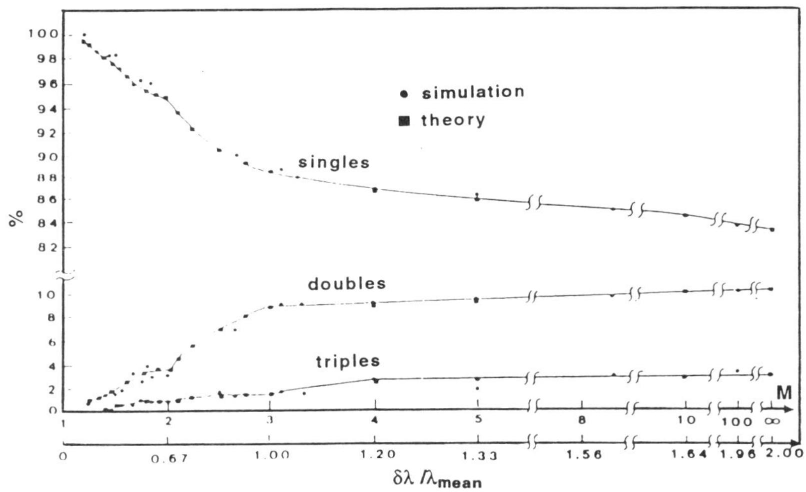

Figure XII.3. — The variation with M (= $\lambda_{max}/\lambda_{min}$) of the proportions p(1), p(2) and p(3) of RLPs lying on single, double and triple rays for the case of $\lambda_{max}<2/d^*_{max}$. After Cruickshank et al. (1987).

A probability map can be constructed for the general case (finite λ_{max} and λ_{min}) giving details of the likelihood of reciprocal lattice points being recorded as single, double, triple, etc; see figures XII.4(a), (b), (c). A striking feature of the probability map for singles, for example, is the region of zero probability between $d^*_{max}/2$, and the origin, on the one hand and $2\lambda_{min}$ and λ_{max} on the other.

Moreover, the number of accessible reciprocal lattice points in a region of revolution bounded by spheres of radii d^*, $1/\lambda_{max}$ and $1/\lambda_{min}$ is given by

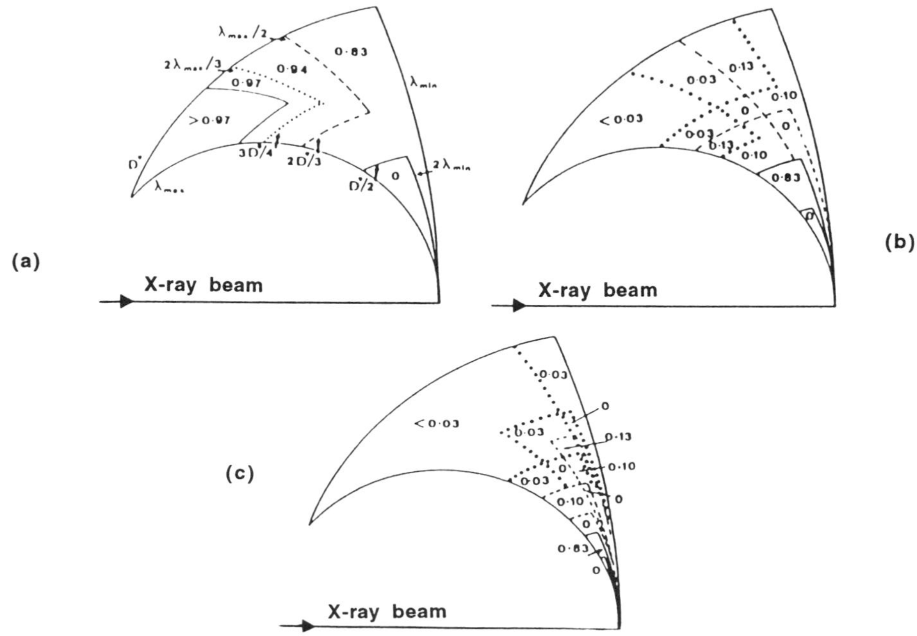

Figure XII.4. — The probability of a reciprocal lattice point being (a) single, (b) double and (c) triple in different regions of reciprocal space. The wavelength labels indicate which Ewald sphere is appropriate. The actual radius of each Ewald sphere is the reciprocal of the label shown. The labels for the resolution spheres e.g. d^*_{max} are true distances on the diagram. After Cruickshank et al. (1987).

$$\frac{\pi d^{*4}_{max}(\lambda_{max} - \lambda_{min})}{4V^*} \text{ for } \lambda_{max} < 2/d^*_{max} \qquad \text{(XII.3)}$$

where V^* is the reciprocal unit cell. The variation of d^* to the fourth power shows that the volume of low resolution data sampled in one exposure is actually quite small. The number of reciprocal lattice points within the resolution sphere is

$$\frac{4}{3}\frac{\pi d^{*3}_{max}}{V^*} \qquad \text{(XII.4)}$$

The Laue method therefore appears very effective at sampling data between d^*_{max} and $d^*_{max}/2$. These reflections are largely recorded as singles but, of course, their intensities need wavelength normalisation if they are to be used for structure analysis (see section **XII.3.**)

The effective sampling of the region between $d^*_{max}/2$ and the origin of reciprocal space needs a method to achieve the separation of the component reflections in a spot. There are several possible methods available to achieve this. The process is often referred to as wavelength or energy deconvolution.

These methods include:

(a) The use of multiple films to record a Laue pattern. The variation of the film absorption factor with wavelength means that the different components of (and wavelengths in) a Laue spot are attenuated at different rates (Zurek, Papiz, Machin and Helliwell (1985); Helliwell et al. (1989a,b)).

(b) In the case of high symmetry it is possible that a reciprocal lattice point recorded in one place as a component of a 'double' spot may also occur elsewhere as a 'single' spot. In this case, the other component of a double may therefore be derived. Alternatively, in the case of multiples a reciprocal lattice point may occur several times, stimulated by different wavelengths.

(c) Direct methods (Hao et al. (1993))

The first method has been used and reasonable merging R factors obtained for doubles although not as good as for singles (Zurek et al. (1985); Helliwell et al. (1989a, b).

XII.2.2. Spatial overlaps

The Laue diffraction spots on a film are of a certain size dependent on factors such as crystal mosaic spread, beam divergence and collimator size (Andrews, Hails, Harding and Cruickshank (1987), Helliwell et al (1989a). If a spot encroaches on its nearest neighbour then these Laue spots may be classed as spatially overlapping. The fraction of spots that are spatially overlapping can be large (e.g. >50%) in the protein crystal case and even higher fractions can be found in the case of virus Laue patterns. Computational approaches to the problem of evaluating their intensities are described as spatial deconvolution. Obvious experimental approaches involve the use of large detector sizes and long crystal to detector distances; this is feasible when the sample is well ordered and because of the tight collimation of the X–ray beam. This, of course, may well not be the case in a kinetic experiment when the sample may disorder slightly at certain stages as has, in fact, been observed (Hajdu et al. (1987)).

It can be shown (Cruickshank, Helliwell, Moffat (1991)) that:

(a) The largest local density of spots occurs along the arcs approaching nodal spots. (Nodal spots are of low h,k,l values or multiples thereof and are associated with principal zones of the lattice).

(b) The nearest neighbours to a nodal spot are single (see also Jeffery (1958)).

(c) Most overlaps occur between singles. This explains why the population of spatial overlaps is essentially distinct from the population of energy overlaps (figure XII.1 (c), (d) and (e)).

The spatial overlap problem is not as fundamental an obstacle as the energy overlap problem once appeared to be. In small molecule X-ray Laue patterns there are virtually no spatial overlaps. However, in neutron diffraction Laue patterns (see later) and in macromolecular X-ray crystallography it is a matter for serious concern (Cruickshank et al. (1991) and figure XII.5 (a)).

For a flat film set at various distances from a crystal, the numbers of Laue spots, reciprocal lattice points and spatial overlaps vary. There is an optimum distance which arises because as the film is moved further back the number of spots lost to spatial overlaps is balanced by those spots that are lost as they pass beyond the

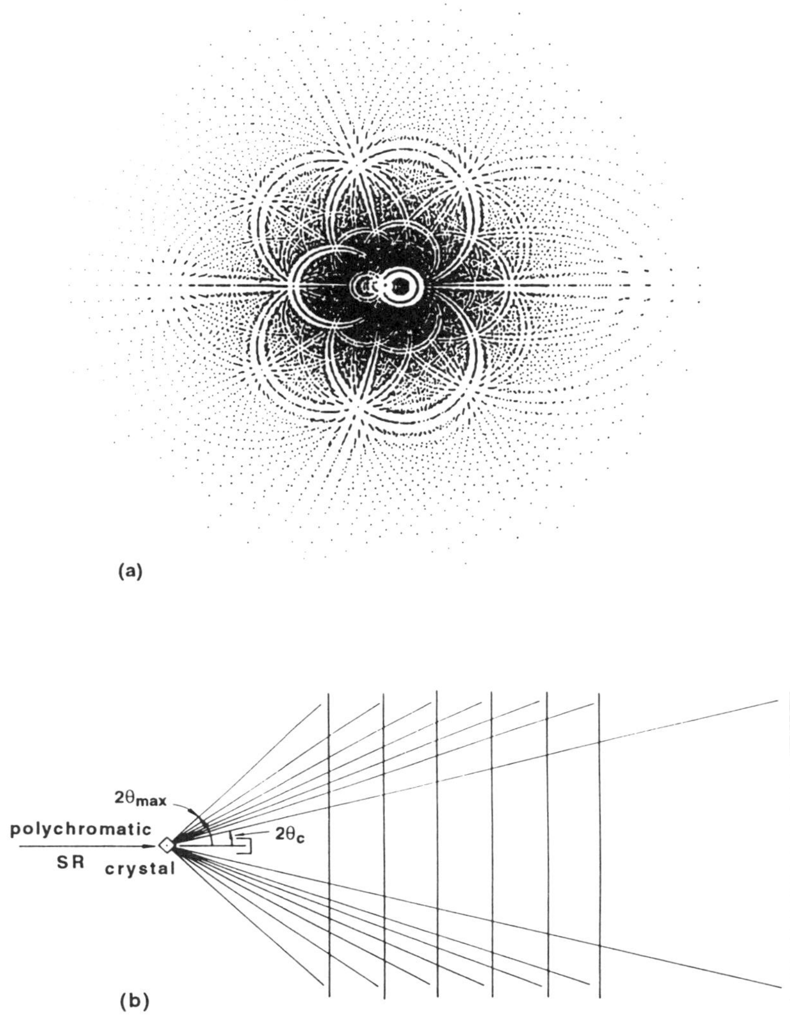

Figure XII.5. — Spatial overlaps dominate a protein crystal Laue pattern when a standard film is set to accept all stimulated spots. The example shown here in (a) is a computer prediction equivalent to figure 1(b) for the pea lectin case but the film is now set at 34mm from the crystal. After Cruickshank et al. (1991).

In (b) the toast-rack arrangement of films is shown schematically. The closest film to the crystal accepts the whole pattern whereby the outer parts (high θ, long λ) contain spots that are not too densely packed. The inner (low θ, short λ) parts of the front film would not be quantified because of too high a density of spots. Instead these would penetrate to the rear films and expand to fill the available aperture but with a greatly reduced density of spots. From Helliwell (1991).

edges of the film. However, an experimental method to solve a significant part of the spatial overlap problem has been proposed (Helliwell (1991)). By arranging films in a special cassette with a pre-set gap between them (e.g. 10mm) the front film allows quantitation of the long λ, high θ spots and on the back film(s) the short λ spots can penetrate and expand to fill the whole aperture. Such an arrangement is referred to as a toast-rack (see figure XII.5(b)).

XII.3. Analysis of Laue Data and Wavelength Normalisation

The wavelength normalisation problem was originally thought (Bragg (1949)) to be more serious than is in fact the case. The measured intensity of a single Laue spot or of an energy deconvoluted spot needs correction or normalisation for a variety of wavelength-dependent factors to obtain structure factor data.

The recorded Laue intensity (strictly, integrated power) is given by Campbell et al. 1986 and references cited therein, for the reflection hkl (abbreviated to the symbol **h**),

$$I_L(\mathbf{h}) = \frac{e^4}{m^2c^4} I_o(\lambda)\, \lambda^4 \frac{1}{2\sin^2\theta\, V_o^2} PAD\,|F(\mathbf{h})|^2 \qquad \text{(XII.5)}$$

Here, $I_o(\lambda)$ denotes the spectral intensity distribution of the incident X-ray beam; V_x is the volume of sample illuminated; V_o is the sample unit cell volume; P is the polarisation factor; A is an absorption correction for the sample in its capillary and D is a detector sensitivity and obliquity factor. Let us consider some of these factors.

(a) $I_o(\lambda)$, the intensity spectrum which is incident at the sample, is modified at the long wavelength end by absorption due to beam line windows or any airpaths in the Laue camera. Also the short wavelengths are cut off if a reflecting mirror is used (and this may be the case for focussing or to aid more precise definition of spot multiplicity).

(b) P, the source polarisation state varies with wavelength (see Papiz and Helliwell (1984) for the SRS example).

(c) Sample absorption A truncates the long wavelengths compared with short wavelengths.

(d) Film absorption increases at long wavelengths. The bromine and silver K absorption edges at 0.92Å and 0.49Å, respectively, enhance the film absorption efficiency D in a wavelength region where it is otherwise decreasing (Clifton et al. (1985)).

(e) Anomalous scatterers in the structure affect the magnitude (and phase) of the structure factors.

Since quantities such as P, A and D vary with any or all of λ, θ and x, the position of the diffracted beam on the detector; the spectral intensity distribution is, in general, not precisely known in advance; and the detector may suffer from spatial distortion and non-uniformity. Thus equation (5) may be written as

$$I_L(\mathbf{h}) = KG(\lambda, \theta, x)\,|\,\mathbf{F}(\mathbf{h})\,|^2 \qquad \text{(XII.6)}$$

where K is a constant. Assuming that all quantities which depend on more than one variable are factorable then $G(\lambda, \theta, x) = f(\lambda)g(\theta)j(x)$, and

$$|F(h)|^2 = [Kg(\theta)j(x))]^{-1}[f(\lambda)]^{-1} I_L(h) \qquad \text{(XII.7)}$$

It is reasonable to assume that K, g(θ) and j(x) are known, to a very good approximation. Thus, the quantifying of Laue diffraction patterns depends almost entirely on determination of f(λ), known as the wavelength normalisation curve or λ–curve (figure XII.6).

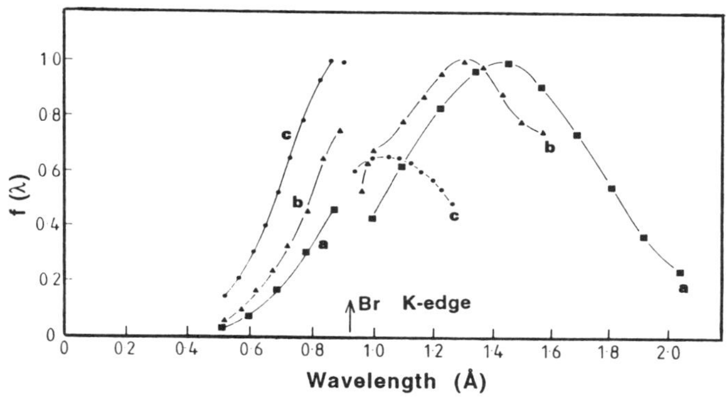

Figure XII.6. — λ–curves determined under various X-ray synchrotron beam line conditions and from protein crystal Laue film data: (a) SRS station 9.6 prior to mirror installation (Helliwell (1984); Helliwell et al. (1989b)) pea lectin. (b) SRS station 9.6 after mirror installation; mercury α–amylase crystal derivative (Helliwell et al. (1989a)). (c) SRS station 9.7, no mirror, with 50μm aluminium filter, cubic concanavalin–A. Each curve on an arbitrary relative level. Also, from Helliwell et al. (1989a).

The methods which have been used to derive a wavelength normalisation f(λ), curve are:

(i) The use of a silicon standard crystal and a film in a pilot study on aluminium phosphate (Wood, Thompson and Matthewman (1983)).

(ii) The use of monochromatic data as a reference set for the same sample (Helliwell et al. (1989b)).

(iii) The use of symmetry equivalent reflection intensities measured at separate wavelengths (Campbell et al. (1986), Helliwell et al. (1989b)). Similar information is also available when the same reflections are measured on several film packs with different crystal orientations.

Methods (ii) and (iii) have generally been the ones used to process protein and small molecule crystal Laue data. The protein crystal processing statistics are almost as good as monochromatic oscillation camera data (Machin and Harding (1985); Helliwell et al. (1986), (1989a); Machin (1987); Temple and Moffat (1987)). Method (iii) breaks down in the presence of strong anomalous scattering; in such a case either method (ii) should be used or a λ–curve derived by method (iii) from a non–anomalously scattering crystal structure; alternatively, in trying to derive a set it should be possible, in principle, to derive the wavelength–independent and –dependent parts of the structure factor separately using the method of Karle (1980).

In the case of monitoring a sample of known structure via the Laue pattern it is possible to use the fractional differences whereby the wavelength–dependent corrections cancel. This difference ratio method can be used with time resolved or perturbation measurements. A simplification, in principle, of the time resolved Laue experiment over its static counterpart now appears: only fractional changes in intensities are needed (Bilderback et al. (1984), Hajdu et al. (1987)).

XII.4. Development of the Synchrotron Laue Method

Synchrotron Laue data were recorded from a crystal of aluminium phosphate and analysed by Wood et al. (1983). The R–factor on I in the Laue data refinement was 19% compared with a conventional monochromatic single crystal study (Thong and Schwarzenbach (1979)) with an R–factor of 2.2% on F. For the Laue data 20% of the measurements were rejected on the basis of bad intensity agreements. Of the eight refined positional parameters six were within 2σ and only one outside 3σ. The temperature factors were somewhat less well determined. This was considered to be due to the significant extinction and multiple diffraction effects for a hard material such as aluminium phosphate.

Laue data from a protein crystal were first recorded by Moffat et al. (1984, 1986) who used a somewhat restricted wavelength range to avoid the multiplicity problem. It was established that rapid exposure times were feasible. Broad bandpass Laue data were recorded from a crystal of the protein pea lectin using the SRS wiggler (Hails et al. (1984), Helliwell (1984), (1985)). These data were used in a pilot study to assess protein crystal Laue data quality. This protein crystal is relatively radiation insensitive. Overall merging statistics for the Laue data after wavelength normalisation (Campbell et al. (1986)) were almost as good as monochromatic data. The merging R–factors on intensity for Laue and monochromatic pea lectin data were respectively 8.2% and 6.3%. The mean fractional difference on F of the Laue data scaled to monochromatic data was 12.9% to 2.6Å resolution. These results are summarised in Helliwell et al. (1989b). The best quality data occur in the λ range $0.59\text{Å}<\lambda<1.57\text{Å}$. At the short wavelength end the photographic films become increasingly transparent and at the long wavelength end sample absorption increases. Both effects weaken the diffraction spot. Sample absorption also introduces systematic errors at long wavelengths, although in principle correction factors for this effect can be calculated.

A difference Fourier map was calculated for the enzyme glycogen phosphorylase b using synchrotron Laue data (Hajdu et al. (1987), 1988)). The map showed electron density at the binding site for bound maltoheptose in phosphorylase b crystals. The sensitivity of the Laue method to weak scattering centres such as hydrogen in a small molecule crystal structure has been shown (M. Helliwell et al. (1989)). A study on a low pH form of carbonic anhydrase (Lindahl et al. (1992)) has shown that a sensitivity to the scattering of one water molecule in 29000Da of enzyme is possible with the synchrotron Laue method. A time resolved synchrotron Laue experiment has been performed by Schlichting et al. (1990) to study GTP hydrolysis in the ras p21 protein. The above studies are clearly establishing the credentials of the X-ray Laue method to yield accurate intensity and structure data.

XII.5. Neutron Laue Diffraction

The mean intensity of Bragg reflections in a single crystal experiment is proportional to the product of the incident beam flux and the volume of the sample, but inversely proportional to the volume of the unit cell of the crystal. As described in the sections above, X-ray techniques have been used to determine the structures of single crystals containing large molecules of biological significance, which by their nature have large unit cells and are often available only in the form of small crystals. Neutrons come in relatively low flux beams, but have the advantage over X-rays that they can easily detect the positions of the hydrogen atoms in a crystal structure. Since protons and water molecules are vital ingredients of all biological processes, it is clear that neutron diffraction can supply biologically useful information which is not available from X-ray diffraction (Savage et al. (1987), Forsyth et al. (1989), Wlodawer et al. (1983), Roth et al (1989)).

In the case of a steady state source, such as the high flux neutron beam reactor at the Institut Laue Langevin (ILL) in Grenoble, in a traditional monochromatic beam experiment ($\Delta\lambda/\lambda\sim$ a few %) on a single crystal of a small protein structure, data collection times are of the order of two to four weeks when the crystal volume is several cubic millimetres. Since neutrons are a scarce resource, one would like to use a greater fraction of those which are available in the beam from the reactor in order to speed up data collection (i.e. use a Laue method). This enables the technique to be extended to smaller volume crystals, or those with larger unit cells. Modified Laue methods, in which attempts were made to separate chromatically overlapped orders, were used in earlier ILL instruments (Maier-Leibnitz (1967), Klar (1972), Hohlwein (1977)) and a white beam, time-of-flight method is the standard technique for neutron spallation source diffraction (Schultz et al. (1984)).

XII.5.1. Choice of experimental parameters

Recent X-ray Laue analysis (**XII.2.1**) shows that chromatic overlap is not as serious a problem as was once thought, but since large samples and beam divergences are used in neutron scattering, spatial overlap can be problematical. The beam divergence from a nickel-coated guide is 0.18 λ degrees (λ in Å) full width at half height. At λ = 4Å, this is 0.72°, which is greater than the crystallite spread of a normal single crystal and approximately the angle subtended at the detector by a sample lmm in size. These are therefore important factors determining the resolution of reflections. The number of reflections on a detector capable of accepting Bragg reflections between θ_1 and θ_2, subject to the condition that $2\sin\theta_2/\lambda_{min} < d^*_{max}$, for wavelengths between λ_1 and λ_2 is given by

$$(4\pi V_o/3)\{(1/\lambda_1)^3 - (1/\lambda_2)^3\}\{\sin^4\theta_2 - \sin^4\theta_1\}$$

Even for a small protein structure like lysozyme with a unit cell volume V_o of 27,000 Å^3, there will be many thousands of reflections simultaneously active on the detector with a white beam. For this reason, and also to reduce incoherent background scattering from the specimen, it is desirable to limit the wavelength range used and use a 'quasi' Laue technique (Wilkinson et al. (1991). Figure XII.7 shows a Laue photograph of a 2mm crystal of tortoise egg white lysozyme taken on a white thermal neutron beam at ILL. The exposure time was 90 minutes and the specimen to film distance was 80mm. Comparison with the X-ray Laue diagram of figure XII.1(a) shows that as expected, the reflections are larger in the neutron case and that the background level is higher.

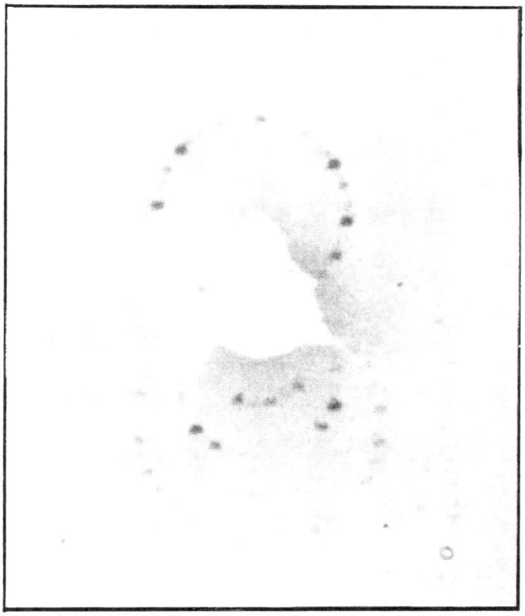

Figure XII.7. — Neutron Laue photograph of 2mm edge crystal of tortoise egg white lysozyme. Exposure time 90mins on thermal beam, specimen to film distance 80mm.

XII.6. Instrument layout and detection systems

The layout of the a proposed quasi-Laue diffractometer (LADI) is shown in figure XII.8. The neutrons arrive along a guide, and a wavelength band is chosen by the velocity selector. In the diagram, scattered neutrons are shown as being detected by two groups of four close-packed position sensitive detectors (PSDs), each with two detectors of 10cm diameter above and below the horizontal plane. An alterative method of detection which is also being investigated is the use of neutron-sensitive image plates which can be made to subtend a large solid angle at the specimen. These detection methods are described below.

XII.6.1. Photomultiplier Tube PSD detector

The detector groups can be moved in the horizontal direction. When placed at 10cm from the specimen, they subtend angles of ~50° in the azimuthal and 120° in the horizontal direction. Reciprocal space is explored by rotating the crystal about a vertical or a horizontal axis. One type of detector currently being investigated is the position sensitive photomultiplier tube. Such detectors have been developed for neutron diffraction topography (Baruchel et al. (1988)), and have been shown to be capable of a resolution of 0.2mm using a ZnS-LiF scintillator. Greater efficiency, although slightly lower resolution, can be obtained with a Li glass scintillator.

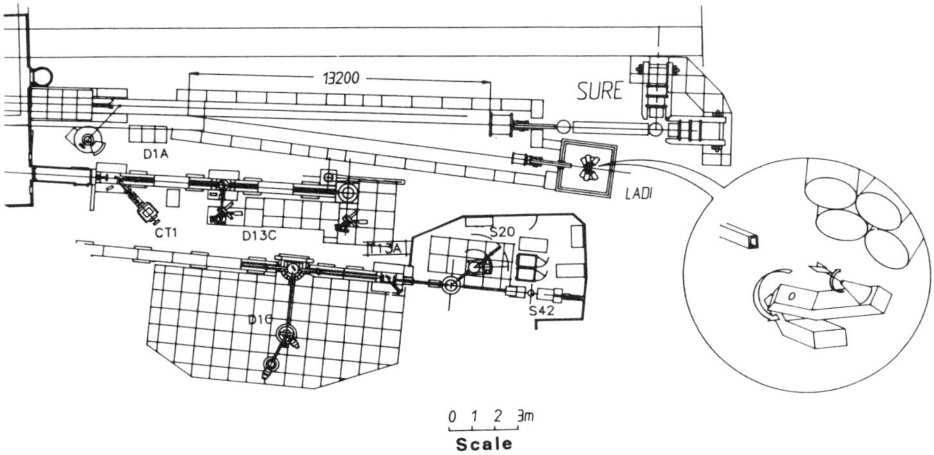

Figure XII.8. — Schematic diagram of the neutron instrumental layout. The neutrons arrive along a guide, and a wavelength band is chosen by the velocity selector. Scattered neutrons are detected by two groups of four close-packed PSDs, each with two detectors of 10cm diameter above and below the horizontal plane. The specimen to detector distance is 10cm.

The resolution of the detector has been demonstrated using Laue reflections from a single crystal of niobium, which gave low background scattering. The specimen was in the form of a 2mm edge cube with 100 faces and was oriented to give a 110 reflection at the centre of the detector (figure XII.9); two other reflections can be seen on the edges. The full width at half maximum of the 110 reflection was 1.2mm x 2.0mm. There is thus little broadening of the reflections due to the finite detector resolution.

XII.6.2. Image plate detector

Image plates are made from flexible plastic sheets about 500 microns thick coated with a 150 micron layer of finely powdered photostimulable phosphor crystals (normally BaFBr doped with Eu^{2+} ions) combined in an organic binder (Miyahara et al. (1986)). When irradiated with X-rays or gamma-rays, electrons which are liberated by ionising Eu^{2+} to Eu^{+} can be trapped in isolated states just below the conduction band, which have been created by substitution of Eu for Ba in the host lattice. These trapped electrons are in a metastable state, with a decay period of many hours, and thus form a stored image of the radiation which has fallen on the plate. The electrons can be released by photostimulation with visible radiation, the resultant light emitted lying in the blue region of the spectrum. The plate can thus be 'read' by scanning with a laser (normally He-Ne) and measuring the intensity of the emitted light as a function of position (Amemiya et al. (1988)). Any signal which remains after reading can be erased by illuminating the plate with an intense source of visible light and the plate can then be re-used. Combined read-out and erasure times are normally of the order of several minutes. Image plates have the advantage over photographic film that the dynamic range which can be

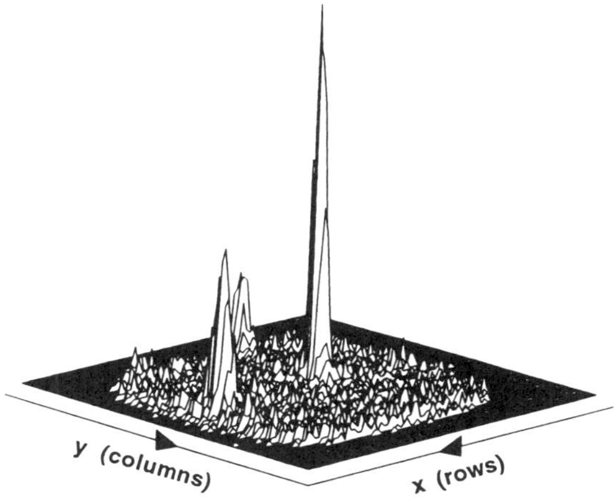

Figure XII.9. — Laue Histogram of the 110 reflection from a niobium single crystal, recorded at $2\theta = 90$ degrees with a photomultiplier tube PSD. The full width at half height of the reflection is of the order of the crystal size.

recorded is three orders of magnitude greater and that a digitised image can be obtained without any intermediate wet-processing stages. The resolution obtainable is usually limited by the dimensions of the laser beam used for read-out, with a pixel size which is typically of the order of 100 microns, giving a full width of about 400 microns at the 1% intensity level.

XII.6.3. Neutron scintillator

In order to use these image plates as neutron detectors, a scintillator is necessary to convert neutrons to electromagnetic radiation. Gadolinium is convenient for this purpose, as it has a very strong n, γ resonance for thermal neutrons giving gamma-rays with kilo-electron volt energies, the range in which the plate is most effective (Rauch et al. (1967), Backlin et al. (1982)). A Kodak storage phosphor screen type S023 which has a sensitive surface area of 200 x 250 mm was placed in a lighttight duralloy cassette, with a gadolinium oxide (Gd203) paint scintillator pressed tightly against the active surface of the plate. Figure XII.10 shows a Laue diagram from a single crystal of lysozyme, approximately 2mm on edge, placed in the well collimated (3mm diameter) beam (Wilkinson et al. (1992)). The plate, which was exposed for 15 minutes, was placed at right angles to the beam with the direct neutrons substantially absorbed by a square B4C beamstop 10mm on edge. The background (due to direct neutron impact on the plate, neutrons incoherently and inelastically scattered from the specimen, and gamma-rays produced by n, γ reactions at the specimen) was significant, and has been subtracted from the image by performing a simple average over square regions 4mm on edge.

XII.6.4. Image plate camera

As an alternative to the photomultiplier PSDs, an on-line cylindrical image plate reader for LADI has ben designed (figure XII.10) and a prototype is presently under construction. The neutrons which are Bragg reflected by the crystal at the centre of a stationary cylinder pass easily through the cylinder wall and are recorded on a specially made phosphor in which gadolinium oxide has been incorporated. After recording the pattern with an exposure of about thirty minutes, the cylinder is spun on its axis at 400rpm while a read head tracks vertically down the outside of the cylinder in about 4 minutes, giving a spiral coverage of the cylinder surface with a pixel size of 250 microns.

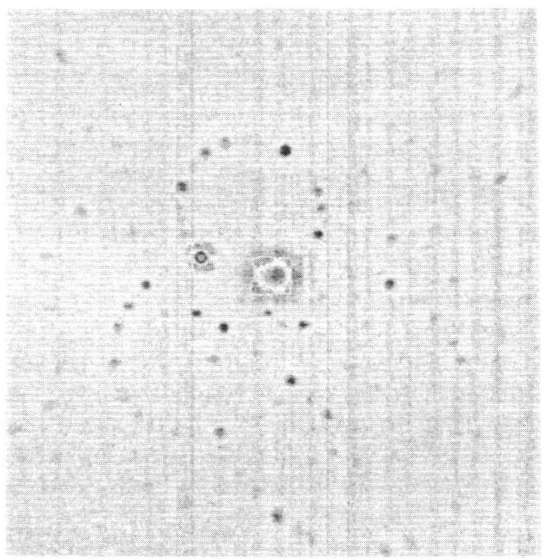

Figure XII.10. — Image plate neutron Laue diagram of a 2mm lysozyme crystal taken with GdO scintillator and cold neutron beam. Exposure time 15mins, specimen to film distance of 44mm.

XII.7. Concluding remarks

This chapter has reviewed the way in which the synchrotron Laue technique has emerged as a viable method for protein crystal structure analysis in specific and important situations where time-resolved studies of structure and reactivity are needed. Indeed, as crystallographically determined protein structures appeared, the criticism was sometimes levelled that these structures were not representative of the in vivo function of these molecules. Since the Laue method has allowed time-resolved studies of structural intermediates in crystals, this objection is clearly invalid. Finally, the extension of the Laue method to neutron diffraction offers the prospect of wider application of neutrons in protein crystallography, with the important benefit of sensitivity to the scattering from hydrogen atoms, which are critical in certain biological reactions.

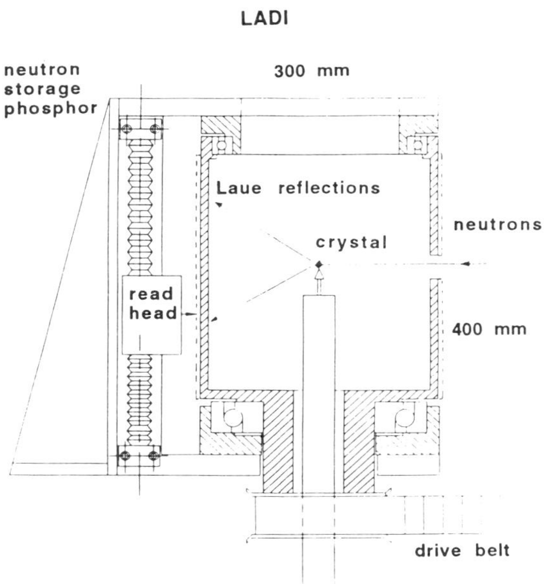

Figure XII.11. — On-line image plate reader for neutron diffractometer. The plate is read after exposure by spinning the cylinder on its axis while tracking the read head vertically.

REFERENCES

AMEMIYA, Y., MATSUSHITA, T., NAKAGAWA, A., SATOW, Y., MIYAHARA, J., CHIKAWA, J., 1988 - Nucl. Instr. and Meth. A266, 645

AMORóS, J.L., BUERGER, M.J., AMORóS, M.C., 1975 - The Laue Method, New York Academic Press.

ANDREWS, S.J., HAILS, J.E., HARDING, M.M., CRUICKSHANK, D.W.J., 1987 - Acta Cryst. A43, 70

BACKLIN, A., HEDIN, G., FOGELBERG, B., SARACENO, M., GREENWOOD, R.C., REICH, C.W., KOCH, H.R., BAADER, H.A., BREITIG, H.D., SCHULT, O.W.B., SCHRECKENBACH, K., VON EGIDY, T., MAMPE, W., 1982 - Nucl. Phys. A380, 189

BARUCHEL. J., KURODA, K., LIAUD, P., MICHALOWICZ, A., SILLOU, D., 1988 - J. Appl. Cryst. 21, 28

BRAGG, W.L., 1949 - "The Crystalline State: A General Survey " in Vol. 1 of "The Crystalline State" edited by W.H. and W.L. Bragg. G. Bell and Sons Ltd (London)

BRAGG, W.L., 1975 - The Development of X-ray Analysis edited by Phillips, D.C. and Lipson, H. G. Bell and Sons Ltd (London)

BILDERBACK, D.H., MOFFAT, K., SZEBENYI, D., 1984 - Nucl. Instrum. and Methods, 222, 245

BURAS, B., GERWARD, L., 1975 - Acta Cryst. A31, 372

CAMPBELL, J., HABASH, J., HELLIWELL, J.R., MOFFAT, K., 1986 - Info., Quarterly for Protein Crystallography, No. 18, Daresbury Laboratory, pp. 23

CLIFTON, I. J., CRUICKSHANK, D.W.J. ,DIAKUN, G.. ELDER, M., HABASH, J., HELLIWELL, J.R., LIDDINGTON, R.C., MACHIN, P.A., PAPIZ, M.Z.. 1985 - J. Appl. Cryst. 18, 296.

CRUICKSHANK, D.W.J., HELLIWELL, J.R., MOFFAT, K., 1987 - Acta Cryst. A43, 656

CRUICKSHANK, D.W.J., HELLIWELL, J.R., MOFFAT, K., 1991 - Acta Cryst. A47, 352

FORSYTH, V.T., MAHENDRASINGHAM, A., PIGRAM, W.J., GREENALL R.J., BELLAMY, K.A., MASON, S.A., FULLER, W., 1989 - Int. J. Biol. Macr. 11, 236

FRIEDRICH, W., KNIPPING, P., VON LAUE, M., 1912 - Bayerische Akademie der Wissenschaften, 303

HAILS, J., HARDING, M.M., HELLIWELL, J.R., LIDDINGTON, R, PAPIZ, M.Z. 1984 - Daresbury Laboratory DL/SCI/TM428E.

HAJDU, J., ACHARYA, K.R., STUART, D.I., BARFORD, D., JOHNSON, L.N., 1988 - Trends in Biochemical Sciences 13, 104

HAO, Q., CAMPBELL, J.W., HARDING, M.M., HELLIWELL, J.R., 1993 - Acta Cryst A (in press).

HAJDU, J., MACHIN, P.A., CAMPBELL, J.W., GREENHOUGH, T.J., CLIFTON, I.J., ZUREK, S., GOVER, S., JOHNSON, L.N., ELDER, M., 1987 - Nature 329, 178

HELLIWELL, J.R., 1984 - Reports on Progress in Physics 47, 1468

HELLIWELL, J.R., 1985 - J. Mol. Struct. 130, 63

HELLIWELL, J.R., 1989 - Physics World, January issue, 29

HELLIWELL, J.R., 1991 - Nucl. Instrum. and Methods A308, 260

HELLIWELL, J.R., 1992 - Macromolecular crystallography with synchrotron radiation, Cambridge Cambridge University Press.

HELLIWELL, J.R., HARROP, S., HABASH, J., MAGORRIAN, B., ALLINSON, N.M., GOMEZ DE ANDEREZ, D., HELLIWELL, M., DEREWENDA, Z., CRUICKSHANK, D.W.J., 1989 a- Rev. Sci. Inst. 60, 1531

HELLIWELL, J.R., HABASH, J., CRUICKSHANK, D.W.J., HARDING, M.M., GREENHOUGH, T.J., CAMPBELL, J.W., CLIFTON, I.J., ELDER, M., MACHIN, P.A., PAPIZ, M.Z., ZUREK, S., 1989 b- J. Appl. Cryst. 22, 483

HELLIWELL, J.R., PAPIZ, M.Z., GLOVER, I.D., HABASH, J., THOMPSON, A.W, MOORE, P.R., HARRIS, N., CROFT, D., PANTOS, E., 1986 - Nucl. Instrum. and Methods A246, 617

HELLIWELL, M., GOMEZ DE ANDEREZ, D., HABASH, J., HELLIWELL, J.R,, VERNON, J., 1989 - Acta Cryst. B45, 591

HOHLWEIN, D., 1977 - Acta Cryst. A33, 649

JEFFERY, J.W., 1958 - Z. Krist. 110, 321

KALMAN, Z.H., 1979 - Acta Cryst. A35, 634

KARLE, J., 1980 - Int. J. Of Quant. Chemistry, 7, 356

KLAR, B., 1972 - Acta Cryst. A28, S250

LINDAHL, M., LILJAS, A., HABASH, J., HARROP, S., HELLIWELL, J.R., 1992 - Acta Cryst. B48, 281.

MACHIN, P.A., HARDING, M.M., 1985 - Info. Quarterly For Protein Crystallography No. 15, Daresbury Laboratory.

MACHIN, P.A., 1987 - in "Computational Aspects of Protein Crystal Data Analysis", Daresbury Laboratory DL/SCI/R25, Edited by Helliwell, J.R., Machin, P.A. and Papiz, M.Z., pp. 75

MAIER-LEIBNITZ, H., 1967 - Ann. Acad. Sc. Fenn. A VI, 267

MIYAHARA, J., TAKAHASHI, K., AMEMIYA, Y., KAMIYA, N., SATOW, Y., 1986 - Nucl. Instr. and Meth. A246, 572

MOFFAT, K., SZEBENYI, D., BILDERBACK, D.H., 1984 - Science 223, 1423

MOFFAT, K., BILDERBACK, D., SCHILDKAMP, W., VOLZ, K., 1986 - Nucl. Instrum. and Methods A246, 627

PAPIZ, M.Z., HELLIWELL, J.R., 1984 - Daresbury Laboratory DL/SCI/TM42E.

RAUCH, H., GRASS, F., FEIGL, B. 1967 - Nucl. Instr. and Meth. 46, 153

ROTH, M., LEWITT-BENTLY, A, MICHEL, J., DEISENHOFER, J., HUBER, R., OESTERHELT, D., 1989 - Nature 340, 659

SAVAGE, H.F.J, P.F. LINDLEY, P.F., J.L. FINNEY, J.L., P.A. TIMMINS, P.A., 1987 - Acta Cryst. B43, 280

SCHLICHTING, I., ALMO, S.C., RAPP, G., WILSON, K., PETRATOS, K., LENTFER, A., WITTINGHOFER, A., KABSH, W., PAI, E.F., PETSKO, G.A., GOODY, R.S., 1990 - Nature 345, 309

SCHULTZ, A.J., SRINIVASAN, K., TELLER, R.G., WILLIAMS, J.M., LUKEHEART, C.M., 1984 - J. Amer. Chem. Soc. 106, 999

STEINBERGER, I.T., BORDAS, J., KALMAN, Z.H., 1977 - Phil. Mag. 35, No. 5, 1257

TEMPLE, B., MOFFAT, K., 1987 - in "Computational Aspects of Protein Crystal Data Analysis", Daresbury Laboratory DL/SCI/R25, Edi

ted by Helliwell, J.R., Machin, P.A. and Papiz, M.Z., pp. 84

THONG, N., SCHWARZENBACH, D., 1979 - Acta Cryst. A35, 658

WILKINSON, C., GABRIEL, A., LEHMANN, M.S., ZEMB, T., NE, F., 1992 - International Society for Optical Engineering Conference Proceedings, 1737, 324

WILKINSON, C., LEHMANN, M.S., 1991 - Nucl. Instr. and Meth. A310, 411

WLODAWER, A., SJOLIN, L., 1983- Biochemistry 22, 2720

WOOD, I.G., THOMPSON, P., MATTHEWMAN, J.C., 1983 - Acta Cryst. B39, 543

WYCKOFF, R.W.G., 1924 - The Structure of Crystals, New York: Chemical Catalog Co.

ZACHARIASEN, W.H., 1945 - Theory of X-ray Diffraction in Crystals, New York: Dover.

ZUREK, S., PAPIZ, M.Z., MACHIN, P.A., HELLIWELL, J.R., 1985 - Daresbury Laboratory, Info., Quart. for Prot. Cryst. No. 16, 37

CHAPTER XIII

NEUTRON CRYSTALLOGRAPHY OF BIOLOGICAL MOLECULES

M.S. LEHMANN

XIII.1. Introduction

The last few years have seen an explosive growth in the number of biomolecular structures determined by single-crystal diffraction techniques. A vast majority of these were obtained using X-ray methods, and the increased output is due to important developments in all the fields required for a successful outcome, starting with the new techniques for protein engineering, followed by better crystallisation procedures and more advanced data collection equipment and methods, helped along by a constant evolution of structure solution techniques employing more and more powerful computers until the final structure is studied with sophisticated, real-time graphic display systems.

X-ray diffraction covers the full range of analysis from the structure determination to the understanding of the biological function. By contrast neutron diffraction techniques are mostly used to analyse particular aspects of the molecular structure. The main reason for this is the scarcity of neutron sources and their low flux, but also that the most common tools for structure determination such as heavy atom and anomalous scattering techniques are strongly limited due to the nature of the scattering factors. The applications of neutron diffraction techniques in biology are thus complementary to the X-ray analysis. Once the overall features are known, neutrons are used to study what cannot be seen by X-rays.

The use of neutrons is to a very large extent related to the scattering power of hydrogen and deuterium. The list in table XIII.1 of scattering lengths shows that while the X-ray scattering from an atom is proportional to the number of electrons such a simple rule does not hold for the scattering of neutrons. Moreover, the scattering of X-rays, being from an extended cloud of electrons, has an angular dependence, while this is not the case for neutrons which are scattered from the much smaller nuclei.

Table XIII.1. — Selected scattering lengths for neutrons and X-rays. For simplicity the X-ray values have been expressed in electrons and are given for zero scattering angle. The unit fm is 10^{-15} m.

Atoms	H	D	C	N	O	S	Au	
Neutrons	-3.7	6.6	6.6	9.4	5.8	2.8	7.6	fm
X-rays	1	1	6	7	8	16	79	electrons

It is clear from this table that it is much easier to observe a hydrogen atom with neutrons than with X-rays, especially when one takes into account that the intensity measured is proportional to the sum of the squares of the scattering lengths. Thus, the signals from deuterium and carbon are nearly equal for neutrons while deuterium or hydrogen is only 3% of the signal from carbon in the X-ray case. In situations where it is possible to reach atomic resolution, neutrons are therefore very useful for the location and detailed study of specific hydrogen atoms, either in the macromolecule or in the solvent.

In addition, the neutron scattering lengths for hydrogen and deuterium are very different. The neutron scattering densities for materials containing hydrogen or deuterium are therefore very different. In cases where atomic resolution cannot be reached, for example when the material is disordered, neutrons can thus help to single out specific features. It is not always straightforward, though, to exchange deuterium for hydrogen, but one case is normally simple, namely the exchange of D_2O for H_2O. This is often used in the study of biological materials which by nature contain much water. In small angle scattering studies of macromolecules in solution the approach is applied extensively as discussed in previous chapters VIII, IX and XIV, and specific labeling with deuterated molecules is discussed in chapter X.

The procedure is called contrast variation. The solvent surrounding the material is made up by mixing H_2O with D_2O in varying proportions, giving a large range of values for the scattering of the solvent. The study is only done at a resolution much less than the inter-atomic distances, and the scattering density can therefore be described as a constant for a given region. It is -0.06 fm/Å^3 for H_2O and 0.63 fm/Å^3 for D_2O. The value for proteins is typically around 0.25 fm/Å^3, while for DNA or RNA the number lies near 0.42 fm/Å^3. By choosing the right mixture the relative signal from one of these components with respect to the solvent can therefore be reduced or enhanced.

The different types of investigations require quite different kinds of experimental conditions. For the high resolution studies of atomic features short wavelengths are needed, while for low resolution analysis of overall features very long wavelengths are more adequate. Thanks to the white spectrum of a neutron source both are easily available and contrary to the X-ray case absorption gives many fewer difficulties for the long wavelength studies. As a consequence radiation damage is not a problem. Indeed, once a suitable crystal is available for a study - which is not always an easy matter - the whole set of measurements can be done on this sample. It is thus - within limits - of less importance how long it takes to record the data. We should always keep in mind that the largest part of the work comes afterwards, in the data analysis and in the understanding.

XIII.2. Methodology

In a diffraction experiment, as described in chapters VI and VII of volume I of this series, we measure intensities $I_\mathbf{h}$, reflected off the different Bragg planes $\mathbf{h}$=(h,k,l) of the crystal. $I_\mathbf{h}$ is related to the so-called structure factor $\mathbf{F_h}$ by

$$I_\mathbf{h} = (\text{experimental constants}) \times |\mathbf{F_h}|^2$$

The structure factors $\mathbf{F_h}$ are obtained as the Fourier transform of the content of the unit cell (and vice versa)

$$F_h = \iiint \rho(x,y,z) \exp(-2\pi i(hx + ky + lz))\, dV$$

where $\rho(x,y,z)$ is the scattering density at the point (x,y,z) in fractional coordinates in the unit cell of the crystal.

It should be kept in mind that the F_hs are complex quantities, and that the measurements only give the absolute values, the structure factor amplitudes. The associated phase angles must therefore be found by other methods, and are normally deduced from a known or partially known structure determined by X-ray diffraction.

The distribution, $\rho(x,y,z)$, is made up of atoms and F_h can therefore also be written as a sum over the atoms (see chapter VI of volume I)

$$F_h = \Sigma_j\, b_j \exp(-2\pi i(hx_j + ky_j + lz_j))\, T_j(\sin\theta/\lambda)$$

where b_j is the scattering length of atom j, (x_j,y_j,z_j) the position, T_j is the temperature factor, which takes into account the distribution due to thermal motion and disorder of the individual atom, λ the wavelength, and 2θ is the angle of the diffracted beam.

The two expressions above for F_h are essential for the analyses of the diffraction data. In one approach the density analysis is used while the other type of study is based on the locations of the atoms.

XIII.2.1. Density analysis

Here most of the structure is known, and the neutrons are used to find a small difference or a small locally induced change, e.g. an additional small guest molecule. In this case the approach is to study the difference density

$$\Delta\rho(x,y,z) = \rho(x,y,z)_{\text{new, partially unknown}} - \rho(x,y,z)_{\text{known}}$$

The two densities are obtained as Fourier transforms of the corresponding sets of structure factors. For the known structure both the amplitude and the phase of the F_hs are reasonably well determined, but for the new, slightly different system only the amplitudes are known. This is what was measured. No phases are available and the best approach is therefore to use the phases from the known structure as a first approximation. This of course has to be done with caution, and techniques have been developed to do this in an optimal way taking into account the relative accuracies of the available phases (see Roth, 1987, and references herein). It is obvious that the smaller the difference, the better the approximation, giving in the best case a signal from the new part of the system which is about half the true value.

XIII.2.2. An atomic model

In this approach the exact positions of the atoms, (x_j,y_j,z_j) and their thermal motion behaviour, T_j, are studied.

In its simplest form T is assumed to describe an isotropic motion

$$T = \exp(-B(\sin\theta/\lambda)^2)$$

in which case B is related to the mean square displacement of the atom, $\langle u^2\rangle$, by

$$B = 8\pi^2 \langle u^2 \rangle .$$

In more general terms an anisotropic motion can be described as

$$T = \exp(-2\pi^2 (u_{11}h^2a^{*2}+.. +2u_{12}hka^*b^*+ ..)$$

where the quantities u_{ij} relate to displacements in the different directions in space. An expression of this kind is at present very rarely used in the analysis of large molecules, mainly due to the high values of B, which limit the accessible data. It might however be appropriate when low temperatures are used in which case the thermal motion can sometimes be reduced.

The calculations to get the best atomic positions and thermal parameters are done by non-linear least-squares methods, minimising a quantity of the type

$$\Sigma_h w_h (| F^{obs}_h | - | F^{calc}_h |)^2$$

where F^{obs}_h and F^{calc}_h are the observed and calculated structure amplitudes, the latter based on some model. The quantity w_h is a weight factor taking into account the quality of the individual observation. A number of different methods are available to obtain the best set of parameters, and the programs used are in all cases based on the equivalent programs developed for X-ray diffraction, having undergone a number of suitable modifications.

XIII.2.3. The measurements

The systems analysed normally contain several thousand atoms. If the structure is studied at atomic resolution the number of data is then typically in the order of tens of thousands of Bragg reflections, while less will do for low resolution work. In all cases, though, a considerable amount of data is required. It is therefore obvious that many Bragg reflections must be recorded quasi-simultaneously.

The basic principle is very simple. The diffraction geometry is described in chapter VII of volume I (for a recent review see also Majkrzak et al., 1990) and the instrument consists of a sample holder and a detector. The detector can be zero, one or two-dimensional (chapter V of volume I. See also Convert and Forsyth, 1983), but in order for the measurement to finish within a reasonable time, two-dimensional detectors are necessary.

Traditionally a neutron beam with only a few percent wavelength spread has been used for these measurements, but methods employing larger $\Delta\lambda/\lambda$ combined with longer wavelengths are now being explored (Wilkinson and Lehmann, 1991 and chapter XII of this volume), as is the use of pulsed neutron sources (Schoenborn, 1992).

XIII.3. Diffraction at or near atomic resolution

XIII.3.1. Looking for a single hydrogen atom

The first neutron diffraction studies of myoglobin (Schoenborn, 1969, 1970; Norvell and Schoenborn, 1975) showed that it is possible to locate ordered hydrogen and deuterium atoms from scattering density maps at a resolution of approximately 2.0 Å. This set off a number of studies where neutron diffraction techniques were used to locate hydrogen atoms that are key parts of an enzymatic process or involved in local dynamics.

An important study of this kind was the identification of the catalytic base in the active site of trypsin (Kossiakoff and Spencer, 1980). This enzyme hydrolyses the peptide bond on the carboxyl side of lysine and arginine, and the catalytic site holds three invariant residues, a histidine, an aspartic acid and a serine. Two models were available for the intermediary complex, both involving histidine 57. They correspond to different protonation states as shown in fig. XIII.1.

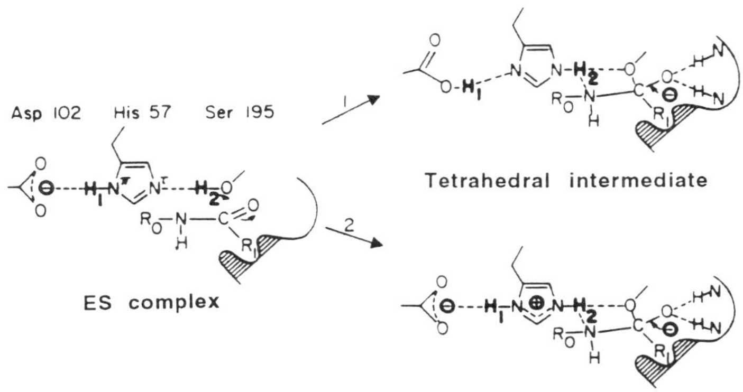

Fig. XIII.1. — Two models for the intermediary state in the protein hydrolysis by trypsin (from Kossiakoff and Spencer, 1980. Reprinted with permission from Nature). The difference is in the protonation state of the histidine residue.

Large crystals (1.5 mm^3) of bovine trypsin could be grown, and the neutron diffraction study gave data to 2.2 Å on a crystal that had been soaked for about one month in D_2O solutions. The H/D exchange aimed both at reducing the background of the non-negligible incoherent scattering from hydrogen, and at enhancing the signal from the hydrogen (i.e. deuterium) atoms under study.

The atoms of interest were determined from different kinds of scattering density and difference maps where the phasing of the structure amplitudes were based on a number of models. In one model the deuterium atom between the aspartic acid and histidine was omitted from the calculation of phases, and a significant peak was found near the histidine. In another calculation excluding the two deuterium atoms bound to this residue a density map showed density at both the locations. Yet another difference map where a deuterium was placed at the aspartic acid location also gave preference for the histidine side as indicated in fig. XIII.2. The combined calculations therefore point to model 2 (fig XIII.1) as the most likely intermediary in the reaction.

Similar studies have been done on lysozyme which cleaves certain saccarides. The crystal structure was solved by D.C. Phillips and his collaborators (Blake et al., 1965), and a mechanism for the function was proposed (see e.g. Phillips, 1966). It involves proton donation from a carboxyl group, and during a neutron diffraction study of the triclinic crystal form of hen egg-white lysozyme (Mason et al., 1984) this was studied in detail. The crystal was grown in normal water, and hydrogen-deuterium exchange was then done in D_2O solution. Data were collected to 1.8 Å, and again difference scattering densities showed clearly the location of the deuterium, in agreement with the earlier proposal.

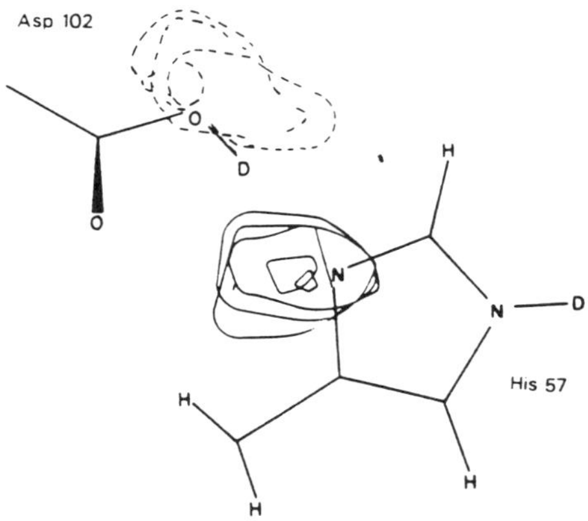

Fig. XIII.2. — Difference density map for the asparagine-histidine interaction (from Kossiakoff and Spencer, 1980. Reprinted with permission). In the model the hydrogen (i.e. deuterium) atom was placed on the carboxylic group of asparagine. The dotted contours corresponding to negative atomic density clearly show that the deuterium was misplaced and should be found in the positive peak near histidine.

XIII.3.2. Looking for water

Obviously water is also a target for study. Water is most often involved in hydrogen bonding (fig XIII.3) and while it is easy enough to find the oxygen atoms, for example using X-rays, it is by no means straightforward to identify a hydrogen bonding network from the oxygen atoms only. This was clearly seen in the studies of vitamin B_{12}.

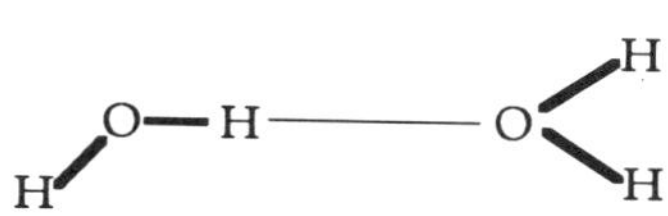

Fig. XIII.3. — Schematic diagram of two water molecules connected by a hydrogen bond. Typically the H---O distance is around 1.8 Å with a large range of values for the angles H---O-H centered around 109.5° and 120°.

Vitamin B_{12} is a relatively small molecule with 209 atoms. In the crystal form it binds about 17 waters per molecule of vitamin B_{12}. Good, large crystals that scatter well are available, and neutron studies to very high resolution (i.e. 1 Å) were carried out at 6 °C (Savage et al., 1987). These can be compared to the high resolution X-ray analysis, and in the solvent region the information from neutron studies is richer than from the X-ray analysis as seen in figure XIII.4.

As in the earlier examples the hydrogen atoms were exchanged for deuterium wherever possible. Indeed, the crystals were grown from D_2O, and the deuterium atoms are visible in the neutron scattering density, while as expected only the oxygen atoms are seen in the X-ray map.

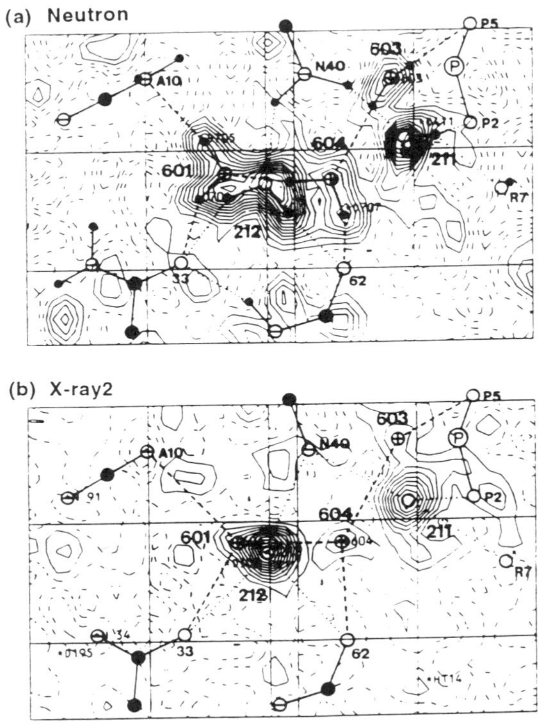

Fig. XIII.4. — Sections through difference densities where the water molecules have been excluded from the model (from Savage, 1986. Reprinted with permission). (a) is for neutrons and gives complete D_2O molecules, while (b) is for X-rays and shows only oxygen atoms.

It is of course also possible to combine X-ray and neutron data (Wlodawer and Hendrickson, 1982), and this approach has been exploited by Wlodawer and Sjölin (1983) in a study of ribonuclease A. Adding neutron data allowed the repositioning of a number of residues from the original model, and it was also very useful in the construction of the solvent map.

An interesting example of disorder was found in a detailed study of one of the water molecules of vitamin B_{12}. The density is shown in fig. XIII.5 and is an example of multi-site disorder, either with a small number of main sites or a continuous distribution. It is of course not possible to judge from this whether the situation is static or dynamic, but from the look of the density in the middle it is tempting to speculate that this represents a sum of observations of a water molecule 'moving by'. It is known from the deuteration that water molecules do exchange, and a good part of the density is therefore likely to be made up by individually moving water molecules temporarily bound in specific locations. When studying the disorder, one should therefore always keep in mind that a good part of this is dynamic.

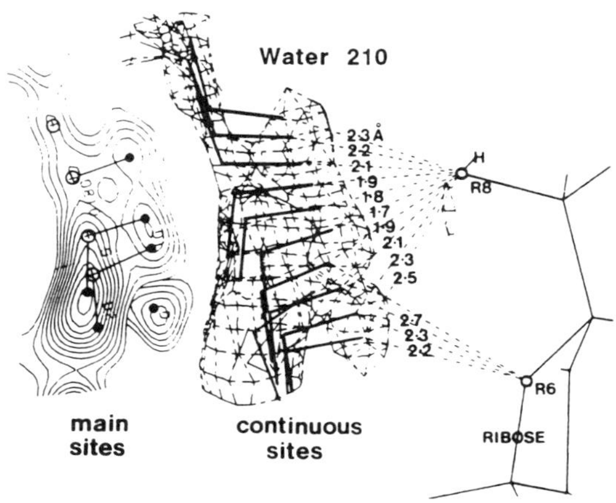

Fig. XIII.5. — Difference density showing a water molecule inside crystals of B12 (from Savage, 1986. Reprinted with permission). The left part shows contour levels and indicates possible main sites for the molecule, while the right hand side gives the three-dimensional representation of high scattering density.

Recently B12 was studied again with neutrons, this time at 15 K, and the first results from this study (Bouquiere et al., 1993) confirm that much of the disorder can be 'frozen out' and is thus of dynamical nature. As a comparison 56 solvent atoms (O or H) were found to be distributed over 23 sites, while at 279 K 113 solvent atoms were found to be distributed over 47 sites. Similarly the hydrogen bond pattern is tightened up at 15 K compared to the 279 K structure, with a very narrow distribution of oxygen-oxygen distances centered around 2.8 Å.

XIII.3.3. Hydration surfaces and hydrophobic cages

Protein crystals contain in general more than 30% water which can be approximately divided into three classes, namely a first hydration shell, a second hydration shell and bulk water. The behaviour of the water in each of the three groups is defined by their environment, ranging from the first hydration shell with bound, well located molecules to the freely moving water further away, typically at a distance of 4 to 5 Å from the macromolecular surface.

The contribution of the water scattering to the structure factor is quite large. It is therefore difficult to obtain a description of the water arrangement using the straightforward difference map calculation outlined in XIII.2.1, and for larger molecules than vitamin B_{12} more sophisticated techniques following the same principles are therefore employed.

Beyond the sheer size of the problem another difficulty is the description of the disorder. A good portion of the water is partially fluid, and this cannot be handled by introduction of more fractional atoms. The fluid structure will only give a signal at low angle, and thus not enough data to allow for such a procedure. In a study of myoglobin Cheng and Schoenborn (1990) therefore divided the solvent into shells around the protein. The shell thickness was chosen to be around 0.3 Å, and the shells were assumed to start 1.6 Å from the known atomic positions of the protein. A grid was set up, and points within each shell gave rise to a structure factor of the type

$$F^{shell}{}_h = \rho \sum_j \exp(-2\pi i(hx_j + ky_j + lz_j)) \ \exp(-B(\sin\theta/\lambda)^2)$$

where ρ is the scattering density within the shell, (x_j,y_j,z_j) the coordinates of the grid point j and B is called a liquidity factor and is equivalent to a temperature factor. The total structure factor is therefore the sum of $F^{shell}{}_h$ and the structure factor for the protein moiety, and crystallographic refinements can now be done. In the present case the protein was held fixed, and the ρ and B for each of 15 shells were obtained by a stepwise minimisation process.

Several sets of protein data were used and fig. XIII.6 gives the result for myoglobin monocarboxide. A minimum in B corresponds to low fluidity and indicate regions where water is bound, and the two minima in the figure at around 2.3 Å and 3.9 Å conform very well with the location of the first and second layer of water.

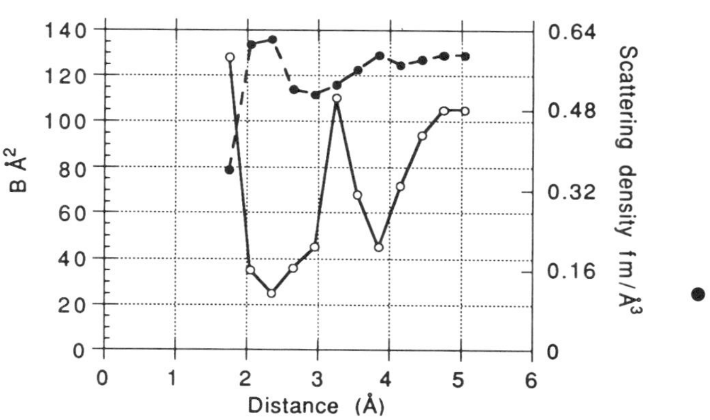

Fig. XIII.6. — The fluidity factor B (open spheres) and the scattering density ρ (dots) for the shells as a function of distance from the surface of the protein (from table 3 of Cheng and Schoenborn, 1990).

The improved description of the scattering density gave better phases for the structure factors, and combined with the knowledge of the mean scattering density in the various regions this was used to locate the water and ion molecules. In all 87 water molecules and 5 ions were found. The majority of these, 67, were placed in the first shell, i.e. attached directly to the protein, while the last 22 were bound to other water molecules only. Binding to polar groups was dominant, and an analysis showed that the order of preference was acidic>basic>polar>nonplar, as expected.

Another interesting approach is the combination of data recorded both for crystals containing H_2O and D_2O, and this has been applied to the structure of bovine trypsin (Kossiakoff et al., 1992). In principle the two structures are indentical except for the exchangeable hydrogens, and the differences between the two set of structure factors will therefore show these atoms due to the large difference between the scattering lengths of deuterium and hydrogen. In fact the average difference in intensity between the two sets of data is 30-40 %, and this excludes the use of a simple difference density maps.

An iterative approach using experimentally determined phases for the solvent structure based on the two sets of data was therefore developed. Its principle is show schematically in the Argand diagram in figure XIII.7.

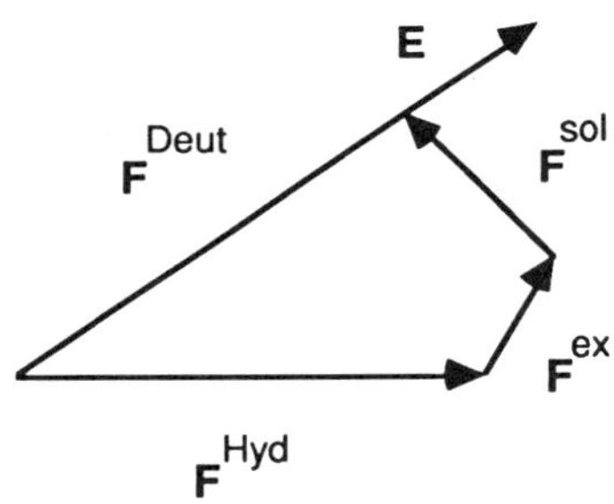

Fig. XIII.7. — Diagram showing the different structure factor components. F^{Deut} and F^{Hyd} are the structure factors for the two structures, F^{ex} is for the H to D exchanged part, F^{sol} contains both ordered water and the bulk solvent and **E**, the closing error, contains the information about the missing part of the solvent model.

The structure was assumed to contain three parts, namely the protein itself which remains fixed, the bulk which is found at a distance of more than 5 Å from the protein and the region for ordered or partially ordered water in the intermediate belt. A good part of the difference density is thus known, notably the inside of the protein which is the same for the two sets of data and the bulk region which should stay constant, and this can be included in the calculation. F^{Hyd} was also kept constant, and its phase was determined by a prior refinement without inclusion of any H_2O molecules. This is possible due to the cancellation effect of the two hydrogens with scattering length of -3.7 fm against oxygen with 5.8 fm. The phase of F^{Deut} can now be found by minimisation of **E**, and water molecules can then be identified and refined from the difference maps. A similar approach was used for the H/D locations.

In a number of iterations 291 solvent peaks were identified. They form clusters with up to 19 members. Many of the larger water cluster are in the form of long chains and are found along the intermolecular boundaries and they are thus characteristic for the crystal form. As in the study of myoglobin most of the sites were associated with the protein surface, leaving 69 peaks to connect only with other water molecules. An interesting aspect is the interactions between the hydrophobic part of the protein surface and the bound water. In trypsin the 53% apolar surface is mainly spread out in small patches, but wherever larger hydrophobic areas occur they are bridged over by one or more water molecules from a cluster. Finally the coverage is thus evenly partitioned between polar and apolar parts.

An exceptional case of water ordering around a hydrophobic group is found in crambin, a small plant protein with 46 residues where leucine 18 is nearly completely embedded in water. The structure has been studied to high resolution with both neutrons and X-rays (Teeter and Whitlow, 1986), and the result for leucine 18 is show in fig. XIII.8.

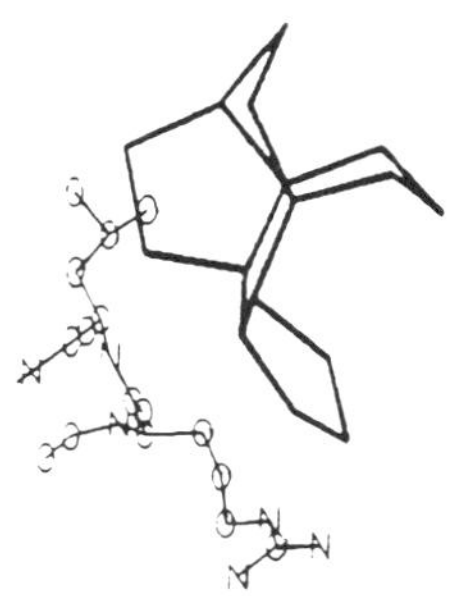

Fig. XIII.8. — Pentagons of water around leucine 18 of crambin (from Teeter and Whitlow, 1986). Only the oxygen atom positions are shown.

Here the hydrophobic part is surrounded by a nearly complete cage of water in a similar manner to what is found in many clathrate hydrates. The case is at present unique. Crambin is very hydrophobic, and a number of other hydrophobic proteins probably have to be studied before similar cages are found.

A fair number of the water molecules in the studies described above are engaged in interprotein connections. A study of the water surface cannot therefore give a direct picture of the hydration of the protein in solution. Nevertheless, no other method gives such detailed structural information about hydration, and it is obviously possible from the crystal structure to propose a model for the individual protein molecule and its water.

XIII.3.4. Hydrogen/Deuterium exchange and the dynamics of molecules

Neutron diffraction can also be used to study H/D exchange. Typically the crystals are grown in H_2O, and are then left to exchange in D_2O for a period ranging from days to years. If there is exchange the scattering factor for the H/D atom changes, and this can be identified by the crystallographic refinement techniques discussed in XIII.2.2. This is clearly seen from the expression for the structure factor when this is rewritten slightly as

$$F_h = \Sigma_j b_j \, occ_j \exp(-2\pi i(hx_j + ky_j + lz_j)) \, T_j(\sin\theta/\lambda)$$

adding in the site occupation factor, occ_j. Suppose that b_j is set to 1. Then occ_j is -3.7 for H only and 6.6 for D only, and for approximately 30 % D it will be zero! Occ_j is adjusted during the structural refinement, and its final value gives the ratio of deuterium to hydrogen for a given hydrogen location.

The H/D exchange in a number of proteins has been studied, and an extensive discussion of these results was given by Kossiakoff (1985).

One example of an H/D exchange comes form the analysis of trypsin (Kossiakoff, 1982). In this case the crystal was soaked for one year, after which neutron diffraction data were measured. The result for the exchangeable amide protons of the peptide units was that 68 % were exchanged, while 24 % were unexchanged. It was observed that the non-exchanged hydrogen atoms were found mainly in the β-sheet regions, which in this structure forms the so-called β-barrels. A similar pattern was observed in pancreatic trypsin inhibitor (Wlodawer et al., 1984) and also a study of ribonuclease A (Wlodawer and Sjölin, 1983) showed that some of the β-sheets were well protected. More recently a mutant of subtilisin has been studied (Kossaikoff et al., 1991). Here it was observed that the amide groups in the secondary elements were most resistant to changes with the largest number of unexchanged sites found in the seven-stranded parallel β-sheet.

Information about local dynamics can also be obtained. The H/D pattern in residues 47-52 of trypsin (fig. XIII.9) shows this.

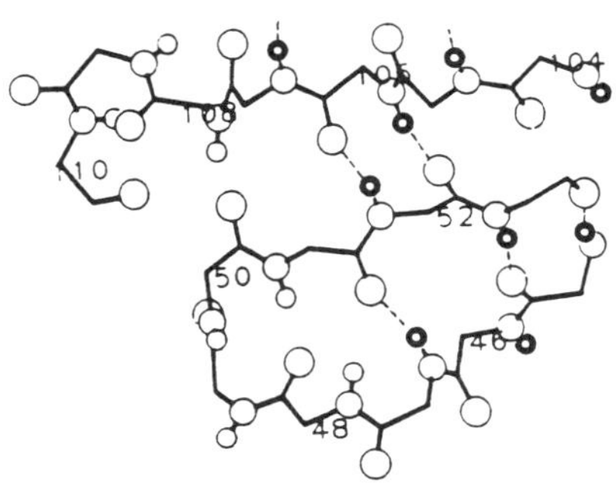

Fig. XIII.9. — Main chain of trypsin from residue 47 to 52, showing the internal hydrogen bonding (from Kossiakoff, 1984. Reprinted with permission). Dark circles are unexchanged hydrogen atoms while the small spheres are deuterium.

The bend formed by the residues can 'open' in several different ways. One possibility would be a 'zipper' motion, that periodically release all the hydrogen bonds along this piece of sheet. This does obviously not happen, as the hydrogen on residue 47 is not exchanged, and the motions that allow for a H/D exchange in the bend are consequently of a more local nature.

XIII.3.5. Small solvent molecules

Specific deuteration can be done in many ways. Different variants of H_2O/D_2O exchange have been discussed above, but it is of course also possible to study other, suitable deuterated small molecules in the surroundings. A very large contrast between the small solvent molecules and their environment is obtained by using normal water for the mother liquor. By difference techniques the location of the molecules can then be found due to the fact that the scattering density of H_2O is near zero. In addition structural information can be obtained from standard crystallographic refinements if sufficient data is available, and this will show the changes in the macromolecule due to the addition of the small molecules.

Lysozyme interactions with ethanol (Lehmann et al., 1985), dimethylsulfoxide (Lehmann and Stansfield, 1989) and tetramethylammonium (Bouquiere et al., 1993) were studied in this way. The experiment consisted in adding the molecules to the crystals and then comparing the structural results with the original structure using both difference density maps and structure refinement techniques. The approximate scattering density of the different parts are given in table XIII.2, and fig. XIII.10 shows a stereo plot for the scattering density of one of the dimethylsulfoxide molecules.

Table XIII.2. — Scattering density for selected molecules. The percentages introduced into the crystal for the scattering experiments are given.

compound	fm/Å^3	comment
H_2O	-0.06	
CD_3CD_2OH	0.50	25 % added to crystal
$(CD_3)_2SO$	0.61	15 % added to crystal
$(CD_3)_4N^+$	0.64	30 % added to crystal
Protein	0.19	non-deuterated

In all cases several molecules were found; for ethanol 13, for dimethylsulfoxide 6 and for tetramethylammonium 4. No significant changes were found in the structure, and for the denaturants ethanol and dimethylsulfoxide this led to a proposition of model for denaturation which proceeds via a dehydration of the protein at high alcohol concentration. Both these molecules seem to be mainly bound to the protein via hydrophobic interactions, while the tetramethylammonium interaction is dominated by polar attractions. This might well be due to the fact that the latter molecule is a protein protectant and therefore would have a very different type of interaction with the macromolecule.

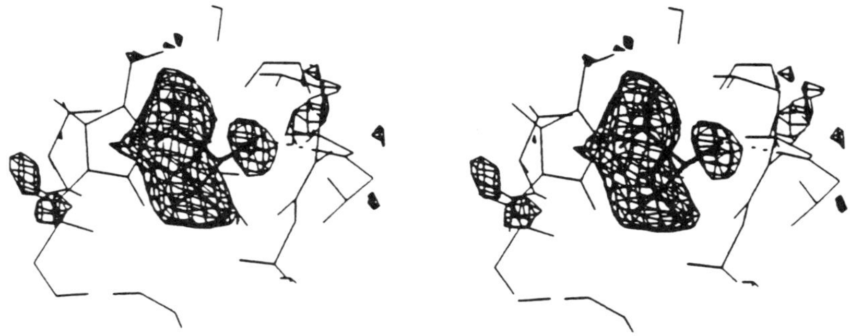

Fig. XIII.10. — Stereo view of a molecule of dimethylsulfoxide bound to lysozyme. The protein density has been removed.

XIII.4. Low resolution studies

XIII.4.1 Phospholipids inside or outside

When part of the biological structure is completely disordered, which can be the case for the DNA inside a virus capsid or for membranes surrounding a macromolecule, it is of great interest to know exactly where the various components are found. Because the parts under study are disordered the goal is only to get very low resolution information on a scale where their scattering density is constant, and the studies therefore do not go much beyond 10 Å d-spacings.

While atomic or near atomic resolution studies are best done with wavelengths between 1 and 4 Å, low resolution studies can be done with much longer wavelengths, typically around 8 Å. These are available with neutrons, and have the advantage of giving much higher reflectivity, thus smaller crystals can be used. Moreover, the angle between individual Bragg-reflections is larger, allowing for bigger unit cells. It is therefore possible to attack larger systems, and two examples are given below.

Lipovitellin, which exists as a dimer with a molecular weight of 352,000 D, binds lipids. The molecular structure is known from X-ray analysis (R. Raag et al., 1988), and has in the middle a large cavity of 68,000 $Å^2$. Fig. XIII.11 shows the polypeptide part of the molecule.

As the individual atoms in a disordered molecule cannot be located by diffraction techniques, the best way to show that the cavity actually holds lipids is by contrast variation using neutrons (Timmins et al., 1992). Data were recorded to about 12 Å resolution with H_2O in the crystal mother liquor as well as for mixtures with 20, 59, 85 and 93 % D_2O in H_2O. From these measurements and from the knowledge of the structure as obtained from X-rays it is then possible to estimate the phase of the additional component. Moreover, it is possible to extrapolate to other mixtures. At 10 % D_2O the lipid has the same scattering density as the solvent, and should thus be invisible, which is the case; the X-ray structure is reproduced at low resolution with neutrons. At 40 % D_2O the protein is invisible, and indeed there is density inside the protein. This density is assigned to the lipid, which seems to be bound as a bilayer. The major interactions of the lipid with the protein cavity are hydrophobic, while the lipid head groups are projected into the solvent or are bound to solvent in the cavity.

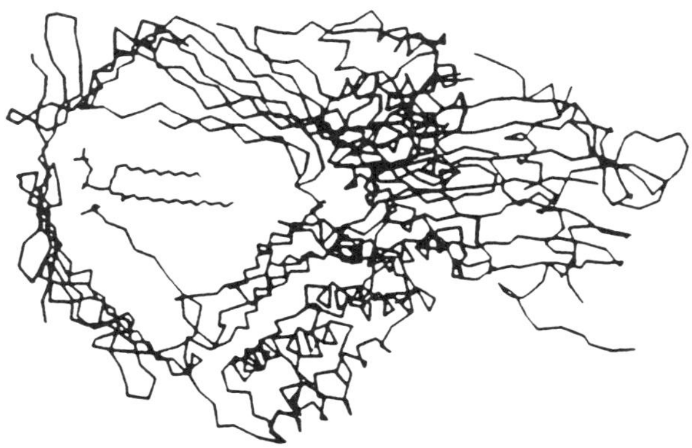

Fig XIII.11. — Polypeptide chain of lipovitellin. From Raag et al. (1988). (Reprinted with permission). A lipid molecule is placed inside the cavity in an arbitrary position.

The photosynthetic reaction center from *Rhodopseudomonas viridis*, which was crystallised and solved recently (Deisenhofer et al., 1985), is embedded in a membrane. Detergent was used for the crystallisation and serves to mimic the membrane. As this is disordered it could not be located as the scattering density for X-rays is very similar to the surrounding water. Again, with neutrons, it was possible to find the detergent (M. Roth et al. 1989) using data with 0, 15, 55 and 87 % D_2O in H_2O. At 40 % D_2O, which can be derived from the other observations, the protein is nearly invisible, while the membrane has a relatively large negative contrast. The detergent is found to be concentrated in a ring around the center. It covers the space along the α-helices, which are hydrophobic in nature, and confirms the hypothesis that these helices constitutes the membrane-spanning part of the complex.

XIII.4.2. Structure solution

Variation of the scattering density for a part of the crystal, e.g. the H_2O/D_2O exchange of the solvent, will lead to large changes in the scattered Bragg intensities, but they do not directly give any phase information. Nevertheless, choosing a given macromolecule/solvent contrast can help in the analysis by simplifying the density. This is for example the case when both RNA and protein are present, and in addition the contrast can be made much larger with neutrons than with X-rays. These features of neutron scattering are presently being developed to study the 50S subunit of ribosomal particles using single crystals (Eisenstein et al, 1991).

Suitable crystals of ribosomes for neutron scattering are available from several species. The first data were measured on ribosomes from *Halobacterium marismortui*, and test measurements were done on samples from *Thermus thermophilus*. Because of the small crystal size and the very large unit cell, diffraction experiments are very time-consuming, but data are now available to 30 Å resolution for 100% D_2O and partially for H_2O.

At present *ab initio* phasing is being developed. In addition to direct methods, maximum entropy and likelihood methods developed by Bricogne and Gilmore (Gilmore et al., 1991) are used. The density maps presently obtained are consistent with the electron-microscope model for the ribosome, and work is being done on the introduction of a two-level model to distinguish between protein and RNA. If this gives results, the method could obviously be extended to deal with normal size protein crystallographic problems.

XIII.5. Concluding remarks

Neutron scattering studies of single crystals of biological molecules cover a wide range of problems. They can give a detailed analysis of the water structure or the position of atoms crucial for the activity of the molecules, but they can also be used to determine the position and nature of the matrix where these reactions take place. The reasons for this range of possibilities lie in the large range of useful wavelengths available, but also in the unique scattering power of hydrogen and deuterium.

In nearly all cases where single crystal neutron diffraction is used it serves either to explain or elucidate points that cannot be resolved by X-ray techniques, and it is therefore very much a complementary technique which is used after the X-ray work is completed.

A central problem in neutron scattering is the flux, which at present limits the applications, and efforts and proposals are therefore made to circumvent this. They go both towards larger detectors - at the ILL e.g. the D19 detector is being enlarged (Mason, 1993) - and towards the use of a wider range of wavelengths either on a steady state reactor (Wilkinson and Lehmann, 1991) or with spallation neutrons (Jauch and Dachs, 1984; Lehmann, 1984; Schoenborn, 1992). These new developments will extend considerably the range of projects that can be tackled.

REFERENCES

BLAKE, C.C.F, KOENIG, D.F., MAIR, G.A., NORTH, A.C.T, PHILLIPS, D.C. and SARMA, V.R., 1965 - Nature, **206**, 757

BOUQUIERE, J.P., FINNEY, J.L. and LEHMANN, M.S., 1993 - Proc. Faraday Soc. (in press)

BOUQUIERE, J.P., FINNEY, J.L., LEHMANN, M.S., LINDLEY, P.F. and SAVAGE, H.F.J., 1993 - Acta Cryst. **B49**, 79

CHENG, X.and SCHOENBORN, B.P., 1990 - Acta Cryst. **B46**, 195

CONVERT, P. and FORSYTH, J.B., Eds., 1983 - Position-Sensitive Detection of Thermal Neutrons, Academic Press.

DEISENHOFER, J., EPP, O., MIKI, K., HUBER, R. and MICHEL, H., 1985 - Nature **318**, 618

EISENSTEIN, M., SHARON, R., BERKOVITCH-YELLIN, Z., GEWITZ, H.S., WEINSTEIN, S., PEBAY-PEYROULA, E., ROTH, M. and YONATH, A., 1991 - Biochimie **73**, 879

GILMORE, C.J., HENDERSON, H.N. and BRICOGNE, G., 1991 - Acta Cryst. **A47**, 842

W. JAUCH and H. DACHS, 1984 - in Proceedings of the Workshop on Neutron Scattering Instrumentation for SNQ. Eds. R. Scherm and H. Stiller, Jülich Report Nr. 1954, 31

KOSSIAKOFF, A.A., 1984 - in Neutrons in Biology, ed B.P. Schoenborn, Plenum Press, 281

KOSSIAKOFF, A.A., 1985 - Ann. Rev. Biochem. **54**, 1195

KOSSIAKOFF, A.A, 1982 - Nature **296**, 713

KOSSIAKOFF, A.A., SINTCHAK, M.D., SHPUNGIN, J. and PRESTA, L.G., 1992 - PROTEINS: Struct., Funct. and Gen. **12**, 223

KOSSIAKOFF, A.A. and SPENCER, S.A., 1980 - Nature **288**, **414**

KOSSIAKOFF, A.A, ULTSCH, M., WHITE S. and EIGENBROT, C., 1991 - Biochemistry **30**, 1211

LEHMANN, M.S., 1984 - in Proceedings of the Workshop on Neutron Scattering Instrumentation for SNQ. Eds. R. Scherm and H. Stiller, Jülich Report Nr. 1954, 53

LEHMANN, M.S., MASON, S.A. and MCINTYRE, G.J., 1985 - Biochemistry, **24**, 5862

LEHMANN, M.S. and STANSFIELD, R.F.D., 1989 - Biochemistry, **28**, 7028

MAJKRZAK, C.F., LEHMANN, M.S. and COX, D.E., 1990, - in Physical Methods of Chemistry, Vol V, Eds. B.W. Rossiter and J.F. Hamilton, John Wiley and Sons, 123

MASON, S.A. 1993 - Personal communication

MASON, S.A., BENTLEY, G.A. and MCINTYRE, G.J., 1984 - in Neutrons in Biology, ed B.P. Schoenborn, Plenum Press, 323

NORVELL, J.C. and SCHOENBORN, B.P., 1975 - Brookhaven Symp. Biol. **27**, **II**, 12

PHILLIPS, D.C.. 1966 - Sci. Amer. **215**, 78

RAAG, R., APPELT, K., XUONG, N.-H. and BANASZAK, L.J., 1988 - J. Mol. Biol. **200**, 553

ROTH M., 1987 - Acta Cryst. **A43**, 780

ROTH, M., LEWIT-BENTLEY, A., MICHEL, H., DEISENHOFER, J., HUBER, R. and OESTERHELT, D., 1989 - Nature **340**, 659

SAVAGE, H.F.J., 1986 - in Water Science review **2**, Ed. F. Franks, Cambridge University Press, 67

SAVAGE, H.F.J, LINDLEY, P.F., FINNEY, J.L. and TIMMINS, P.A., 1987 - Acta Cryst. **B43**, 280

SCHOENBORN, B.P., 1969 - Nature **224**, 143

SCHOENBORN, B.P., 1971 - Cold Spring Harb. Symp. quant. Biol. **36**, 569

SCHOENBORN, B.P., 1992 - SPIE **1737**, 235

TEETER, M.M. and WHITLOW, M.D., 1986 - in Transactions of Am. Cryst. Ass. **22**, 75

TIMMINS, P.A., POLIKS, B. and BANASZAK, L., 1992 - Science, **257**, 652

WILKINSON, C. and LEHMANN, M.S., 1991 - Nucl. Inst. and Meth. **A310**, 411

WLODAWER, A. and HENDRICKSON, W.A., 1982 - Acta Cryst. **A38**, 239

WLODAWER, A. and SJÖLIN, L., 1983 - Biochemistry **22**, 2720

WLODAWER, A., WALTER, J., HUBER, R. and SJÖLIN, L., 1984 - J. Mol. Biol. **180**, 301

GENERAL READING

WLODAWER, A., 1982 - Neutron Diffraction of Crystalline Proteins. Prog. Biophys. molec. Biol. **40**, 115

SCHOENBORN, B.P., Ed. 1984 - Neutrons in Biology, chapters VI and VII. Plenum Press.

SAVAGE, H. and WLODAWER, A., 1986 - Meth. in Enzym. **127**, 285

CHAPTER XIV

THE STRUCTURE AND ASSEMBLY OF VIRUSES: CONTRIBUTIONS FROM X-RAY SYNCHROTRON AND NEUTRON RADIATION

P.A. TIMMINS

XIV.1. Introduction

The aim of this chapter is to give a survey of the role played by neutrons and synchrotron X-rays in the elucidation of the structures of viruses. In order to fully appreciate the importance of structure it is first necessary to know what a virus is and what are its biological functions.

Viruses are parasitic particles which can self-replicate only in the presence of a cellular replication mechanism. Hence on entering the cell they take over the machinery of the cell in order to replicate and in doing so eventually kill the cell.

The infective agent in a virus particle is a piece (or pieces) of RNA or DNA. This nucleic acid is encapsulated in a protein coat which protects it from being degraded. The virus may in some cases be surrounded by a lipid membrane which is derived from the membrane of the host cell. The RNA or DNA may contain a few or several genes which encode the information for a number of proteins. In the simplest viruses the nucleic acid contains information only for the capsid protein, an enzyme for replicating the nucleic acid and perhaps one or two other proteins concerned with cell recognition or release. In more complicated viruses there may be several different capsid proteins which are encoded in addition to enzymes necessary for the replication of the nucleic acid such as DNA polymerase or reverse transcriptase.

Viruses can also transform cells by inserting their nucleic acid into the cellular DNA thus modifying its replication and hence the control of protein synthesis.

A knowledge of virus structure is essential for an understanding of assembly and disassembly as well as for the design of anti-viral drugs. In addition all the processes from receptor recognition to cell penetration, assembly, disassembly and release from the cell are paradigms for normal cellular processes. For this reason the interest in viral processes exceeds merely the understanding of the viral infection *per se* , but extends to the study of a large number of cellular processes. In this chapter we show how neutron and synchrotron radiation along with other biophysical techniques have contributed to these goals.

XIV.2. Principles of Virus Structure

The first virus to be crystallised was tomato bushy stunt virus (TBSV) in 1938. From this date onwards both x-ray diffraction and electron microscopic studies as well as biochemical characterisations were carried out on a number of viruses. Two observations were of fundamental importance:

(i) It was observed that viruses are always composed of nucleic acid and protein and sometimes of lipid. The amount of protein in a complete virus particle is much greater than can be coded for in the nucleic acid genome of the particle.

(ii) Two very different kinds of structure are observed by electron microscopy. In the earliest observations these could be described as rod-like and spherical particles of which typical examples are tobacco mosaic virus (TMV) and polio virus respectively (Fig.XIV.1).

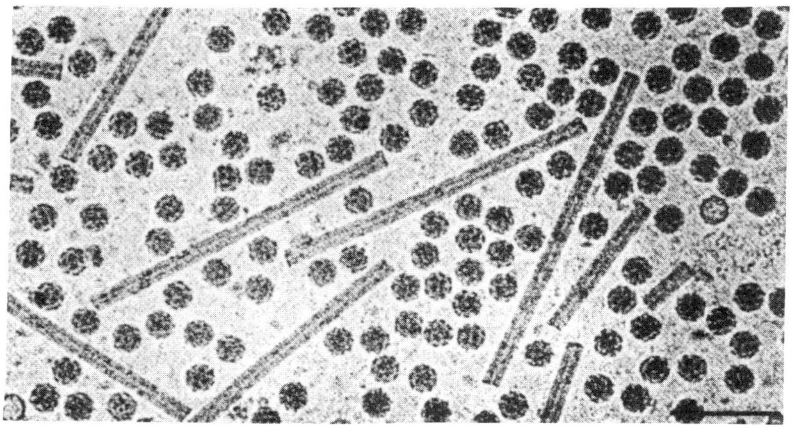

Fig.XIV.1. — Electron micrograph of a typical rod-shaped virus, tobacco mosaic virus (TMV) and a typical T=3 spherical virus, turnip yellow mosaic virus (TYMV), in amorphous ice. The bar represents 1000Å. From Adrian et al. (1992) with permission.

From these observations Crick and Watson (1956) suggested that the shells of all simple viruses must be composed of multiple copies of identical sub-units packed together in a regular way. They showed that in tobacco mosaic virus (TMV), the archetypal rod-shaped virus , that the symmetry was helical allowing the particle to be formed by repetition of a single kind of protein interaction, the ultimate length of the particle being limited by the length of its constituent RNA. For the so-called spherical viruses they showed that the most efficient way to build a closed shell from multiple copies of the same sub-unit was to arrange them with cubic symmetry. There are three different forms of cubic symmetry, 23, 432 and 532 symmetry which require 12, 24 or 60 sub-units respectively to form a closed shell. They were unable to suggest which of the cubic symmetries may be most likely but Caspar (1956) showed that the X-ray diffraction pattern of TBSV crystals was consistent with a particle having 532, icosahedral symmetry (Fig.XIV.2). Later chemical analysis showed that the number of sub-units in a TBSV particle was 180, three times the number predicted from the icosahedral symmetry and inconsistent with the idea of multiple *exactly* repeating contacts. This dilemma was resolved by the theory of quasi-equivalence of Caspar and Klug (1962) who showed that the lowest energy state of a particle would still be produced when almost identical - *quasi-equivalent* - interactions were used.

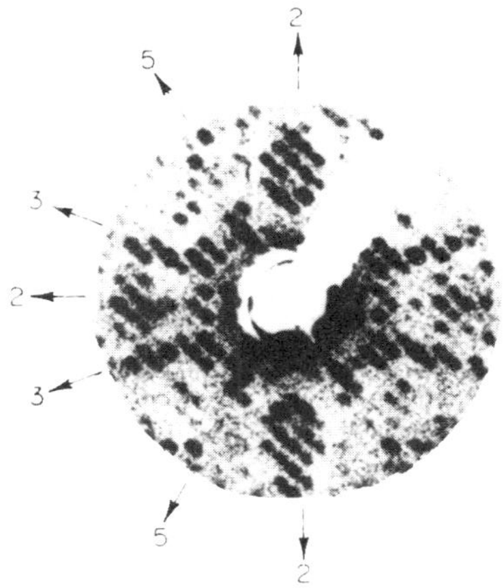

Fig. XIV.2. — Diffraction pattern from a single crystal of tomato bushy stunt virus (TBSV). The spikes of high intensity indicate the directions of the particle symmetry axes. From Girard et al. (1989) with permission.

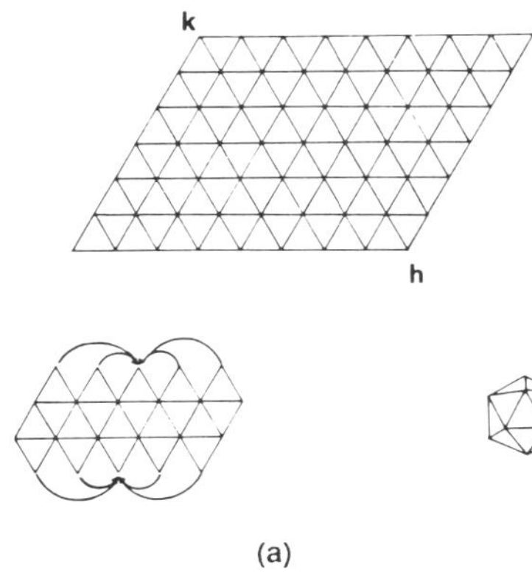

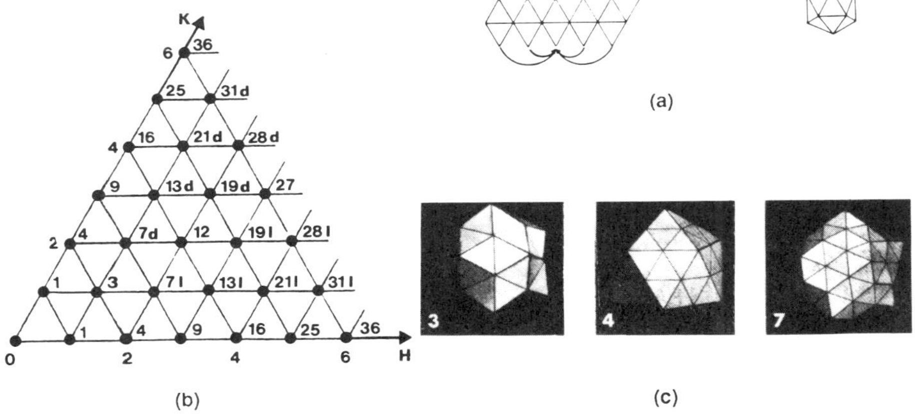

Fig.XIV.3. — (a) Construction of a T=1 icosahedron from an hexagonal net by elimination of 6-fold vertices (b) 6-fold vertices to be eliminated in order to produce shells of differing triangulation number. Adapted from Girard et al. (1989) with permission. (c) Shells of triangulation number T=3, 4 and 7.

The way in which an icosahedron can be constructed from a hexagonal net is shown in Fig.XIV.3. In this case the solid figure obtained contains 20 identical equilateral faces and any point placed upon one of these faces will be repeated exactly 60 times by the symmetry elements to give 60 points with identical environments. We can however construct other shells, also having icosahedral symmetry by cutting the hexagonal net in different ways. In the first case every hexagonal point of the planar net became a five-fold symmetry apex of the icosahedron. If now we only select further neighbours from the origin as five-fold symmetry apices then we will produce bigger shells but still having icosahedral symmetry. The different kinds of shells are defined by their triangulation number T, defined as :

$$T = h^2 + hk + k^2$$

where h and k are any integers including 0.

XIV.3. Neutron Scattering

Neutron scattering has been a most useful technique in the study of virus structure largely through the exploitation of the D_2O/H_2O contrast variation method. This is described in detail in Chapter II, but here we recall some of the principles relevant to viruses which are composed of protein, RNA or DNA and sometimes containing lipid. The scattering length densities of viral components are very different hence their contributions to the scattering are minimised in different D_2O/H_2O mixtures. Thus proteins become "invisible" in a mixture of ~42% D_2O/H_2O, RNA in 65% D_2O/H_2O, DNA in 70% D_2O/H_2O and lipid in 10 - 20% D_2O/H_2O depending on the exact chemical composition. Neutron scattering studies have been devoted almost exclusively to icosahedral viruses.

XIV.3.1. Small Angle Scattering

The scattering of spherical or near spherical particles is characterised by a series of alternating maxima and minima. Fig. XIV.4 shows the scattering from solutions of TBSV in various mixtures of D_2O/H_2O. The positions of the maxima are characteristic of the particle as viewed at a given contrast. We see in particular in this example that the curve from particles in 40% D_2O/H_2O is very different from the

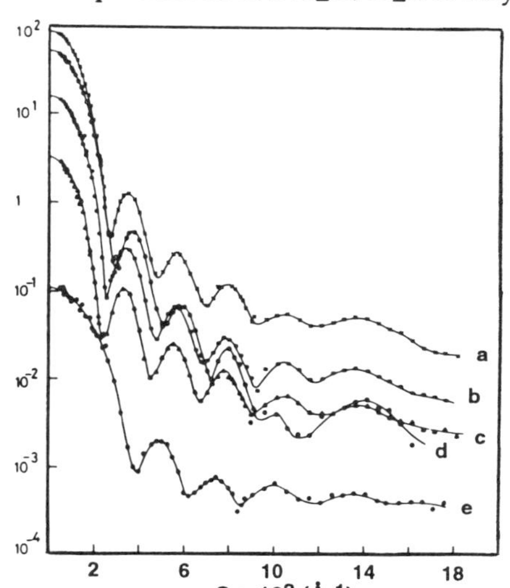

Fig.XIV.4 . — Neutron scattering curves from TBSV solutions of various D_2O / H_2O content (A) 100%, (B) 68%, (C) 55%, (D) 0%, (E) 41%. From Jacrot, (1981) with permission.

other curves. This indicates that the RNA which dominates the scattering at this contrast is located at low radii and that the protein forms an external shell. This is the result to be expected from previous studies by other techniques but we may put this on a much more quantitative basis by determining, on an absolute scale, the radial distribution of matter in the particle.

$$\rho(r)-\rho_s=\frac{1}{\pi}\int_0^{\infty} Q^2F(Q)\frac{\sin Qr}{Qr}dQ \qquad \text{(XIV.1)}$$

F(Q) for a spherical particle is always real and may therefore be computed from the measured quantity I(Q):

$$F(Q)=\pm\sqrt{I(Q)} \qquad \text{(XIV.2)}$$

The only uncertainty then in F(Q) is its sign but as it is a well-behaved function, changing sign at each zero, we can usually determine its sign everywhere if it is known at Q=0. This may be calculated as I(0) depends only on the chemical composition of the virus and solvent and is invariably +ve in pure H_2O and -ve in pure D_2O.

In practice there are a number of more or less important drawbacks to this approach:

1.) The most important limitation is that the integral in equation XIV.1 is calculated for Q from 0 to ∞, whereas the data is only measured to some finite value of Q. The consequence of this is to produce series termination errors resulting in oscillations in the density have no physical significance.

2.) A second limitation, closely related to the first, is the departure from spherical symmetry. We know that the virus particle in fact has icosahedral symmetry and depending on the exact clustering of sub-units this will be manifested in the scattering at Q values greater than about $0.15Å^{-1}$, corresponding to a resolution of about 40Å. Thus beyond this limit the scattering cannot be considered as a spherical particle and eqn. XIV.1 is no longer valid.

3.) A third but much less serious limitation is the influence of instrumental parameters on the scattering (smearing). For neutrons scattering this is the beam divergence, most important at low Q, and the polychromatic nature of the beam, most important at high Q. Modern deconvolution techniques are however capable of correcting for these effects (see for example, Glatter, 1982).

Due to the above effects the most widely used method for interpreting small angle scattering from spherical viruses has been the model-fitting approach. Here the structure is described by a series of concentric shells of varying scattering length density each having a form factor:

$$F_i=4\pi\int_{S_i}^{R_i} r^2(\rho(r)-\rho_s)\frac{\sin Qr}{Qr}dr \qquad \text{(XIV.3)}$$

where R_i and S_i are the external and internal radii respectively.

Thus for a model composed of n shells the predicted scattered intensity I(Q) is given by:

$$I(Q) = \left(\Sigma^{n} F_i(Q)\right)^2 \qquad \text{(XIV.4)}$$

A model described by n shells, each having an inner and outer radius and an excess scattering length density is fit to the data for each contrast. A number of physical constraints can be applied:

1.) The geometrical parameters (radii) must remain the same at all contrasts.

2.) The scattering length densities obtained should be a linear function of the contrast.

The instrumental smearing may be taken into account either by desmearing the data or by including smearing parameters in the model.

In the case of TBSV this led to a structure which was described by a model of 4 concentric shells of different scattering length densities. The manner in which the scattering length density (on an absolute scale) in each shell varied as a function of D_2O content enabled the amount of protein, RNA and water in each shell to be estimated.

An important flexibility in solution scattering is the possibility of varying the solution conditions. For TBSV studies were also carried out on virus particles which had been swollen by removal of calcium and change in pH. Fig XIV.5 shows the compact and swollen structures schematically and demonstrates that the swelling process can be explained by a simple radial translation of all viral components, which must of course undergo reorganisation in the plane of the surface to maintain inter sub-unit contacts.

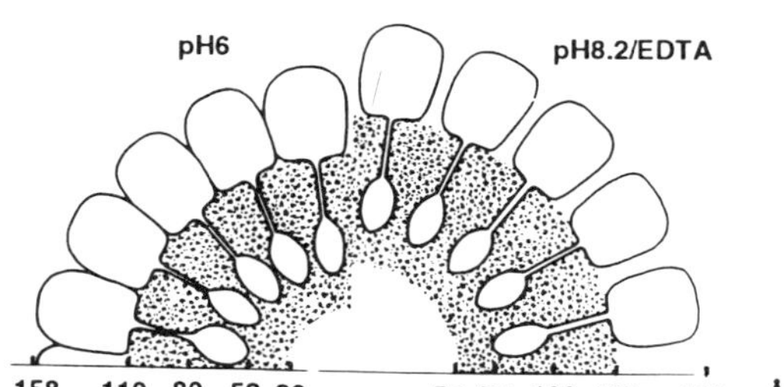

Fig. XIV.5. — Schematic representation of the radial distribution of protein and RNA in compact and swollen TBSV. From Jacrot (1981) with permission.

Another striking example of the success of contrast variation in solution was in the case of influenza virus. In this case the virus is much larger and more complicated and the analysis therefore less straightforward. Influenza virus contains an RNA genome, a protein coat and a lipid bilayer envelope derived from its host cell (Cusack, 1982). Scattering curves at five different contrasts are shown in Fig. XIV.6 along with the model derived for the virus structure. It contains an outer glyco-protein coat extending from a radius of 450Å to 590Å, which constitutes some 45% of the viral mass, much higher than previously believed. The lipid bilayer was found to be located between 400 and 450Å.

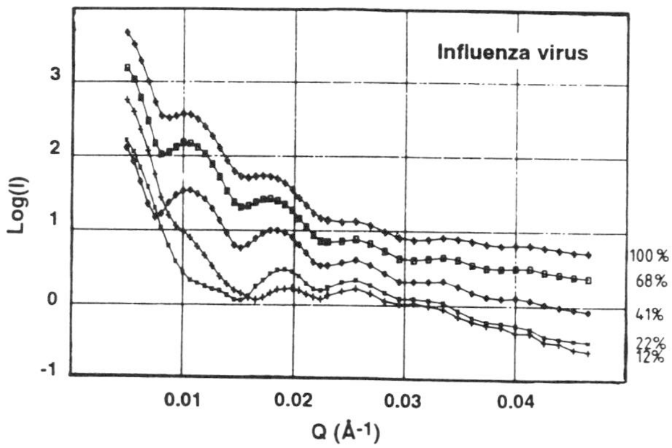

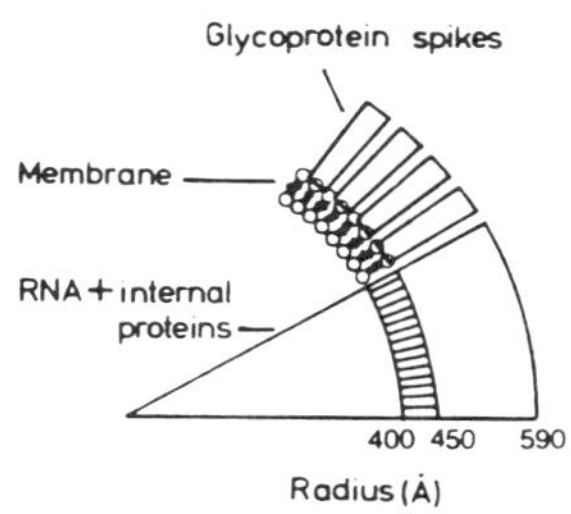

Fig. XIV.6. — (a) Neutron scattering curves from solutions of influenza virus in different D_2O/H_2O mixtures as indicated. (b) Low resolution model of the influenza virus as deduced from the small angle scattering curves. From Cusack (1982) with permission.

XIV.3.1.2. Assembly/Disassembly

Due to the low flux of a neutron source it is usually not possible to follow real-time kinetics as is the case for synchrotron radiation . However, neutron solution scattering may in favourable cases be used to follow the assembly/disassembly of a virus in a stopped flow system. This was the case for Brome Mosaic Virus (BMV), a small plant virus which can be disassembled and re-assembled as a function of pH and ionic strength. Using a stopped-flow method it was possible, through the use of contrast variation, to show that reassembly occurs through a co-condensation of protein and RNA rather than by formation of empty capsids with subsequent insertion of the RNA (Cuillel et al., 1987).

XIV.3.2. Low Resolution Crystallography

In all the virus structures determined to date by high resolution X-ray crystallography a part or all of the RNA and/or Protein has been invisible in the electron density maps. This is due primarily to the fact that these components are disordered with respect to the crystal lattice and that their smeared out density has very low contrast with respect to the solvent. As crystals of biological macromolecules contain large amounts of water the principles of H_2O/D_2O contrast variation can be applied to resolve these problems in the same way as for solutions. This technique has been described in its application to non-viral structures in

Chapter XIII. As in all crystallographic problems it is necessary to obtain the phases of the reflections. In this case, however, contrast variation can be a direct aid in solving the phase problem for a crystal in a given D_2O/H_2O mixture so long as part of the structure is known. This is usually the case as the structure will normally have been studied and in large part solved by X-ray diffraction. The ordered part of the structure solved with X-rays serves as a starting point for phasing the neutron data used to locate the more disordered part of the structure. Two examples are the case of Satellite Tobacco Necrosis Virus (STNV) where the nucleic acid was located by neutron scattering (Bentley et al., 1986; Fig. XIV.7) and TBSV where both the nucleic acid and 25% of the protein which is not visible in X-ray maps were located (Timmins, Wild and Witz, unpublished).

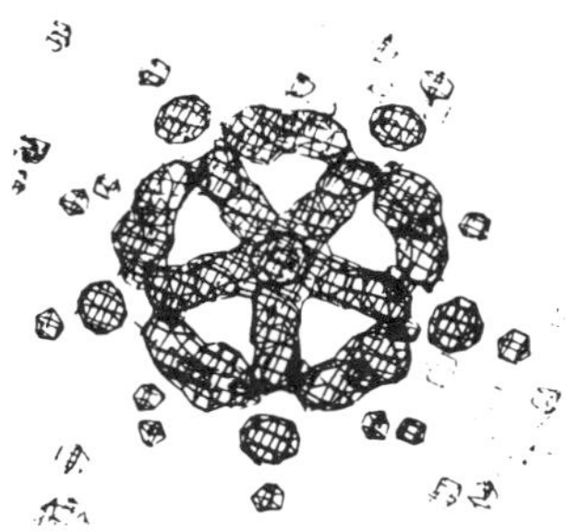

Fig. XIV.7. — The RNA structure in particles of STNV. The tubes of density have a diameter similar to that of an RNA duplex and run along the internal edges of the viral icosahedron. From Bentley et al. (1987) with permission.

XIV.4. Synchrotron X-Ray Scattering

XIV.4.1. Small Angle Scattering

The power of synchrotron radiation in small angle scattering lies mainly in the high X-ray flux available.

XIV.4.1.1. Radial density distributions

X-ray small-angle scattering is in general less important than neutron small angle scattering for the determination of the radial distribution of scattering density due to the difficulty of contrast variation. It is straightforward to measure the scattering curve and obtain a nominal spherical dimension on the assumption that the particle is homogeneous but to obtain a measure of the internal structure is much less easy.

XIV.4.1.2. Assembly/Disassembly

The high flux of synchrotron radiation opens up a field for the investigation of the assembly/disassembly processes that occur with viruses on a time scale not accessible to neutron scattering. The most extensive studies to date have been carried out on Tobacco Mosaic Virus (TMV) a helical rod-like virus whose structure was solved by X-ray fibre diffraction (Namba and Stubbs,1986). The coat protein of TMV can be disassembled and reassembled in the absence of viral RNA. Various intermediate states including the so-called 20S disk have been characterised by analytical centrifugation and electron microscopy. Synchrotron radiation has been used to follow the assembly process induced by a temperature jump (Hiragi et al., 1990). Scattering curves were measured in a minimum time of 7.5 seconds which defines the effective time resolution of the experiment.

XIV.4.2. High Resolution Crystal Structure Analysis

The advent of synchrotron radiation has revolutionised the high resolution structure determination of spherical viruses. Until the early 1980s only three small plant virus structures had been solved, those of TBSV, Southern Bean Mosaic Virus (SBMV) and Satellite Tobacco Necrosis Virus (STNV), each using laboratory based X-ray generators. This early work has been reviewed in depth by Harrison (1983). These structures were solved using "classical" methods of multiple isomorphous replacement (MIR) combined with the use of non-crystallographic symmetry averaging. Each particle in a virus crystal has 532 symmetry but the 5-fold axis can never be expressed as a symmetry axis of the crystal. It may however be used for improving phases by averaging the structure obtained from MIR, back-transforming the structure and combining the calculated phases with those obtained from MIR. Poliovirus was the first example of a virus where only one heavy atom derivative was available but, these single isomorphous replacement (SIR) phases in conjunction with non-crystallographic symmetry averaging, were sufficient to determine the structure.

Apart from the problems of phasing of structure amplitudes from virus crystals there were also many practical problems associated with data collection from virus crystals. The most widespread problem was their sensitivity to radiation which was often such that on a laboratory source the crystal was often destroyed before any useful data could be obtained. Synchrotron radiation now allows this problem to be largely overcome, even though in some cases it is still possible to obtain only one or very few exposures per crystal. For example in the case of Mengo Virus (Luo et al., 1987) more than ~600 crystals were used, one for each 0.3° oscillation photograph. Another major advantage of synchrotron radiation is the collimation obtained which allows spots from unit cells of up to ~1000Å to be easily resolved.

The improved data obtained from synchrotron radiation prompted further improvements in phase determination, refinement and extension. A major advance was made by Rossmann who showed that it was indeed possible to solve virus structures without any isomorphous derivatives. He showed that in many cases the structure at medium resolution (5 - 6Å) was sufficiently similar to that of an already known virus structure that this model could be used to obtain starting phases which could be refined in the usual way using non-crystallographic symmetry averaging and solvent flattening. This however is often insufficient to obtain correct phases in the high resolution range. Rossmann et al. (1985) showed that this could be done simply by extension from the medium resolution structure. In this method the phases at medium resolution are refined, the model obtained is back transformed to calculate structure factors to a slightly higher resolution and the new phases applied to the observed structure factors to produce a slightly higher resolution model. After non crystallographic symmetry averaging the procedure is repeated cyclically, adding data in shells of 0.1 - 0.2Å resolution until the high resolution structure is

Table XIV.1

Virus structures determined by high resolution X-ray crystallography

Name (Acronym)	Resolution (Å)	No. of Reflections per data set	X-Ray Source	Cell dimensions (Å)	Method of structure analysis*	Reference
Tomato Bushy Stunt Virus (TBSV)	2.9	199 000	Rotating anode	383 x 383 x 383	MIR + NC	Harrison et al., 1978
Southern Bean Mosaic Virus (SBMV)	2.8	321 600	Rotating anode	a=334.5 c=755.8 γ=120°	MIR + NC	Abad-Zapatero et al., 1980
Satellite Tobacco Necrosis Virus (STNV)	3.0 2.5	299 000 503 000	Rotating anode	317 x 304 x 185 β=94.4°	MIR + NC	Liljas et al., 1982
Swollen TBSV	8.0	36 100	Rotating anode	546 x 433 x 383 β=134°	(TBSV) + MIR + NC	Robinson & Harrison, 1982
Polio Virus	3.0	893 900	Rotating anode	323 x 358 x 380	SIR + NC + PE	Hogle et al., 1985
Human Rhino virus 14 (HRV14)	3.0	509 900	Synchrotron	445 x 445 x 445	MIR + NC + PE	Rossmann et al., 1985
Turnip Crinkle Virus (TCV)	3.2	310 000	Rotating anode	349 x 379 x 397	(TBSV) +NC	Hogle et al., 1986
Tobacco Mosaic Virus (TMV)	2.9	-	Various lab. X-ray sources		Fibre diffraction	Namba & Stubbs, 1986
Mengo Virus	3.0	995 000	Synchrotron	441 x 427 x 422	(HRV14) + NC	Luo et al., 1987
Cow-pea Mosaic Virus (CpMV)	3.5	38 600	Rotating anode + Synchrotron	317 x 317 x317	SIR + NC + PE	Stauffacher et al., 1987
Black Beetle Virus (BBV)	3.0	110 700	Synchrotron	362 x 362 x 362	MIR + NC + PE	Hosur et al., 1987

Rhino Virus Serotype 1a (HRV1a)	3.2	135 600	Synchrotron	a=341 x c=466 γ=120°	(HRV14) + NC + PE	Kim et al., 1987
Bean-pod Mottle Virus (BPMV)	3.0	698 000	Synchrotron	311 x 284 x 350	(CpMV) + PE + NC	Chen et al., 1989
Foot and Mouth Disease Virus (FMDV)	2.9	129 000	Synchrotron	345 x 345 x 345	(Mengo + HRV) + NC + PE	Acharya et al., 1989
MS2 (bacteriophage)	3.3	540 500	Synchrotron	466 x 288 x 274 β=121.7	MIR + NC + PE	Valegård et al.,1990
Canine Parvovirus (CPV)	3.25	583 700	Synchrotron	263 x 349 x 267 β=90.8°	Shell model + SIR + NC + PE	Tsao et al., 1990
Simian Virus 40 (SV40)	3.8	262 000	Synchrotron	558 x 558 x 558	SIR + NC + PE	Liddington et al., 1991
ΦX174 (bacteriophage)	2.7	1 000 300	Synchrotron	306 x 361 x 300 β=92.9°	(CpMV) + NC + PE	McKenna et al., 1992
Flock House Virus (FHV)	3.0	408 700	Synchrotron	a=332 x c=782 γ=120°	(BBV) + NC + PE	Fisher & Johnson, 1993
Satellite Tobacco Mosaic Virus (STMV)	3.0	68 000	Rotating anode	174 x 192 x 203	MIR + NC	Larson et al., 1993

* SIR single isomorphous replacement
MIR multiple isomorphous replacement
NC non-crystallographic symmetry averaging + solvent flattening
PE phase extension
(XXX) used structure of XXX as initial phasing model

obtained. These improvements in methodology have gone hand in hand with the improved data quality obtained from synchrotron radiation.

Since 1985 a large number of structures has been solved, using synchrotron radiation, which could not be envisaged with a laboratory source. The magnitude of the task involved is summarised in Table XIV.1. It will be seen that in most cases data collection involves several hundreds of thousands of reflections measured for the native virus plus as many again for derivatives.

Some of the later virus structure determinations have been reviewed by Liljas (1986) and Rossmann and Johnson (1988). Recent progress is summarised in Harrison (1991,1992). Below we consider some of the landmark structures that have been determined.

XIV.4.2.1. Human Rhino Virus (HRV)

Human Rhino Virus serotype 14 (HRV14) was the first virus structure to be solved using synchrotron radiation data. The picorna virions, of which HRV is a member along with polio, foot and mouth disease, hepatitis A, and several other serious pathogens, contain 4 different structural proteins designated VP1, VP2, VP3 and VP4. These encapsidate a single positive strand of RNA. Data were collected on the Cornell High Energy Synchrotron Source (Chess) and the structure solved to 5Å resolution by multiple isomorphous replacement and non-crystallographic symmetry averaging. The resolution was then extended to 3Å using the technique described above.

The structure of HRV14 was very surprising in its similarity with the plant viruses TBSV and SBMV. These latter two viruses each contain 180 copies of a single protein sub-unit and are hence T=3 particles. Quasi-equivalence is achieved by the polypeptide having three slightly different conformations in the particle. In the case of HRV it was immediately clear that the coat was formed only by the proteins VP1, VP2 and VP3 but that, although they were chemically distinct, their basic folding pattern was essentially the same and, moreover, it was the same as in the sub-units of TBSV and SBMV. It should be noted that the virus is technically a T=1 particle with the repeating unit being the VP1-VP2-VP3 trimer. Rossmann has termed this pseudo-symmetry with a P rather than a T number. Another characteristic of the structure is the presence of a deep depression or 'canyon' some 25Å deep and 12-30Å running around the 5 fold vertices. Rossmann suggested that this may be the site for receptor cell binding (Fig. XIV.8). As the canyon is too narrow to be accessed by the

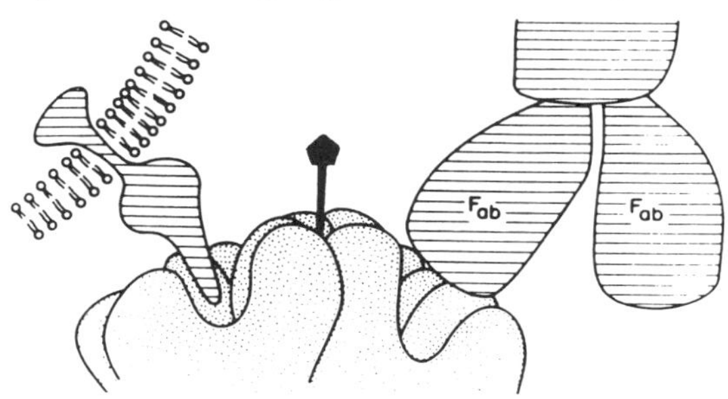

Fig XIV.8. — The canyon hypothesis of Rossmann. The receptor binding site is found at the bottom of the canyon which sterically impedes the binding of the antibody FAB fragment. From Luo et al. (1987) with permission. Copyright 1987 by the AAAS.

FAB fragment of an antibody the residues on the canyon floor would not be under immune selective pressure and could retain there specificity for the same receptor. Recently experiments have been carried out to investigate the conformational changes which take place on exposing HRV14 to acid vapour (Giranda et al., 1992). The acid-induced changes are believed to reflect processes that trigger uncoating of the virus. These experiments involved measuring data from crystals exposed to acid and which became disordered within 15 minutes of exposure. Thus only through the use of synchrotron radiation was it possible to measure data sufficiently rapidly for a structure analysis.

XIV.4.2.2. Foot and mouth disease virus (FMDV)

Foot and mouth disease virus is another member of the picorna virus family. As with HRV14 the crystals available were very radiation sensitive such that only one 0.4-0.5° oscillation photograph could be obtained from each crystal on the Daresbury synchrotron (Acharya et al., 1989). An important factor in the limitation of radiation damage was the use of relatively short (~0.9Å) wavelength radiation as well as the high collimation of the beam. Initial phases were obtained in this case from a model which was a hybrid of two known structures, HRV14 and Mengo virus (Luo et al, 1987), and phases refined and extended by non-crystallographic symmetry averaging and solvent flattening. An important factor in the success of the phase extension procedure was the use of a very tight envelope around the viral coat, within which averaging and back transformation was carried out. Phase extension, nevertheless, required 519 cycles of calculation!

The structure of FMDV showed once again that even though the subunits are considerably smaller, the basic folding and packing of the coat proteins is very similar to that of other picorna viruses as well as to the small T=3 plant viruses. However, the surface is much smoother than that of HRV14 and does not seem to possess any major canyons. As a result an hypothesis more general than the canyon hypothesis was proposed in which the constant receptor binding site is concealed by 'camouflaging' this region within a sea of considerable variability.

XIV.4.2.3. Bean pod mottle virus (BPMV) and cow-pea mosaic virus (CpMV)

Bean pod mottle virus and cowpea mosaic virus are members of the comovirus group and are very similar in properties, including structure. BPMV has been solved to 3.0Å resolution (Chen et al., 1989) and CpMV to 3.5Å resolution (Stauffacher et al., 1987). One of the particularities of both of these viruses is that the particles contain 60 copies each of two coat proteins, one large (L) and the other small (S). The solution of the structure showed that despite this different stoichiometry the coat structure was again very similar to the T=3 simple plant viruses and also to the picorna viruses described above. In fact the large sub-unit contains two domains each having the canonical ß-barrel shape. These two domains plus the small sub-unit thus form the three quasi-equivalent units of the T=3 shell.

Another important feature of this structure was the observation of ordered nucleic acid (see XIV.4.3).

XIV.4.2.4. Simian Virus 40 (SV40) and Murine Polyoma Virus.

The structure of murine polyoma virus was solved to low resolution (22Å) by Rayment et al., (1982) and showed a remarkably different quaternary organisation from that observed in all other icosahedral viruses. It consists of 360 copies of a major structural protein VP1 arranged as 72 pentameric building blocks on the viral surface. Twelve of the pentameric building blocks are situated on the icosahedral 5-fold axes each surrounded by 5 other pentamers. The other 60 pentamers are, however, not situated on symmetry axes and are surrounded by six other pentamers. This produces a structural puzzle. How do pentameric structures fit into hexameric holes? The resolution of the murine polyoma structure was insufficient to unravel this but the structure of a related virus SV40 allowed the puzzle to be solved. Liddington et al. (1991) showed from the 3.8Å resolution structures that the mismatch between 5-fold symmetry and 6-fold coordination was mediated by a system of flexible arms which both maintain the integrity of the constituent pentamers and form specific interactions with neighbouring pentamers.

XIV.4.3. Nucleic acid structure and protein/nucleic acid interactions

The methods used in determination of virus structure from protein crystallography preclude the determination of the structure of the internal nucleic acid at high resolution, even if such a structure exists. This is due to the icosahedral symmetry of the protein coat, a symmetry which the nucleic acid cannot possess. Even if there were a unique conformation the experiment would itself smear out the structure (a) due to the symmetry averaging used in the structure determination process and (b) due to the fact that the exact icosahedral symmetry of the outer surface means that the crystal is "unaware" of the structure of the internal RNA and therefore, even if one could obtain phases without using the icosahedral symmetry, the structure would probably be averaged by the crystal itself. As mentioned earlier RNA has been located at low resolution using neutron diffraction but at high resolution no RNA was visible in the first virus structures. Recently, however, some nucleic acid has been observed by X-ray crystallography implying that parts of the nucleic acid are at least ordered with approximate icosahedral symmetry. In BPMV some 20% of the RNA binds to the interior of the capsid. It is single stranded and has well-defined polarity and stereo-chemistry. In satellite tobacco mosaic virus (STMV) a T=1 particle similar to STNV, a segment of RNA double helix comprising 7 nucleotides is found associated with each coat protein dimer and lying along the icosahedral edges. This is reminiscent of the ordered RNA observed by low resolution neutron scattering in STNV (Bentley et al., 1987). An ordered RNA duplex of 20 nucleotides, interacting with a helical protein domain inside the capsid shell, is found in Flock House Virus (FHV; Fisher and Johnson, 1993). The structure is very similar to that of Black Beetle virus (BBV; Hosur et al., 1987) which was used as a starting point for the data analysis of FHV. Some of these results have been reviewed by Stockley (1992).

XIV.5. Conclusions

XIV.5.1. Advances in the understanding of virus structure

The most striking results concerning the structure of icosahedral viruses have come from the very recent crystallographic studies. The tertiary and quaternary structures of all the viruses so far studied, with the exceptions of the bacteriophage MS2 (Valegård et al., 1990), and SV40 (Liddington et al., 1991) have shown a remarkable similarity. MS2 is the only virus to display a significantly different fold for its coat protein whereas SV40 appears to depart from the basic principles of quasi-equivalence in its quaternary structure. Viral sub-units (except in MS2) always

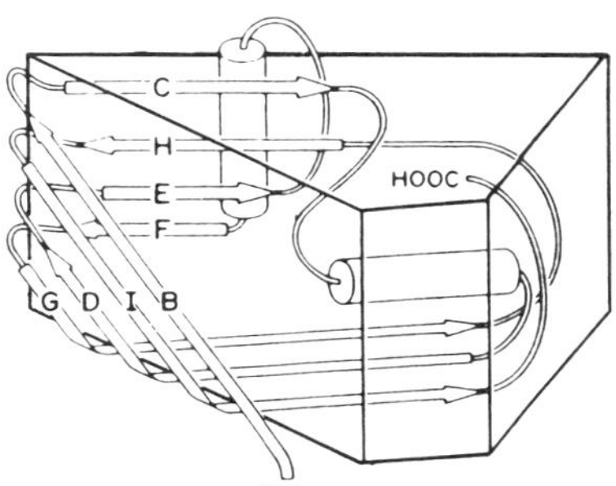

Fig. XIV.9. — The canonical eight-stranded ß-barrel of which almost all viral coat proteins are constructed. The differences between different proteins lie in the connecting loops and in the termini. From Hogle et al. (1985) with permission. Copyright 1985 by the AAAS.

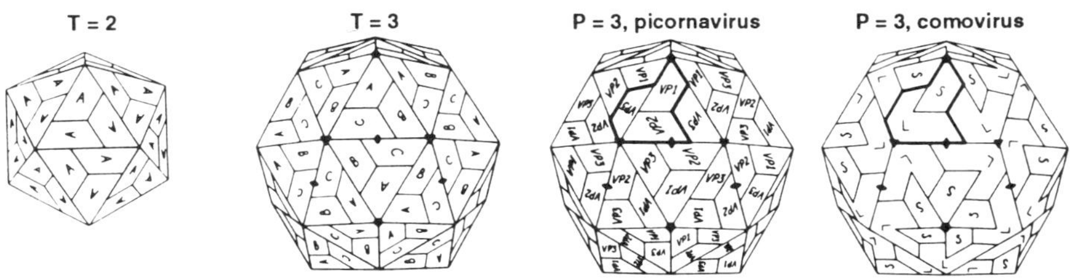

Fig. XIV.10. — The packing of sub-units in different classes of icosahedral viruses. All of the shells have exact icosahedral symmetry. Each sub-unit is represented by a trapezoid which has the approximate shape of the ß-barrel viewed from the top. The icosahedral asymmetric unit of T=1 capsids is one sub-unit whereas for the T=3 particle it is the central triangle containing the sub-units A, B and C. These three sub-units have the same amino-acid sequence but slightly different environments and hence slightly different conformations. The P=3 picorna virus shell is technically a T=1 particle and is closely related to the T=3 shell being composed of 180 ß-barrel domains. The three sub-units, VP1, VP2 and VP3 are however distinct proteins. In the case of the como viruses there are two chemically distinct proteins, the large (L) and small (S) sub-units, but the L protein contains two ß-barrel domains thus producing again a 180 domain shell. From Rossmann and Johnson (1989). Reproduced with permission, from the Annual Review of Biochemistry, Vol. 58, (c) 1989) by Annual Reviews Inc..

fold in such a way as to conserve the canonical ß-barrel (Fig. XIV.9) structure even in the absence of any detectable amino acid sequence homology. The packing of the sub-units on the surface usually follows a T=3 or a pseudo-T=3 surface lattice irrespective of the number of chemically different units making up the shell Fig. XIV.10).

XIV.5.2. Future prospects

In this chapter we have barely discussed the very large, often enveloped, viruses for which there is little or no hope of producing crystals for X-ray analysis. In several of these systems it has been possible to isolate and solve the structure of the coat proteins and with the help of electron microscopy or neutron scattering we may be able to produce models of high resolution by assembling together the components. Increasingly powerful synchrotron sources will allow the use of smaller crystals and allow structures to be solved to higher resolution. The prospect of drug-binding studies is a particularly exciting domain. Finally with the most powerful synchrotron sources we can hope to look in more detail at the kinetics of assembly/disassembly processes if suitable model systems can be found.

REFERENCES

ABAD-ZAPATERO, C., ABDEL-MEGUID, S.S., JOHNSON, S.E., LESLIE, A.G.W., RAYMENT, I., ROSSMANN, M.G., SUCK, D., TSUKIHARA, T. 1980 - Nature 286, 33

ACHARYA ,R., FRY, E., STUART, D., FOX, G., ROWLANDS, D., BROWN, F. 1989 - Nature 337, 709

ADRIAN, M., TIMMINS, P., WITZ, J. 1992 - J.Gen. Virology, 73, 2079

BENTLEY, G.A., LEWIT-BENTLEY, A., LILJAS, L., SKOGLUND, U., ROTH, M. AND UNGE, T. 1987 - J. Mol. Biol. 194, 129

CASPAR, D.L.D. 1956 - Nature 177,475

CASPAR D.L.D. AND KLUG, A. 1962 - Cold Spring Harbor Symposium on Quantitative Biology, 27, 1

CHEN, Z., STAUFFACHER, C., LI Y., SCMIDT, T., BOMU, W., KAMER ,G.,

SHANKS, M., LOMONOSSOF, G., JOHNSON, J.E. 1989 - Science 245, 154

CRICK, F.H.C., WATSON, J.D. 1956 - Nature 177, 473

CUSACK, S. 1982 - Inst. Phys. Conf. Ser. No. 64:Section 4, 351

CUSACK S., RUIGROK R.W.H., KRYGSMAN, P.C.J. AND MELLEMA, J.E.1985 - J. Mol. Biol. 186, 565

CUILLEL, M., BERTHET-COLOMINAS, C., TIMMINS, P.A. AND ZULAUF, M. 1987 - Eur. Biophys. J. 15, 169

FISHER, A., JOHNSON, J.E. 1993 - Nature 361, 176

GLATTER, O. 1982 in *"Small Angle X-ray Scattering"* Editors O.Glatter & O.Kratky, Academic press, London, New York.

GIRANDA, V.L., HEINZ, B.A., OLIVEIRA, M.A., MINOR, I., KIM, K.H., KOLATKAR, P.R., ROSSMANN, M.G., RUECKERT, R.R. 1992 - Proc. Natl. Acad. Sci. USA 89, 10213

GIRARD, M., HIRTH, L., LEBURIER, G., WITZ, J. 1989 in "Virologie Moléculaire" Doin Editeurs, Paris.

HARRISON S.C., OLSON A.J., SCHUTT, C.E., WINKLER, F.K. AND BRICOGNE G. 1978 - Nature 276, 368

HARRISON, S.C. 1983 - Adv. virus Res. 28, 175

HARRISON, S.C. 1991 - Current Opinion in Structural Biology 1, 288

HARRISON, S.C. 1992 - Current Opinion in Structural Biology 2, 293

HELLIWELL, J.R. 1992 - Macromolecular Crystallography with Synchrotron Radiation - Cambridge University Press (Cambridge, UK)

HIRAGI, Y., INOUE, H., SANO, Y., KAJIWARA, K., UEKI, T. and NAKATANI, H. 1990 - J. Mol. Biol. 213, 495

HOGLE, J.M., CHOW, M., FILMAN, D.J. 1985 - Science 229, 1358

HOGLE, J.M., MAEDA, A., HARRISON, S.C. 1986 - J. Mol. Biol. 191, 625

HOSUR, M.V., SCHMIDT, T., TUCKER, R.C., JOHNSON, J.E., GALLAGHER, T.M., SELLING, B.H., RUECKERT, R.R. 1987 - Proteins: Struct. Funct. Genet. 2, 167

JACROT, B. 1981 in Comprehensive Virology, 17, (Eds. Fraenkel-Conrat, H. and Wagner, R.R.) Plenum Press, New York and London.

KIM, S., SMITH, T.J., CHAPMAN, M.S., ROSSMANN, M.G. 1989 - J. Mol. Biol. 210, 91

LARSON, S.B.,KOSZELAK, S., DAY, J., GREENWOOD, A., DODDS, J.A., McPHERSON, A. 1993 - Nature 361, 179

LIDDINGTON, R.C., YAN, Y., MOULAI, J., SAHLI, R., BENJAMIN, T.L., HARRISON, S.C. 1991 - Nature 354, 278

LILJAS, L. , UNGE, T., JONES, T.A., FRIDBORG, K., LÖVGREN, S., SKOGLUND, U., STRANDBERG, B. 1982 - J. Mol. Biol. 159, 93

LILJAS, L. 1986 - Prog. Biophys. Molec. Biol. 48, 1

LUO, M., VRIEND, G., KAMER, G., MINOR, I., ARNOLD, E., ROSSMANN, M.G., BOEGE, U., SCRABA, D.G., DUKE, G.M., PALMENBERG, A.C. 1987 - Science 235, 182

MCKENNA, R., XIA, D., WILLINGMANN, P., ILAG, L.L., KRISHNASWAMY, S., ROSSMANN, M.G., OLSON, N.H., BAKER, S. INCARDONA, N.L. 1992 - Nature 355, 137

NAMBA, K., STUBBS, G. 1986 - Science, 231, 1401

RAYMENT, I., BAKER, T.S., CASPAR, D.L.D., MURAKAMI, W.T. 1982 - Nature 295, 110

ROBINSON, I.K., HARRISON, S.C. 1982 - Nature 297, 563

ROSSMANN, M.G., ARNOLD, E., ERICKSON, J.W., FRANKENBERGER, E.A, GRIFFITH, J.P., HECHT, H-J., JOHNSON, J.E., KAMER, G., LUO, M.,MOSSER, A.G., RUECKERT, R.R., SHERRY, B. AND VRIEND, G. 1985 - Nature 317, 145

ROSSMANN, M.G., JOHNSON, J.E. 1989 Ann. Rev. Biochem 58, 533

STAUFFACHER, C.V., USHA, R., HARRINGTON, M., SCHMIDT, T., HOSUR, M.V., JOHNSON, J.E 1987 - In *Crystallography of Molecular Biology, a NATO Advanced Study Course*, Plenum Press, New York.

STOCKLEY P.G.1992 - Current Opinion in Structural Biology 2:143

TSAO, J., CHAPMAN, M.S., AGBANDJE, M.,KELLER, W.,SMITH, K., WU, H., LUO, M., SMITH, D.J., ROSSMANN, M.G., COMPANS, R.W., PARRISH, C.R. 1991 - Science 251,1456

VALEGÅRD, LILJAS, L.,FRIDBORG, K.,UNGE, T. 1990 Nature 345, 36

CHAPTER XV

COHERENT DIFFUSE SCATTERING IN BIOLOGY AND SOFT-CONDENSED MATERIALS

J. DOUCET

Foreword

In the sense that we shall define in the general presentation, coherent diffuse scattering is a complementary phenomenon to diffraction: **diffraction gives the average microscopic structure of crystals, coherent diffuse scattering gives the deviations from this average structure,** deviations which are often the key to the understanding of their macroscopic properties. Diffuse scattering is a universal phenomenon that occurs for all crystals. This lecture is restricted to the analysis of biological and soft condensed materials by this technique.

In the first part we shall present and explain the topic in more detail; in the second part an overview of the various types of scattering by imperfect crystals and their general mathematical expression will be given. The third part will be devoted to a short description of the experimental devices that are used to collect and analyse coherent diffuse scattering (mainly X-ray and neutron radiation), and the fourth and last part will be dedicated to the description of a few examples of coherent diffuse scattering studies.

It is important to note that the lecture "Large amplitude motions in molecular crystals, polymers and biological systems (incoherent neutron scattering)" has parts in common with ours; we will not of course approach this topic, but it must be taken into account that the two lectures partly cover the same topic.

XV.1. General Presentation

XV.1.1. What is coherent diffuse scattering?

Scattering is one of the interactions of radiation with matter which is characterized by a deviation of the incident radiation without (elastic process) or with (inelastic process) energy change. We will be interested in X-ray and neutron scattering but similar phenomena occur in electron microscopy.

Diffuse means opposite to localized; diffuse scattering is the scattering which is spread out of the Bragg reflexions. Its origin lies in the structure deviations from the three-dimensional packing of a perfect crystal. This scattering has been much less studied than the Bragg scattering, although its existence was proved a few years only after Bragg's discovery (around 1917). This is due to the fact that the analysis of diffuse scattering needs a knowledge of the average crystalline structure, which is deduced from the Bragg scattering; in addition, its study is more complicated and gives less spectacular results than that of Bragg scattering. Figures XV.1-a and XV.1-b give examples of X-ray films exhibiting diffuse and Bragg scattering.

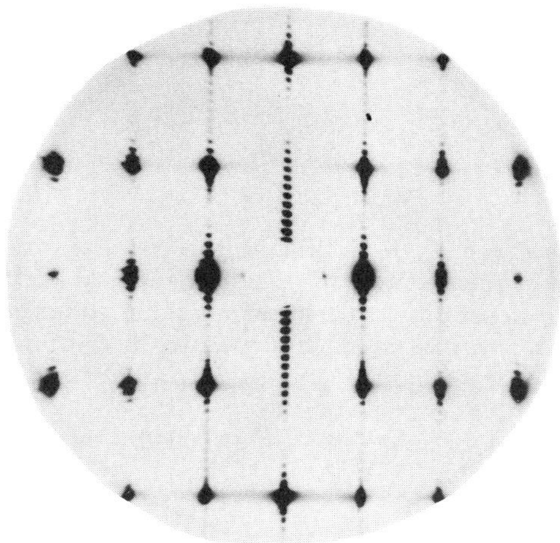

Figure XV.1-a. — X-ray pattern of the crystalline phase of $C_{23}H_{48}$ which exhibits diffuse scattering

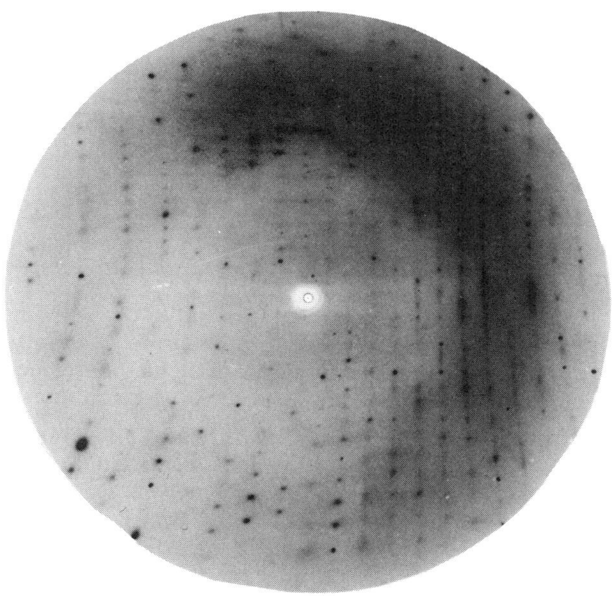

Figure XV.1-b. — X-ray pattern of hen egg-white lysozyme in the orthorhombic form

coherent signifies, for us, the existence of correlations between the deviations from the perfect three-dimensional structure. The extent of the correlations can vary from intracell or intramolecular distances up to the entire crystal, with the crystal periodicity. According to the type of correlation, diffuse scattering is more or less spread out within reciprocal space: highly correlated deviations lead to more condensed diffuse scattering than poorly correlated ones.

It must be emphasized that the notions of coherence and incoherence have not the same significance for the X-ray and for the neutron communities. The above definition corresponds to the X-ray community definition. For the neutron community, coherence is related to the spin coherence: scattering by neutrons is a nuclear process which depends on the spin of the nucleus. Neutron scattering always displays two components: a "coherent" one and an "incoherent" one which is related to the autocorrelation of a given nucleus. "Coherent" neutron scattering corresponds both to correlated and uncorrelated phenomena. Only "coherent" <u>and</u> correlated neutron scattering corresponds to the "coherence" of the X-ray community.

Let us present here two simple images to illustrate the notion of correlated structural deviations and their significance.

1) During a tennis game, the heads of the spectators move alternately and simultaneously to the right and to the left. The average orientation of their heads is the same as if nothing were occurring on the tennis court, but the existence of correlation between the displacements is highly significant: it indicates that a ball is moving from one side to the other. The diffuse scattering features of the two possibilities, random or correlated displacements, are quite different.

2) During a rowing race, all the oars move in a correlated (or coherent) way; if they did not, the boat velocity would be highly reduced. Here also, the diffuse scattering features of the correlated and uncorrelated situations would be different. We note that a knowledge of the average position of the oars is unable to inform us about the origin of the motion of the boat; only the deviations from the average structure and their correlations give us the key to this problem. This simple image can be translated to the atomic structure and properties of the crystals.

XV.1.2. What are soft condensed and biological materials?

We call ***"soft condensed matter"*** all materials that are characterized by a crystallographic order that is intermediate between that of "hard crystals" and liquids.

Liquid crystal phases are classified as soft condensed matter. They are characterized by an orientational disorder of the molecules around their long axes. For instance, terephthal-bis-butyl-aniline (TBBA) exhibits the following phase sequence (with six liquid crystal phases):

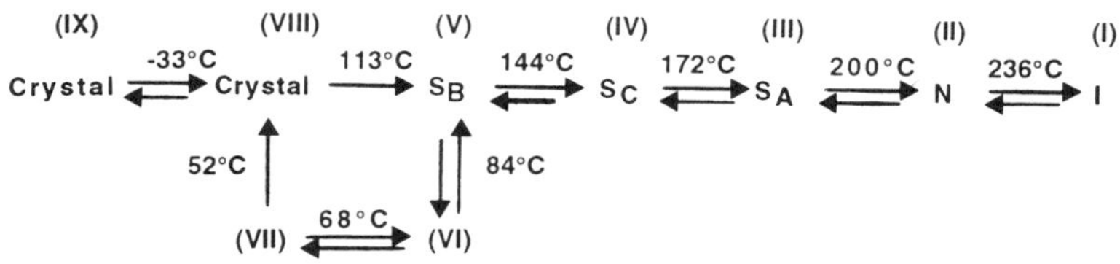

Plastic crystals can be considered as soft condensed matter since in the plastic phase the molecules (globular shape) undergo rotational displacements around their centres.

Polymers also belong to the class of soft materials, whatever their type: semi-crystalline, amorphous or rubbery.

Biological materials consist mostly of three types of compound which play key roles in biology:

Proteins and ***DNA*** can be crystalised into high quality crystals containing between 30% and 95% water.

Membrane structure is analogous to the structure of liquid crystals; the main difference lies in their bilayer structure, which is due to the two different parts of the molecules: a hydrophilic head and hydrophobic aliphatic chains.

XV.1.2.3. Why this lecture ?

Soft condensed and biological materials are characterized by crystalline imperfections which are in general displacements of rather large amplitude. These displacements, which lead to the possibility of deformation at the macroscopic scale, very often play key roles in the biological functions of the molecules: their properties are related to the deviations from the average structure. For instance, the catalytic properties of hæmoglobin cannot be understood from its average structure. It is necessary to consider correlated atomic motions to understand how the oxygen molecules enter within the protein down to the active site.

Diffuse coherent scattering can provide valuable information to study the correlated atomic displacements. The geometric characteristics are deduced from the localization of the diffuse features, whereas the energy characteristics can be obtained through frequency analysis of the diffuse scattering (if performed with neutrons).

XV.2. Diffuse scattering by imperfect crystals

XV.2.1. Total scattered intensity

It is well known that a perfect crystal does not exist. Nevertheless the crystals investigated are generally assumed to be to perfect since X-ray diffraction is not very sensitive to imperfections. Patterns obtained after very long time exposures reveal diffuse scattering, which is due to imperfections, i.e. to deviations from the ideal perfect structure. We describe here the main results without attempting to be exhaustive. A complete analysis can be found in A. Guinier's book: "X-ray Diffraction", W.H. Freeman, San Francisco, London, 1963.

We will only examine crystalline imperfections involving atomic displacements (excluding for instance substitution disorders), which are the most common imperfections for soft condensed materials.

The notation is the following: the crystal contains N unit cells (index n), each cell contains l atoms (index j); the positions notations are shown on figure XV.2.

For a perfect crystal we have : $\mathbf{r}_n^j = \mathbf{r}_n + \mathbf{r}^j$. Let us now assume that each atom is displaced from its theoretical position in the perfect crystal by a vector $\mathbf{u}_n^j$. For this imperfect crystal we have: $\mathbf{r}_n^j = \mathbf{r}_n + \mathbf{r}^j + \mathbf{u}_n^j$. The structure factor of the cell n becomes $F_n(S) = \sum f_j \exp(-2i\pi \mathbf{S}.(\mathbf{r}^j + \mathbf{u}_n^j))$ and the total scattered amplitude is: $A(S) = \sum_n F_n \exp(-2i\pi \mathbf{S}.\mathbf{r}_n)$

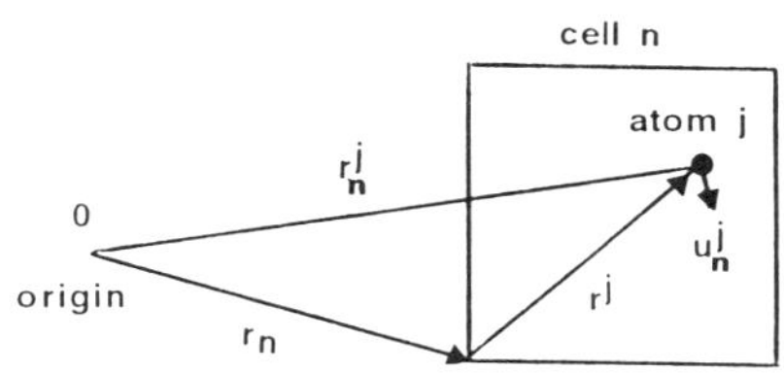

Figure XV.2. — Notations

(note that for a perfect crystal $\mathbf{u}_n^j = 0$, and thus: $A(S) = F \sum_n \exp(-2i\pi \mathbf{S}.\mathbf{r}_n)$)

To get constructive interferences we must have $\mathbf{S}.\mathbf{r}_n = p \in \mathbb{N}$, which implies : A=nF and for the intensity: $I(S) = AA^* = N^2 |F(S)|^2$ with A* complex conjugate of A.

Setting $\mathbf{r}_m = \mathbf{r}_{n'} - \mathbf{r}_n$, we find : $I(S) = \sum_m (\sum_n F_n(S) F^*_{n+m}(S) \exp(2i\pi \mathbf{S}.\mathbf{r}_m))$

The number of terms involved in $\sum_m$ is greater than n but a lot of terms are equal to 0 because $\mathbf{r}_{n+m}$ corresponds to a cell out of the crystal. This is taken into account introducing a term $V(\mathbf{r}_m)$ which is equal to the volume fraction in superposition with itself after a translation of $\mathbf{r}_m$.

We can now relate F_n to its mean value $\langle F \rangle$ introducing φ_n : $F_n = \langle F \rangle + \varphi_n$

with $\langle\varphi_n\rangle = 0$. Thus: $\langle F_n(\mathbf{S})F^*_{n+m}(\mathbf{S})\rangle = |\langle F\rangle|^2 + \langle\varphi_n.\varphi^*_{n+m}\rangle$

and finally the total scattered intensity can be written:

$$I(\mathbf{S}) = \underbrace{N\sum_{\mathbf{m}} V(\mathbf{r_m})\,|\langle F\rangle|^2 \exp(2i\pi\mathbf{S}.\mathbf{r}_m)}_{I_1} + \underbrace{N\sum_{\mathbf{m}} V(\mathbf{r_m})\Phi_{\mathbf{m}} \exp(2i\pi\mathbf{S}.\mathbf{r}_m)}_{I_2}$$

with $\Phi_m = \langle\varphi_n.\varphi^*_{n+m}\rangle$

XV.2.2. Bragg scattering (coherent non diffuse)

The total scattered intensity can be split into two terms I_1 and I_2. Let us first look at I_1.

As $V(\mathbf{r}_m)$ is a real quantity we have $\sum_m V(r_m)\exp(2i\pi\mathbf{S}.\mathbf{r}_m) \approx \exp(2i\pi\mathbf{S}.\mathbf{r}_m) = N$

This means that the intensity I_1 is localized at the nodes of the reciprocal lattice; we have: $I_1 = N^2|\langle F\rangle|^2$

This intensity term constitutes the set of well known Bragg reflexions. It is easy to show that for small amplitude displacements we have:

$$\langle F\rangle = \left\langle \sum_j f_j \exp(-2i\pi\mathbf{S}.(\mathbf{r}^j + \mathbf{u}^j_n))\right\rangle = \sum_j f_j \exp(-2\pi^2 \langle(\mathbf{S}.\mathbf{u}^j_\mathbf{n})^2\rangle) \exp(-2i\pi\mathbf{S}.\mathbf{r}^j)$$

which indicates that the effect of the disorder is to multiply each atomic scattering factor f_j by the quantity $\exp(-2\pi^2 \langle(\mathbf{S}.\mathbf{u}^j_\mathbf{n})^2\rangle)$.

The effect on the Bragg reflexions produced by the atomic displacements from the average perfect crystalline structure is to decrease the intensity by a factor which varies as $\exp(-2\pi^2 S^2\langle\mathbf{u}^2\rangle)$ (Debye-Waller factor or thermal factor); it is important to note that the angular width of the reflexions is not affected.

XV.2.2. Diffuse scattering

Let us look at term I_2. As $\mathbf{r}_m$ increases, the distance between the cell n and n' increases and the correlation between the atomic displacements decreases, which means: $\langle\varphi_n\varphi^*_{n+m}\rangle \approx \langle\varphi_n\rangle\langle\varphi^*_{n+m}\rangle = 0$

Thus we only consider small vectors for which $V(\mathbf{r}_m) \approx 1$, finally giving:

$$I_2(\mathbf{S}) = N\sum_m \Phi_m \exp(2i\pi\mathbf{S}.\mathbf{r}_m)$$

Since Φ_m is a complex quantity, the intensity I_2 is spread out of the reciprocal nodes: it is the diffuse scattering. Its value depends on Φ_m i.e. on the nature of the correlations of the atomic motions from one cell to another. Roughly speaking, diffuse scattering is the Fourier transform of the function Φ_m .

XV.2.3. Incoherent diffuse scattering

Assuming no correlation between the atomic displacements from one unit cell to another (incoherent disorder), all the terms $\Phi_m = 0$ except Φ_0, which gives $I_2 = N\Phi_0$. Since $\Phi_0 = \langle \varphi_0 \varphi_0^* \rangle$, we deduce:

$$\boxed{I_2 = N\left(\langle |F|^2 \rangle - |\langle F \rangle|^2\right)}$$

The diffuse scattering is then continuously spread out in reciprocal space. This incoherent diffuse scattering intensity is generally weak and invisible in normal conditions. It can be simply perceived that I_2 is continuous remembering that the Fourier transform of the correlation function Φ_m, which is in this case a δ function, is a continuous function.

This case is important since it serves also as a basis for the analysis of diffuse scattering: the quantity F can be extended to the entire group of atoms or molecules for which the displacements are correlated.

XV.2.5. Coherent diffuse scattering : thermal diffuse scattering

There nearly always exists some kind of correlation between the displacements. The localization of the diffuse scattering will depend on the type of correlation.The case of thermal motion is found in all crystals. It is well known that atomic displacements can be analyzed into a set of sinusoidal waves (phonons) rather than by individual atomic motions.

Let us first look at the diffuse scattering produced by one sinusoidal wave of wave vector **k** and amplitude **a**. The displacements are for example:

$$\mathbf{u}_n^j = \mathbf{a} \cos 2\pi(\nu t - \mathbf{k}.\mathbf{r}_n^j)$$

For X-ray scattering the atoms can be considered as fixed and displaced with respect to their theoretical position by $\mathbf{u}_n^j = \mathbf{a} \cos 2\pi \mathbf{k}.\mathbf{r}_n^j$ since their frequency ν is about 10^7 times smaller than the X-photon frequency; the energy shifts of the X-photons are negligible. In that case it is simple to deduce the correlation function $\Phi_m = 2\pi^2 |F|^2 (\mathbf{a}.\mathbf{S})^2 \cos 2\pi \mathbf{k}.\mathbf{r}_m$ which leads to :

$$I_2 = N\pi^2 |F|^2 (\mathbf{a}.\mathbf{S})^2 \left(\sum_m V(\mathbf{r}_m) \exp(2i\pi(\mathbf{S}+\mathbf{k}).\mathbf{r}_m) + V(\mathbf{r}_m) \exp(2i\pi(\mathbf{S}-\mathbf{k}).\mathbf{r}_m)\right)$$

The diffuse scattering is concentrated into two peaks, located symmetrically with respect to each node of the reciprocal lattice, at the extremities of the wave vectors $\pm$ **k**. The intensity of these peaks is equal to $I_2 = N\pi^2 |F|^2 (\mathbf{a}.\mathbf{S})^2$. A more detailed analysis shows that the harmonic peaks also exists at $\pm 2\mathbf{k}$, $\pm 3\mathbf{k}$ etc.

Qualitatively this result can be immediately found if one remembers that the Fourier transform of a sinusoidal wave (correlation function) consists of two peaks located at $\pm$ **k**.

Let us turn to the effect produced by the various modes propagating within the crystal. We know that each wave vector corresponds to three independent waves: one purely or nearly longitudinal and two purely or nearly transverse.

Taking into account the dispersion relations between the energy and the wave vector **k** it can be shown that I_2 is proportional to $F^2 S^2/q^2 v^2$ where q is the distance of the nearest reciprocal node to the considered node, v the wave velocity (sound velocity for acoustic modes, which is assumed equal for all modes).

Physically there is a great number of waves propagating through the crystal (3N) which correspond to N vectors **k** regularly located around each reciprocal node. This indicates that each node is surrounded by a diffuse zone proceeding from the juxtaposition of the intensities I_2 scattered by each wave; in a first approximation I_2 decreases as $1/q^2$ around each reciprocal node. (see figure XV.3).

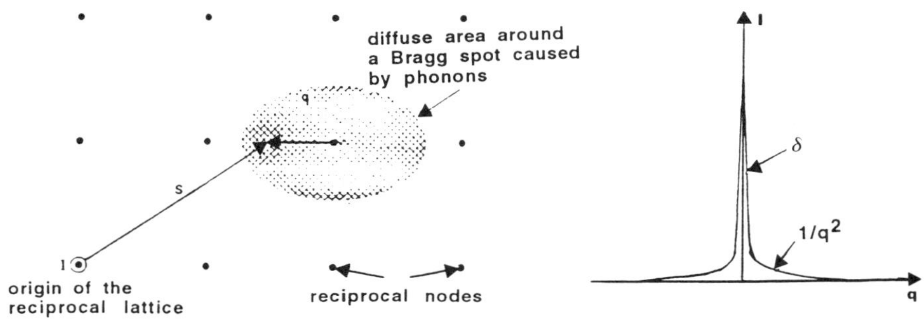

Figure XV.3. — Diffuse scattering by acoustical modes (phonons)

The same conclusions are found for neutron diffuse scattering, but as the neutron energies are of the same order of magnitude as the wave energies, the energy analysis of the coherent diffuse scattering due to the thermal motion can be performed. At each point **k** around a given node, it is possible to observe the three waves and to measure their energy, using a triple-axis-spectrometer. Thus it is possible to get the whole dispersion curves and to measure the sound velocities (slope of the dispersion curves) for the various types of waves (transverse or longitudinal) into the various directions with respect to the crystal axes; examples of dispersion curves are shown below.

It is very important to note that thermal diffuse scattering (TDS) is a quite general phenomenon displayed by all crystals : each Bragg reflexion is surrounded by a diffuse scattering area which is about 10^3 weaker than the corresponding Bragg intensity.

XV.2.6. Other types of diffuse scattering due to correlated displacements

We present here a few types of coherent diffuse scattering which are rather often exhibited by soft condensed and biological materials.

XV.2.6.1. Linear correlation

This type of correlation is characterized by the existence of parallel periodic atomic or molecular rows which are displaced from their theoretical position in the perfect crystal and which remain all parallel (see figure XV.4). The effects of such a disorder on the scattering patterns are the following:

- qualitatively the type of correlation is a periodic row. Its Fourier transform is formed by a family of equidistant reciprocal planes perpendicular to the rows. If **a** is the period along the row, the distance between the reciprocal planes is $1/|\mathbf{a}|$.

- quantitatively the expression for the diffuse scattering is the following:

$$I_2 = |F|^2 \frac{\sin^2(N\pi \mathbf{S}.\mathbf{a})}{\sin^2(\pi \mathbf{S}.\mathbf{a})} \sin^2(2\pi \mathbf{S}.\mathbf{u})$$

in the case of rows containing N molecules of scattering factor F, which are displaced either by +**u** or -**u** without correlation from one to another.

This type of partially coherent disorder is quite common in liquid crystals and proteins.

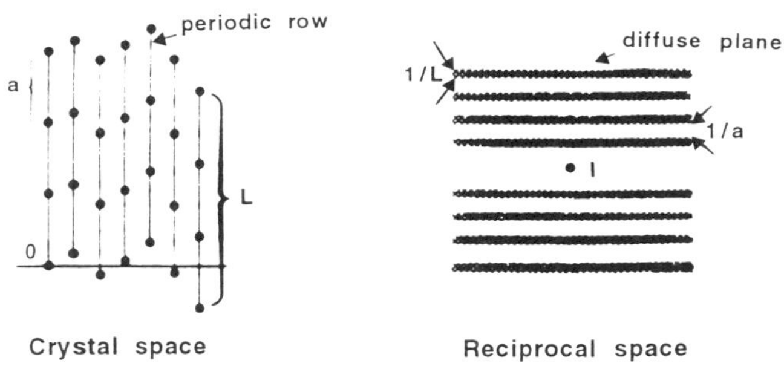

Figure XV.4. — Linear disorder with longitudinal displacements

XV.2.7.2. Planar correlation

The principle is the same as for the linear correlation, but this time we have periodic planes that are translated with respect to their theoretical positions in the perfect crystal (figure XV.5).

The type of correlation is a two-dimensional periodic plane (period **a,b**). Its Fourier transform is a family of reciprocal rows standing perpendicularly to the planes and regularly located according to an array of parameters $1/|\mathbf{a}|$ and $1/|\mathbf{b}|$ If the disorder were complete and the planes infinitely thin, then the rows would be infinitely long. If there exists a partial order or if the planes have a non-zero thickness, the rows are more or less long.

This kind of disorder is exhibited by lamellar structures in which the correlations between neighbouring layers are weak (for instance in liquid crystals, lipid layers and other biological compounds). Each Bragg spot is then transformed into a linear rod along **c*** that is perpendicular to the layers, except for the *00l* reflexions which are not sensitive to such disorder.

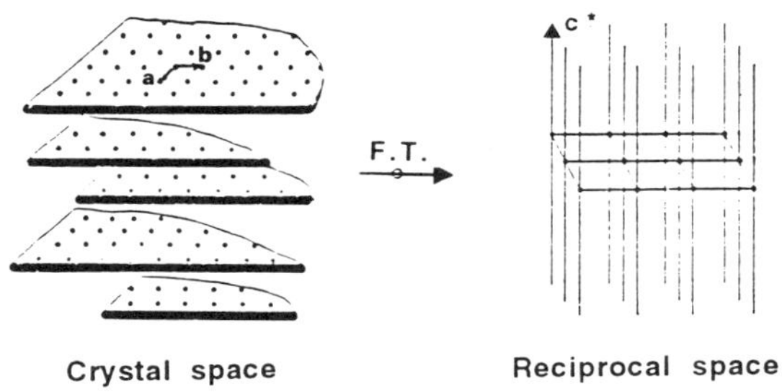

Figure XV.5. — planar correlation

XV.2.6.3 . Local order

Let us imagine a periodic lattice at the nodes of which are located vectors symbolizing any given structural property. If all the vectors stand parallel, we have a perfect order. If the vectors are randomly oriented, we have a perfect disorder superimposed on the periodic lattice. The intermediate state also exists, i.e. the local order that corresponds to small domains with the vectors oriented in the same direction (see figure XV.6).

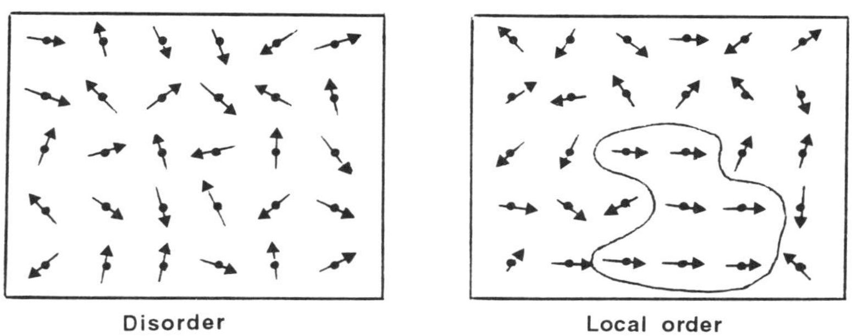

Figure XV.6. — schematic representation of local order; the arrows symbolize any given property

The scattering pattern will then both exhibits Bragg reflexions, corresponding to the periodic lattice, plus diffuse reflexions originating in the small domains where some degree of coherence is preserved in the orientations of the vectors. In other words, it can be said that there is diffraction by very small crystallites, which is spread out in reciprocal space according to the Scherrer formula which gives the angular width Δ of a peak as a function of the size L of the crystallites and of the wavelength λ: $\Delta(2\theta) = \lambda/L \cos\theta$

Such disorder has, for instance, been observed in liquid crystals.

XV.3. Experimental approach

Diffuse scattering intensities are generally 10^2 to 10^5 weaker than Bragg reflexions. Consequently it is necessary to take a few precautions to perform correct and unambiguous observations.

XV.3.1. General considerations

XV.3.1.1. Use of a single crystal

For a given sample mass, a single crystal provides more intensity than a powder. In addition three-dimensional information can be obtained, which is quite important for the interpretation of the diffuse scattering, whereas the powder methods yield an integral at constant scattering vector.

XV.3.1.2. Use of low background radiation

It is essential to use a low background radiation monochromator in order to eliminate the maximum of parasitic contributions (for X-rays).

XV.3.1.3. Use of two-dimensional detection

At least in a first step, it is better to record the patterns on two-dimensional rather than to one-dimensional detectors, since the diffuse scattering is not restricted to precise sites but is spread out over large parts of reciprocal space. Image plates, which are sensitive and exhibit both high spatial resolution and linearity over four decades, seem to be the best type of detector for diffuse scattering detection.

XV.3.1.4. Use of high intensity beams

Since diffuse scattering intensity is much weaker than Bragg intensity, data collection very often requires long exposures. It is then recommended to use high flux radiation.

XV.3.1.5. Spatial resolution

In order to measure correctly the intensities between the Bragg reflexions it is necessary to work with a high spatial resolution.

XV.3.2. X-ray diffuse scattering

From the above considerations, it is clear that the best diffuse scattering collection is performed with synchrotron radiation using image plate detectors. Nevertheless, conventional sources can be used, but it is necessary to work with a monochromator in order to eliminate the continuous scattering.

XV.3.3. Neutron scattering

In normal conditions X-ray techniques do not allow energy analysis of diffuse scattering since the X-rays photons are about 10^7 times more energetic than the phonons. It is therefore impossible to distinguish between static and dynamic disorder. The technique that permits the dynamical study is neutron scattering. Thermal neutron wavelengths are of the same order of magnitude as the lattice parameter lengths and of the same energy as vibrational energies in condensed matter. In terms of scattering, this means that both the direction and energy of the scattered neutron may be affected. The Bragg diffraction and diffuse scattering processes are the same for neutron scattering as for X-rays, but only neutron scattering permits the determination of the energy transfer associated with the diffuse scattering. For instance it is possible with neutron techniques to determine the dispersion curves which are characteristic of the dynamic of the samples.

Such work can be performed using a triple-axis spectrometer; this apparatus differs from an X-ray goniometer in the addition of an analyzer crystal, which provides the energy spectrum of the scattered neutrons (figure XV.7).

The conditions reviewed for X-rays : monochromatic beam, low background, high intensity beams, spatial resolution and use of single crystal, remain of course valid for neutrons.

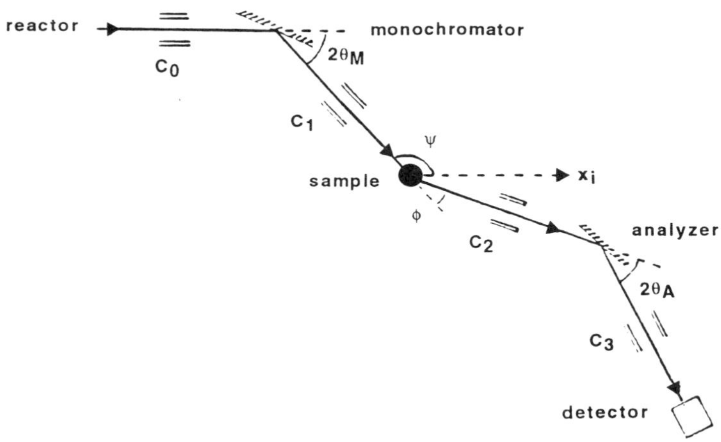

Figure XV.7. — principle of a triple-axis spectrometer for inelastic coherent neutron measurements

XV.4. Examples of coherent diffuse scattering studies

XV.4.1. Liquid crystals (Doucet et al.,1974 - Doucet et al.,1978 - Levelut et al.,1978)

TBBA is an interesting example of compound which presents a very rich polymorphism (see phase sequence in § XV.1.2). Valuable structural information on ordered smectic phases (phases V and VI) have been reached through diffuse scattering analysis given by monodomains (obtained by melting of a single crystal of phase VIII).

XV.4.1.1. Longitudinal order

In figure XV.8-a (phase V), the X-ray beam is perpendicular to the long molecular axes and a set of parallel diffuse sheets appears on the photograph. These sheets correspond to reciprocal diffuse planes perpendicular to the molecules, and their spacing is the inverse of the molecular length (1/29 $Å^{-1}$). As we have seen previously in § XV.2.7.1., such a feature indicates the existence of longitudinal displacements of parallel strings of molecules which are uncorrelated from one string to another. The correlation length along the molecular direction can be measured from the thickness of the diffuse sheets; it is of the order of 200 Å. The intense spots appearing on the diffuse sheets are the Bragg reflexions originating from the smectic layers. The tilt angle of the molecules with respect to the normal to the smectic layers can be directly measured from the photograph and is found to be 32° at 115°C.

In this example coherent diffuse scattering provides structural information (direction of the long molecular axes for instance) which cannot be deduced from the Bragg reflexions.

XV.4.1.2. Lateral order

The photograph in figure XV.8-b (phase V) represents the equatorial plane (**a***,**b***) of the reciprocal lattice (perpendicular to the molecular direction). The six Bragg spots forming a pseudo-hexagon are indicative of a periodic order within the layers (the photograph is over-exposed in order to show the diffuse scattering,

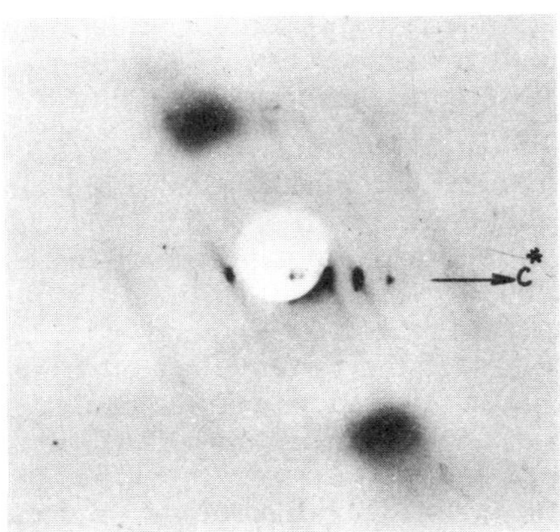

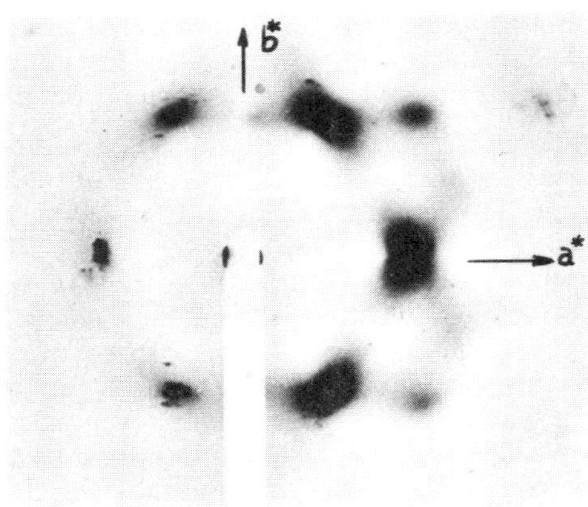

Figure XV.8-a. — X-ray pattern of TBBA in the smectic V phase (beam parallel to the layers)

Figure XV.8-b. — X-ray pattern of TBBA in the smectic V phase (beam perpendicular to the layers)

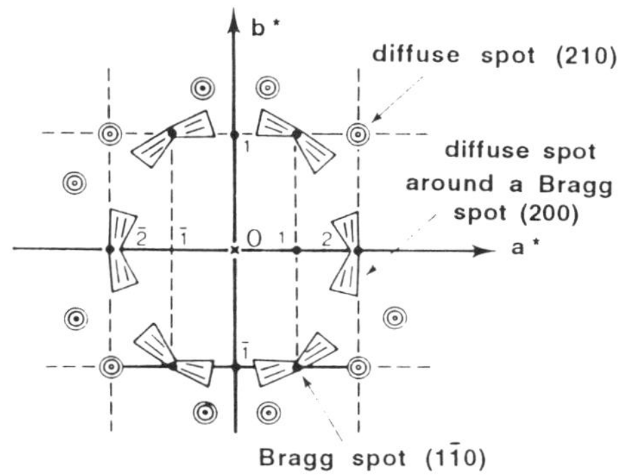

Figure XV.9. — Schematic representation of the X-ray pattern of the equatorial plane **(a*,b*)**

and so the sharpness of the spots is concealed); the lattice is orthorhombic with the C face centred.

In fact, such a description of the lateral structure of the smectic V phase is incomplete since it does not take into account additional scattering features observed on the equatorial plane. These consist of twelve diffuse spots which are schematically represented in figure XV.9. The presence of these spots indicates that the local symmetry is lower than the global symmetry. Let us discuss this point more precisely. If **a** and **b** are the two axes lying in the smectic layers, the diffuse spots have the following indices: (±2 ±1 0), (±5/2 ±1/2 0) and (±1/2 ±3/2 0). Only the spots of the first group are located on reciprocal nodes; the others, with half-integral indices can be deduced from the first group by rotations of 120°.

What is the significance of the four spots (±2 ±1 0)? If we consider them as Bragg spots, then the C face is not centred (the molecular positions at the centre and at the apices of the rectangular lattice (**a,b**) are not equivalent). However this conclusion is valid only for small domains because of the width of the diffuse spots : the size of the domains is given by the Scherrer formula, and is about 30Å. Within a small domain, it can be shown that the molecular lateral packing parallel to the smectic layers is of the herring-bone type (figure XV.10-a where each molecular section is represented by an ellipse). The existence of the two other families of four diffuse spots is explained by assuming that there are three possible orientations for such domains which are related to one another by 120° rotations, corresponding to three orientations of the rectangular cell (figure XV.10-b). Thus, on the whole, all the molecular orientations become equivalent on each node. In this example the existence of a local order, which is different from the average order, is indicated by the coherent diffuse scattering.It has been shown by inelastic neutron scattering that the orientational disorder is a dynamic phenomenon with a relaxation time of the order of 10^{-11} s.

The X-ray patterns of the phase VI of TBBA are very similar to those of the phase V, except for one major difference: four of the twelve diffuse spots are now condensed into Bragg spots and the other eight have disappeared. This indicates that one type of small domain has grown at the expense of the two others. The molecular positions are no longer equivalent, the rotational disorder of the molecules around their long axes being frozen in the phase VI.

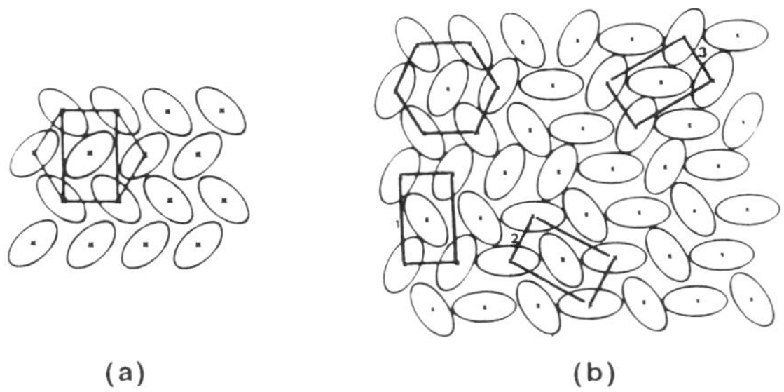

Figure XV.10-a. — small domains with herring-bone packing)

Figure XV.10-b. — the three types of herring-bone packing domains in the phase V of TBBA

c) **phonons**

Examples of dispersion curves are shown on figure XV.11; these curves are obtained by inelastic coherent neutron scattering measurements on a triple-axis spectrometer: the energy shift of the neutron before and after the sample is measured as a function of the distance to a given reciprocal node.

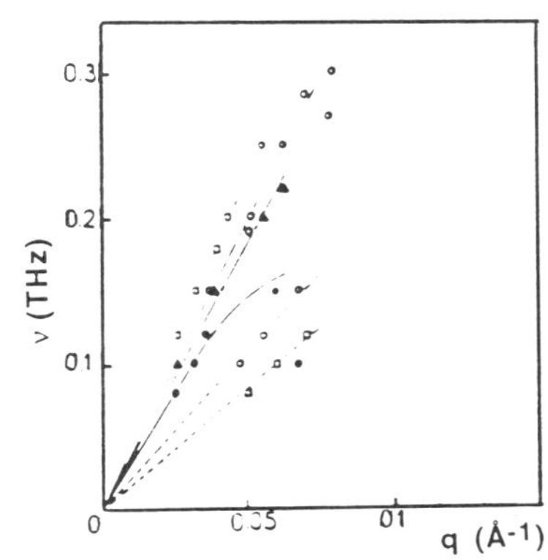

Figure XV.11. — examples of dispersion curves observed in TBBA. The solid line denotes the longitudinal acoustic modes while the transverse modes are indicated by the dashed lines; crystal at 20°C (o), crystal at 104°C (▲), smectic VI (□) and smectic V (●).

The comparative study of the lattice dynamics of the solid and ordered smectic phases shows that the phonons can propagate in the smectic phases. Longitudinal and transverse modes have been clearly detected in the phases V and VI. Except for the phonons propagating along the direction perpendicular to the layers for which a huge quasi-elastic component prevents detection of the modes, the dispersion curves in the ordered smectic phases are similar to those of the crystalline phase.

The slope at the origin is equal to the velocity of the sound for the studied mode. The velocities measured for TBBA (500 m/s to 1500m/s) are analogous to those of lead; this gives some idea of the rigidity of the ordered smectic phases. From these neutron measurements it can be said that the ordered smectic phases are closer to a crystalline phase than to a liquid phase; this conclusion is reversed for the smectic phases A and C which are characterized by a liquid order within the layers.

XV.4.2. Proteins (Glover et al., 1991)

The coherent diffuse scattering given by protein crystals represents a new and promising field of investigation; up to now only half a dozen of studies have been performed, mostly because of the difficulty of such work and of the low number of crystallographic structure determinations. Let us look at two examples: lysozyme and insulin.

XV.4.2.1. lysozyme (Doucet et al.,1987)

One example of X-ray scattering photograph is shown on figure XV.12-a . It was obtained using the synchrotron radiation in LURE (synchrotron radiation source), in fixed geometry (distance crystal-film 75mm, wavelength 1.61Å), on a crystal of hen egg-white lysozyme in the orthorhombic form. In order to collect the diffuse scattering, the exposure time is of the order of 30 minutes whereas the recording of the Bragg spots only requires a few seconds. For this figure the "vertical" rows are the traces of the reciprocal planes perpendicular to **c*** and the horizontal ones are perpendicular to **a***.

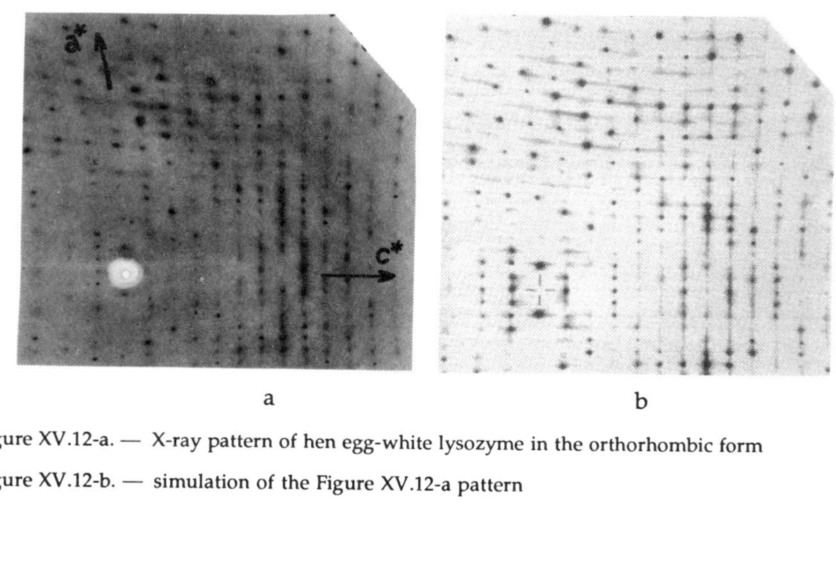

a b

Figure XV.12-a. — X-ray pattern of hen egg-white lysozyme in the orthorhombic form

Figure XV.12-b. — simulation of the Figure XV.12-a pattern

Four intensity components can be observed: Bragg reflexions, thermal diffuse scattering around Bragg reflexions, a large diffuse ring around 0.3 $Å^{-1}$ which arises from the sample holder (capillary glass) and water content of the crystal (about 33%), and finally diffuse scattering features which give evidence of correlated specific disorder.

These diffuse scattering features are located on the photographs along rows perpendicular to **a*** and **c***, and are in fact mainly located in the reciprocal space along two families of planes : (100)* and (001)*. We will not enter here into details of the analysis of this diffuse scattering, which is of the linear correlation type (see § XV.2.7.1) with rows along the **a** and **c** directions corresponding respectively to |**a**| and |**c**| periodicities. We just indicate that the qualitative characteristics are deduced from the localization of the diffuse scattering, whereas the quantitative parameters are derived from the intensity distribution. The quantitative analysis was performed by comparing the experimental films with simulated films (one example of comparison is given on figures XV.12-a and XV.12-b).

Briefly, the analysis is performed in the following way. As explained in § XV.2.3., the diffuse scattering intensity for a given correlation is proportional to: $\boxed{\langle|\Psi|^2\rangle - |\langle\Psi\rangle|^2}$ where Ψ represents the structure factor of one independent unit, i.e in our case a row of molecules along the **a** or **c** direction with the period |**a**| or |**c**|.

Assuming rigid-body translations **u**, it can be easily shown that :

$$\boxed{I_2 = |F|^2 \; (1 - \exp(-4\pi^2(\mathbf{S}.\mathbf{u})^2)}$$

where F is the structure factor of the molecular content which is repeated along the row. Simulations are performed with various molecular contents and **u** directions.

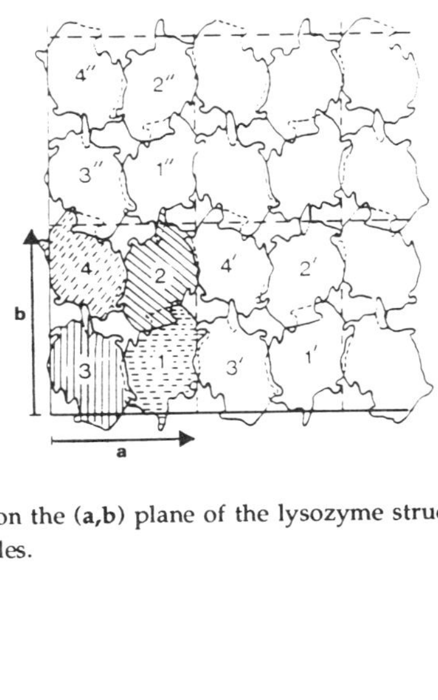

Figure XV.13. — projection on the (**a**,**b**) plane of the lysozyme structure (orthorhombic form); the unit cell contains four molecules.

The main results are the following: molecular displacements, correlated over distances of about 200Å have been detected. These rigid-body displacements are polarized along **a** for the correlations in the **a**-direction and along **c** for those in the **c**-direction; their mean square amplitude is of the order of 5.10^{-2} Å^2. Looking at the structure (figure XV.13) the molecules which are correlated along the **a** direction form two types of rows : (3-1-3'-1'...) and (4-2-4'-2'...), the displacements between the various rows being independent. Similar behaviour is found for the **c**-correlations.

What are the relations between the correlated molecular displacements, the intermolecular contacts and the intramolecular mobility? The correlations are in agreement with the intermolecular contacts, which are ensured by a dozen contacts shorter than 4Å in the **a** and **c** directions, and which do not exist in the **b**-directions for which no correlations are detected. The correlated displacements could be related to intramolecular displacements, which are the most interesting from the biological point of view, but no such information can be deduced from these diffuse scattering features. Nevertheless this type of linear correlation seems interesting since it is displayed by nearly all protein crystals.

XV.4.2.2. insulin (Caspar et al.,1988, Clarage et al., 1992)

A further step in the interpretation of the diffuse scattering features has been recently performed on insulin crystals. After removing the diffuse scattering components due to the solvent, the sample holder and the thermal diffuse scattering, there remains a diffuse background. This diffuse scattering component has been interpreted as short-range correlated displacements (with r.m.s. amplitude 0.4Å) of side chains or parts of the protein over a range 4-8Å. This kind of displacement is important since it is indicative of a high flexibility of the protein, which does not appear from the structure determination. The structure determination provides an idealized average structure that gives a misleading rigid and uniform image of the protein. These displacements have been postulated to explain various processes in biological macromolecules, for example, to explain how oxygen can reach the haem group in myoglobin through a seemingly impenetrable mass of closely packed side-chain atoms.

Another study has been performed on the TDS areas around the Bragg diffraction peaks. It seems that, contrary to a classical crystal where lattice modes propagate over long range, the TDS haloes in insulin correspond to damped molecular displacements over a range of 20-30Å. This conclusion is obtained by the analysis of the intensity decrease of the TDS zone as a function of the distance q to the reciprocal node: it can be shown that for a "classical" crystal it decreases as q^{-2} (harmonic elastic modes) whereas it varies as $q^{-\alpha}$ with $\alpha < 2$ for damped mode propagation.

XV.4.3. DNA

It is well known that DNA double helixes can adopt three main conformations, A,B and Z. The stability of these various conformations depends on many parameters, mainly humidity, nature of the solvent and base sequence. The understanding of DNA polymorphism is important since it is believed that the recognition between proteins and DNA is governed by the local conformation of DNA. Various cases of transition between two forms have been observed; the most studied one is the transition from the B form to the A form when the humidity is lowered. Let us show how information on the stability of the A and B forms has been obtained from a diffuse scattering study performed on crystals of the oligonucleotide d(GGTATACC).

On the X-ray pattern of this nucleotide (figure XV.14) intense diffuse features (DM and DC) are clearly visible in addition to the Bragg spots; they recall the

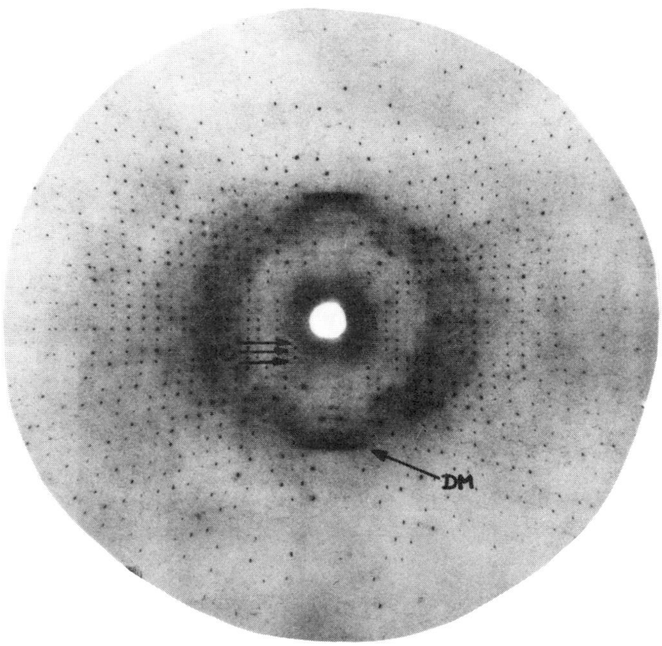

Figure XV.14. — X-ray pattern of the oligonucleotide d(GGTATACC); the **c**-axis (direction of the channels) is vertical.

scattering pattern of a helix with its long axis parallel to **c**. The crystal structure determination shows that the molecules adopt the A form and are located along the walls of hollow cylinder (figure XV.15). The diffuse scattering features prove that the inner part of the cylinder do not only contains solvent, as previously believed, but also extra DNA molecules which do not belong to the crystalline matrix. The interesting point lies in the fact that these extra molecules adopt the B and not the A form like the molecules forming the crystalline matrix. Such a result proves the possibility of coexistence of DNA molecules in the A and B forms; it would never have been obtained without taking into account the diffuse scattering features.

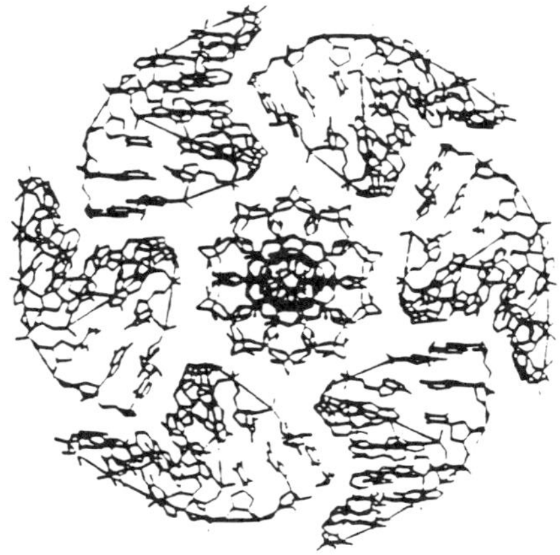

Figure XV.15. — projection of one cylinder of the crystalline structure of d(GGTATACC) parallel to the c-axis; in the central part of the cylinder is represented one additional molecule in the B-conformation, which is responsible for the diffuse features of pattern XV.14.

XV.5. Conclusion

Coherent diffuse scattering analysis of condensed materials (liquid crystals, polymers, plastic crystals...) has been performed for a rather long time, both by X-ray and neutron measurements; much important information has been obtained.

In contrast, the analysis of biological materials by this technique is relatively new and it represents a promising field of investigation for the understanding of intramolecular flexibility and intermolecular interactions.

In addition to the experimental difficulties of diffuse scattering measurements and analysis, the main limitation for the X-ray diffuse scattering method lies in the fact that it is not possible to distinguish static from dynamic phenomena, i.e. displacements from motions. Nevertheless two approaches are possible to overcome that problem.

1) Neutron diffuse scattering. Up to now the results on protein crystals have been quite disappointing because this technique requires very big crystals (at least 0.2 cm^3), beyond the normal volume range for protein crystals, and also requires deuterated compounds to eliminate the incoherent scattering from the hydrogen atoms.

2) Mössbauer spectroscopy. The principle of these technique is the same, but in addition it is possible to measure the energy shift of the gamma photons since the spectral distribution of the incident beam is narrow enough. Up to now the main limitation of Mössbauer measurements lies in the very low beam intensity, but new and important developments will probably occur soon with the intense Mössbauer sources which are scheduled at the ESRF.

Last but not least, it must be emphasized that the coherent diffuse scattering analysis can be performed only on crystallized materials, which constitutes a severe limitation to the interest of the method, mostly for biological compounds which are not naturally in that state. But even in that case, where motions are at least over-damped, original and significant information will be probably obtained soon.

GENERAL READING

GUINIER A. "X-ray diffraction" 1956, Ed W.H. FREEMAN San Francisco, London, 1963

REFERENCES

DOUCET J., LEVELUT A.M., LAMBERT M., 1974 - Mol Cryst. & Liq. Cryst., 24, 317

DOUCET J., LEVELUT A.M., LAMBERT M., 1974 - Phys. Rev. Lett., 32(6), 301

DOUCET J., LAMBERT M., LEVELUT A.M., PORQUET P., 1978, - J. de Phys. 39, 173

LEVELUT A.M., MOUSSA F., DOUCET J., BENATTAR J.J., LAMBERT M., DORNER B., 1981, - J. de Phys. 42, 1651

GLOVER I.D., HARRIS G.W., HELLIWELL J.R., MOSS D.S., 1991, - Acta Cryst B47, 960

DOUCET J., BENOIT J.P., 1987 - Nature, 325, 643

CASPAR D.L.D., CLARAGE J., SALUNKE D.M., CLARAGE M.,1988 - Nature, 332, 659

CLARAGE J.B., CLARAGE M.B., PHILLIPS W.C., SWEET R.M., CASPAR D.L.D., 1992, - Proteins, 12, 145

DOUCET J., BENOIT J.P., CRUSE W.B.T., PRANGE T. KENNARD O., 1989, - Nature, 337, 190

NIENHAUS G.U., HEINZ J., HUENGES E., PARAK F., 1989, - Nature, 338, 665

CHAPTER XVI

LARGE-AMPLITUDE MOTIONS IN BIOLOGICAL SYSTEMS.

S. CUSACK

XVI.1. Introduction

Proteins fold up into well defined three dimensional structures in order to perform their biological functions. However these structures are by no means static or rigid and it is essential to study the dynamics or conformational flexibility of proteins in order to appreciate more fully how they work. This chapter will first give some striking examples of where protein (and nucleic acid) conformational flexibility is an intrinsic part of protein function. This is followed by a discussion of the nature of protein dynamics and theoretical methods for studying and simulating protein dynamics. Finally an example is given, using the protein myoglobin, of how inelastic neutron scattering can give an insight into the nature of atomic fluctuations in proteins and can challenge the ability of molecular dynamics simulations to correctly describe protein dynamics.

XVI.2. Examples of large scale motions in proteins and nucleic acids

XVI.2.1. Triose-phosphate isomerase (TIM)

TIM is the enzyme which catalyses the conversion of dihydroxyacetone phosphate into glyceraldehyde 3-phosphate (i.e. an aldose-ketose isomerisation) as one step of the glycolytic pathway (the conversion of glucose to pyruvate with the production of ATP). The yeast TIM is a dimeric molecule with 247 residues per subunit. The X-ray crystal structure is known of the native enzyme and also of a complex with a transition state analogue. Comparison of these two structures shows that an 11-residue loop (residues 166-176) which is in an open conformation in the native enzyme, closes over the bound ligand in the complex (Figure XVI.1). This motion appears to be of the rigid-body type with conservation of the loop conformation except in the hinge regions (166-168 and 174-177). The motion of this 'lid' has been simulated by molecular dynamics (Joseph et al., 1990) as well as Brownian dynamics methods (Wade et al. 1993). The ultimate aim is to simulate the whole process of substrate binding, including diffusion of the substrate through the solvent towards the active site and closing of the loop. The role of this lid closure is presumably to create an active site environment where solvent is excluded and the isomerisation reaction can occur with high efficiency.

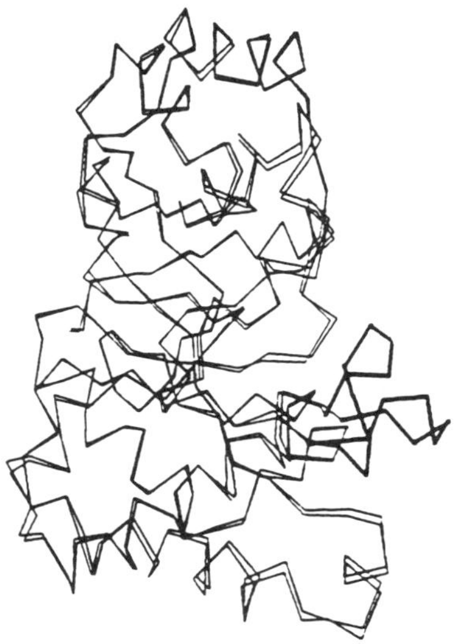

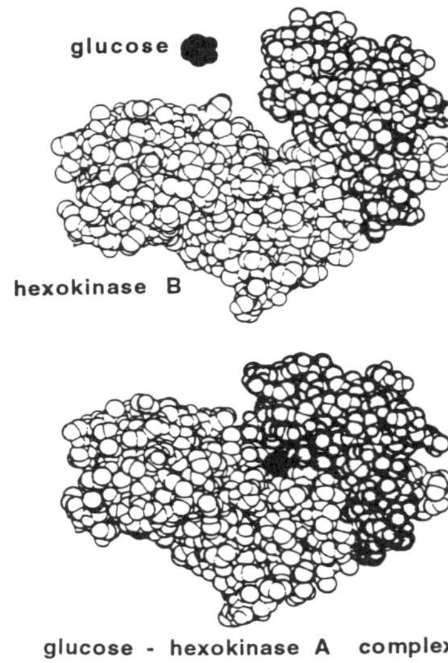

Figure XVI.1. — (Left) Superposition of the backbones of the closed and open forms of yeast triose phosphate isomerase. The two positions of the flexible loop are shown in bold on the right of the molecule. From Joseph et al. 1990.

Figure XVI.2. — (Right) Top - Space-filling model of yeast hexokinase in its native unliganded conformation. Also shown is the substrate glucose. Bottom - structure of the complex of hexokinase with glucose. The active site cleft is closed by a motion that is essentially a rotation of the upper lobe by 12° relative to the lower lobe. From Anderson et al. 1979.

XVI.2.2. Hexokinase

Hexokinase is another enzyme in the glycolytic pathway and catalyses the ATP mediated phosphorylation of glucose to glucose-6-phosphate. The crystal structure of the yeast enzyme (a monomer of 51K daltons molecular weight and 458 residues) shows that the enzyme has two domains, the small domain comprising residues 59-186 and the large domain comprising residues 1-58 and 187-458. Comparison of two crystal structures (Figure XVI.2), the native enzyme and the enzyme complexed with glucose, shows that a substantial conformational change occurs upon substrate binding. This can be described as a rotation of the small domain by 12 degrees relative to the large domain leading to displacements of up to 8Å in the protein backbone. The occurrence of such a considerable conformational change has been verified in solution by a small decrease in the radius of gyration of the molecule upon addition of glucose as measured by small angle X-ray scattering. As with TIM, it is presumed that the purpose of the conformational change is to exclude solvent from the active site during catalysis, perhaps with the additional aim of avoiding hydrolysis of the ATP in the case of hexokinase.

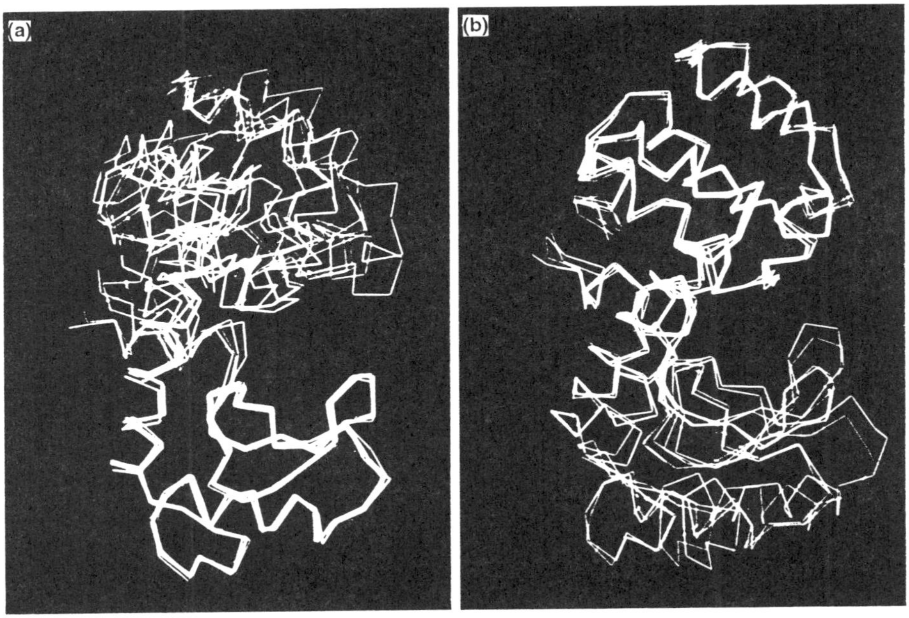

Figure XVI.3. — Superposition of the backbone of five crystal structures of T4 lysozyme. (a) superposition done by aligning the N-terminal domain (residues 15-60), (b) superposition done by aligning the C-terminal domain (residues 80-160). From Faber and Matthews 1990.

Several other enzymes have been shown to undergo relative domain movements comparable or even more dramatic than in hexokinase. Other examples are citrate synthase, phosphoglycerate kinase and adenylate kinase. The latter case has recently been analysed to determine the nature of the 'joints' which allow a small domain to rotate 90 degrees in order to bury the AMP and ATP substrates (Gerstein et al., 1993).

XVI.2.3. T4 lysozyme

A particularly interesting example of protein flexibility has been revealed by crystal structures of T4 lysozyme (Faber and Matthews, 1990). Lysozymes catalyse the hydrolysis of polysaccharides and this particular one is used by the bacteriophage (i.e. virus infecting bacteria) T4 to break down bacterial cell walls to release progeny phage. It is an enzyme of 164 residues, again in two lobes (residues 12-76 and 77-164) with the active site at the interface between the two lobes and it has long been imagined that the two lobes could open and close to allow access and release of the substrate. A mutant enzyme in which methionine 6 is replaced by isoleucine crystallises in the standard trigonal form as well as an orthorhombic form with four molecules in the asymmetric unit. Thus five independent molecular structures have been determined and each shows a different hinge-bending angle between the two lobes with a maximum opening of 32.3 degrees from the closed, native form (Figure XVI.3). The interpretation of this result is that only relatively low energy deformations are required to move the two lobes relative to each other, corresponding to torsional degrees of freedom and that the exact hinge-angle observed in a crystal can be influence by crystal packing forces.

XVI.2.4. Seryl-tRNA synthetase

Another spectacular example of domain motion involves the recognition by seryl-tRNA synthetase of its nucleic acid substrate. This enzyme is one of the twenty members of the family of aminoacyl-tRNA synthetase (one for each amino acid) which play a vital role in protein synthesis in all living cells. Seryl-tRNA synthetase specifically attaches the amino acid serine to the tRNA specific for serine, $tRNA^{ser}$, thus permitting the correct incoporation of this amino acid into a polypeptide chain being synthesised on a ribosome. The structure of this dimeric enzyme shows a globular domain (residues 101-430) containing the active site and an extraordinary N-terminal arm domain (residues 1-100) consisting long antiparallel coiled coil which projects 60Å into the solution (Cusack et al., 1990). Various crystal structures in the absence of bound tRNA show that this helical arm is flexibly hinged and can take up different orientations, as well as twist, depending upon the crystal packing (Figure XVI.4). The crystal structure of the complex with $tRNA^{ser}$ shows the arm in yet a new orientation, which, this time, is determined by the need to specifically recognise the shape of the $tRNA^{ser}$ and to discriminate against incorrect tRNAs.

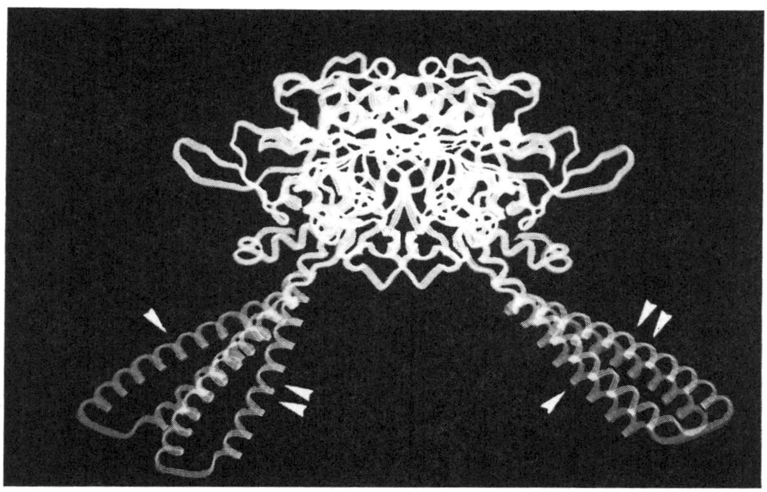

Figure XVI.4. — Two different crystal structures of seryl-tRNA synthetase from *E. coli*. Arrows indicate the different helical arm orientations in the two structures.

XVI.2.5. Transfer RNA (tRNA)

As a corollary to the preceeding example, it should be pointed out that recognition of two macromolecules can involve a mutual conformational adaptation of each partner. A good example of this is provided by crystal structures of another aminoacyl-tRNA synthetase, that specific for aspartic acid (Ruff et al., 1991). In this case, the structure of the $tRNA^{asp}$ is known both uncomplexed and complexed with the aspartyl-tRNA synthetase (the structure of the uncomplexed *enzyme* is not yet known). Comparison of the two tRNA structures shows very considerable distortion of the structure upon interaction with the enzyme, in a way which ensures specific recognition of the correct tRNA (Figure XVI.5).

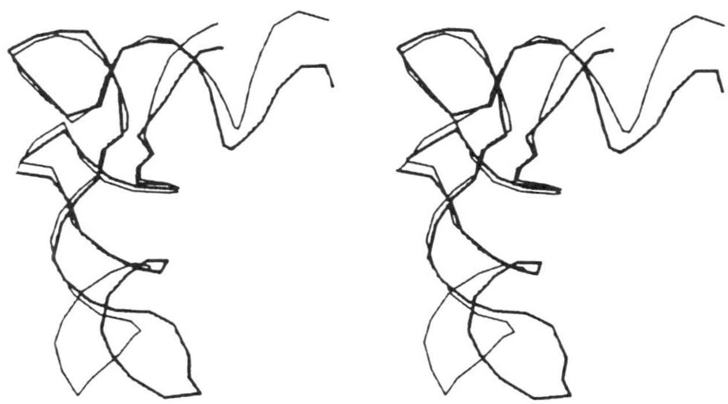

Figure XVI.5. — Stereo diagram showing the superposition of the backbone confromation of $tRNA^{Asp}$ in its native form (thin line) and in its complex with yeast aspartyl-tRNA synthetase. From Ruff et al. 1991.

XVI.3. The nature of protein dynamics: conformational change and equilibrium fluctuations in proteins

In the above examples of biologically important motions in proteins, it is necessary to distinguish between conformational changes and equilibrium fluctuations. In order to perform their biological function many proteins need to be able to adopt two or more different conformations and are induced to change from one to another by ligand or substrate binding or by a change in environmental conditions such as pH. The above examples (and many others not described) show that this may range from subtle reorientations of a few sidechains, to displacements of loops, pseudo-rigid body inter-domain motions, subunit reorientations in oligomeric proteins, to drastic rearrangements of the whole protein (Figure XVI.6).

In whichever conformation a protein might be, it will also be undergoing continual random thermal fluctuations. Indeed from fundamental thermodynamic arguments it is possible to estimate the overall magnitude of internal energy and volume fluctuations in proteins (Cooper 1988). What is the

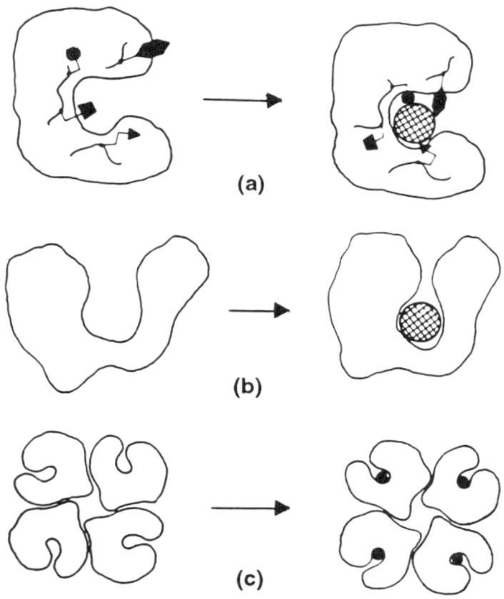

Figure XVI.6. — Schematic diagram of types of ligand induced conformational changes in proteins. (a) local re-arrangements, (b) relative domain movements, (c) subunit re-orientation. From Cooper, A. 1980- Sci. Prog. Oxf. 66, 473.

nature of these fluctuations? Proteins have a complex heirarchichal structure in which a covalently linked polypetide chain is folded into a compact three-dimensional structure which is only marginally stable as a result of the interplay of many weaker interactions, such as hydrogen bonds, and van de Waals and electrostatic non-bonded interactions (see below). As a result proteins possess a wide spectrum of possible equilibrium fluctuations covering different time-scales and different distances (Table XVI.1).

They range from very fast covalent bond vibrations (10^{-13}-10^{-14}s) to slower more global vibrational motion (10^{-11}-10^{-12}s) and surface side-chain motion (10^{-10}-10^{-11}s), inter-domain motion (10^{-7}-10^{-11}s) and on down to internal aromatic ring flips (1-10^{-4}s) and partial denaturing (10-10^{-5}s). Whereas it is clear that conformational change is crucial for biological function, the role of thermal fluctuations in protein function is often less obvious. Are they just an inevitable background jiggling of the protein atoms which permit or lubricate functionally important motions? Or is the pattern of thermal fluctuations tailored by evolution to enhance protein function, for instance in favouring motions which are required along a catalytic reaction pathway? In other words, do proteins function because of or in spite of thermal fluctuations? In some cases the situation is apparently clear: without thermal fluctuations, oxygen would not be able to penetrate into the heme pocket in myoglobin (see section XVI.6). Furthermore, the experiment on freezing ribonuclease described below in section XVI.7 suggests that a certain level of atomic mobility is required in enzymes to enable them to function. On a more general level, it is intuitively obvious that the distinction between conformational change in response to some change in

Motion	Spatial extent(nm)	Amplitude (nm)	Log of characteristic time (s)
Relative vibration of bonded atoms	0.2-0.5	0.001-0.01	-14 to -13
Elastic vibration of globular region	1-2	0.005-0.05	-12 to -11
Rotation of sidechain at surface	0.5-1	0.5-1	-11 to -10
Torsional libration of buried groups	0.5-1	0.05	-11 to -9
Relative motion of different globular regions ("hinge-bending")	1-2	0.1-0.5	-11 to -7
Rotation of medium sized sidechains in interior	0.5	0.5	-4 to 0
Allosteric transitions	0.5-4	0.1-0.5	-5 to 0
Local denaturation	0.5-1	0.5-1	-5 to +1

Table XVI.1. — Typical features of some internal motions of proteins (modified from McCammon and Harvey, 1987).

environment and thermal fluctuations is somewhat artificial since what is at play in both cases is the nature of the underlying atomic interactions and the constraints of the protein structure. In simpler physical systems this connection is generally known as the fluctuation-dissipation theorem. A more sophisticated view of protein dynamics involving the concept of confromational substates will be introduced in section XVI.6.

XVI.4. Theoretical models of protein dynamics

Conformational changes, enzyme kinetics and mechanisms, the notorious protein folding problem and the rational design of engineered proteins all involve understanding of the interplay between protein structure, dynamics and thermodynamics. One way of tackling this problem from the theoretical point of view is to attempt to give as complete as possible a microscopic description of protein systems. This involves knowing the fundamental interactions at play.

XVI.4.1. Empirical potential energy functions

The normal theoretical approach to studying the dynamics of an atomic system is to start with an expression for the configurational energy of the system $V\{\mathbf{R}_L\}$ as a function of each atomic co-ordinate $\{\mathbf{R}_L\}$. The gradient of this potential energy can be used to calculate the force on one part of the system as a result of displacements of another part and hence the dynamical properties can be obtained by integration of the equations of motion (classical or quantum, as appropriate). Despite the complex, many-body nature of protein systems, this is now a routine approach to analysing protein dynamics (McCammon and Harvey 1987, Karplus and Petsko, 1990). However to simplify the problem, an approximate *empirical potential energy function* (or *force-field*) is used containing a sum of parameterised terms corresponding to physically understandable interactions. A typical protein potential energy function is of the form:

$$V(\{\mathbf{R}_L\}) = \sum_{\text{bonds}} k_b(b-b_0)^2 + \sum_{\text{angles}} k_\theta(\theta-\theta_0)^2 + \sum_{\text{dihedrals}} k_\phi(1-\cos[n\phi-\delta])$$
$$+ \sum_{\text{impropers}} k_\omega(\omega-\omega_0)^2 + \sum_{i,j}[4\varepsilon_{ij}(\frac{\sigma_{ij}}{r_{ij}})^{12} - (\frac{\sigma_{ij}}{r_{ij}})^6] + \sum_{ij}\frac{332 q_i q_j}{r_{ij}} \qquad \text{(XVI.1)}$$

In this expression, b, θ, ϕ, ω are the bond lengths, bond angles, dihedral and improper torsional angles and b_0, θ_0, ϕ_0, ω_0 are the equilibrium values for these variables. k_b, k_θ, k_ϕ and k_ω are the associated force constants. The non-bonded terms are pairwise additive and consist of a 12 (repulsive)-6(attractive) Lennard-Jones van der Waals term and a Coulombic term to account for electrostatic interactions between atomic charges. To this must be added, in a full treatment, interaction terms of the protein atoms with the solvent (usually water) and other solutes (e.g. ions). A widely used potential energy function is CHARMM (Brooks et al, 1983). The reader is referred to the book by McCammon and Harvey (1987) for a full discussion of protein potential energy functions.

XVI.4.2. Molecular dynamics simulations

Given such a force-field and the existence of todays powerful computers, it is possible to perform a *molecular dynamics simulation* of the protein dynamics. This involves numerically solving the classical equation of motion simulataneously for each atom in the protein as it evolves under the influence of the potential energy function V. The result is a time series $\{\mathbf{R}_L(t_n)\}$ giving for each atom at discrete times t = nt, n= 0,N where t is typically 0.01ps and Nt is up to 100ps or more. From these trajectories it is possible to calculate directly dynamic properties to compare with experimental measurements, e.g. mean-square atomic displacements, correlation functions or the inelastic neutron scattering spectrum (see below). Molecular dynamics methods have also been extremely important in assisting refinement of protein structures derived by X-ray crystallographic or NMR measurements (Brunger, Kuriyan and Karplus, 1987).

Molecular dynamics simulations give an extremely detailed description of the motion of each atom in a protein (although the accuracy of this detail is another matter!) and ultimately this is what is important in understanding what goes on in an enzyme active site. However without considerable analysis it is difficult to get from them an overall picture of protein dynamics and furthermore few experimental techniques are capable of furnishing equivalently detailed data for comparison. Thus it is of interest to look at simpler models of specific kinds of motion that may occur in proteins.

XVI.4.3. Vibrational motion in proteins

Proteins like all molecular systems are expected to undergo vibrational motion and indeed optical spectroscopic studies of high frequency (100-3000 cm^{-1}) local vibrations characteristic of the particular chemical groups in proteins have for a longtime been used as a probes and monitors of both structure and function [30]. However because of their large size and the importance of soft forces in protein structure they also have lower frequency, more global modes of vibration which were originally conceived of as 'breathing' modes or other distortions of a protein considered as an elastic body. These low frequency modes would be themally excited at room temperature ($h\nu < k_B T = 200$ cm^{-1} at 300K) and would contribute significantly to the atomic mean square displacements. Collective motions of this type have been identified by analysis of molecular dynamics simulations but a more direct way to look theoretically at vibrational motion is to perform a *normal mode analysis* . In this procedure we begin as before with a protein and its empirical potential energy function, but make the further assumption that the protein conformation is at a minimum of the energy function and that motion corresponds to small displacements $\delta \mathbf{R}_L$ of each atom about its equilibrium position $\mathbf{R}_L^0$. We can then expand the potential energy in a Taylor series about the minimum energy configuration giving:

$$V(\{\mathbf{R}_L\}) = V(\{\mathbf{R}_L^0\}) + \frac{1}{2}\sum_{ij}^{3N} \left(\frac{\delta^2 V}{\delta \mathbf{R}_i \delta \mathbf{R}_j}\right)_{\{\mathbf{R}_L^0\}} \delta \mathbf{R}_i \delta \mathbf{R}_j + O(\delta \mathbf{R}^3) \qquad \text{(XVI.2)}$$

If the displacements are small enough we can make the harmonic approximation and only consider the terms quadratic in the displacements. Then using the force constants given by the second derivative matrix in the equations of motion it can be shown that the resultant vibrational motion of each atom can be expressed as a superposition of 3N-6 harmonic oscillations or *normal modes* . Each normal mode is characterised by a nonzero angular frequency ω_λ and a set of N orthonormal eigenvectors $\mathbf{c}_L^\lambda$ giving the relative magnitude and direction of displacement of each atom according to the normal mode λ. The number of modes per unit frequency range is called the density of vibrational states.

Normal mode analyses have been performed on a number of small proteins. In addition, the technique of inelastic neutron scattering (section XVI.5) can be used to experimentally measure the vibrational density of states in the interesting low frequency range. In the case of the small protein, bovine pancreatic trypsin inhibitor (BPTI) a comparison of the experimental vibrational density of states with the results of normal mode analyses using different empirical force-fields (Cusack et al. 1988) has shown the importance of the treatment of long-range electrostatic interactions in correctly predicting the number of very low frequency modes.

Attempts have also been made to describe in terms of normal modes of motion the inter-domain motion of the type described above in lysozyme (paragraph XVI.2.3), this time however with human lysozyme (Gibrat and Go, 1990). One problem is that very low frequency modes well be damped by frictional forces in the system (e.g. viscous damping due to the solvent).

XVI.5. Inelastic neutron scattering and protein dynamics

Inelastic neutron scattering (INS) is a spectroscopic technique which can be used to study internal protein motions on exactly the same timescale (0.1-100ps) that is now widely accessible by computer simulation (Cusack 1988). It can thus be used to test whether for instance current protein potential energy functions are capable of simulating observed dynamic properties of proteins. Indeed molecular dynamics trajectories or a normal mode analysis can both be used to directly calculate the corresponding inelastic neutron scattering spectrum (Bee, 1988, Cusack 1988 and Smith, 1991).

In the case of inelastic neutron scattering, the energy changes occurring when the incident neutrons interact with moving nuclei in the sample are also measured such that an energy spectrum of scattered neutrons is obtained at each scattering angle. The fundamental quantity measured is the *dynamic structure factor* $S(\mathbf{q},\omega)$, where $\hbar\mathbf{q}$ is the momentum transfer between neutron and system (for elastic scattering, $q=4\pi/\lambda\sin(\theta_S/2)$ where λ is the neutron wavelength and θ_S is the scattering angle) and $\hbar\omega$ is the energy transfer from neutron to system. In inelastic neutron scattering a variety of energy/frequency units are employed, $1\text{meV} = 8.07\ \text{cm}^{-1} = 0.24\text{THz} = 11.61^{\circ}\text{K}$.

Since 50% of the atoms in normal (i.e. undeuterated) proteins are hydrogens (^{1}H) and these have an anomalously high incoherent cross-section, the incoherent scattering from them is the overwhelming contribution to the total cross-section. Due to the low intensity of neutron beams, samples for INS need to be of several hundred milligrams, which limits the numbers of biological systems that can be studied. In order to focus on the protein dynamics rather than that of the solvent, the sample is usually hydrated with D_2O. The signal thus arises principally from $-CH_3$,$-CH_2$,-CH and a few unexchanged -NH groups in the protein.

The diagram below shows a schematic representation of the main features of an inelastic neutron scattering spectrum obtained in the case of macromolecules such as proteins.

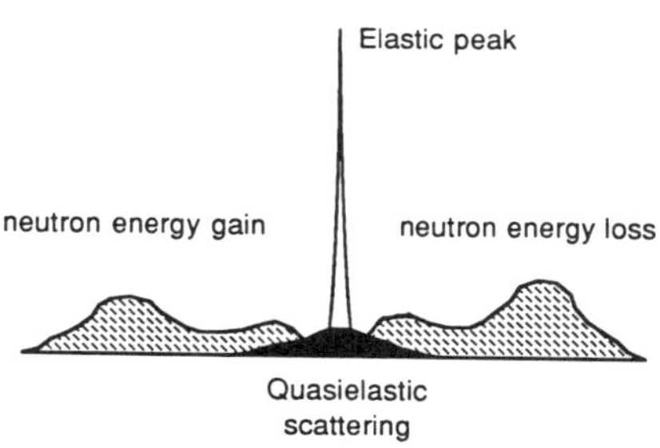

In general, three types of scattering may be distinguished in an inelastic neutron spectrum. These are (a) the *elastic peak* $S(\mathbf{q},\omega=0)$ which arises from neutrons which are scattered with no change in energy and is generally the most intense feature in the spectrum in the case of solid (as opposed to liquid or gaseous) samples, (b) the *quasielastic scattering* which appears as a more or less broad, continuous feature centred at $\omega=0$. This arises from rotational or translational diffusive motion (for example the self diffusion of a particle in a liquid gives rise to a Lorentzian quasielastic scattering with width Dq^2 where D is the translational diffusion coefficient) or activated motions, for example translational or rotational motion of a particle between discrete sites requiring the surmounting of an energy barrier. Transitions between conformational substates of proteins should manifest themselves in the quasielastic spectrum. (c) the *inelastic scattering* which arises from transitions of the system between well-defined energy levels and thus gives rise to features shifted from ω=0. This is the case for a system with harmonic normal modes of vibration in which for instance an incident neutron can excite or de-excite a normal mode of angular frequency ω, giving rise to peaks in the inelastic neutron spectrum at $+\omega$ or $-\omega$ respectively. An analysis of the inelastic spectrum enables the density of vibrational states to be derived (Cusack and Doster, 1990). The separation between low frequency inelastic scattering and quasielastic scattering is sometimes difficult to make in practice as the features due to different kinds of motion may overlap.

XVI.6. The temperature dependence of the dynamics of myoglobin

XVI.6.1. Myoglobin: model system for protein dynamics studies

Sperm-whale myoglobin is a much studied, alpha-helical protein of 153 residues which is found in muscle and its role is to store oxygen. This is done by reversibly binding oxygen to an iron atom in the ferrous state, the central atom in the heme group which is embedded in the protein structure (Figure XVI.7). Since according to the 'static' X-ray structure of myoglobin there is apparently no path between the iron and the solvent it has been imagined that local structural adjustments due to thermal motion must control the rate of oxygen binding and release form the protein (see Karplus and Petsko 1990).

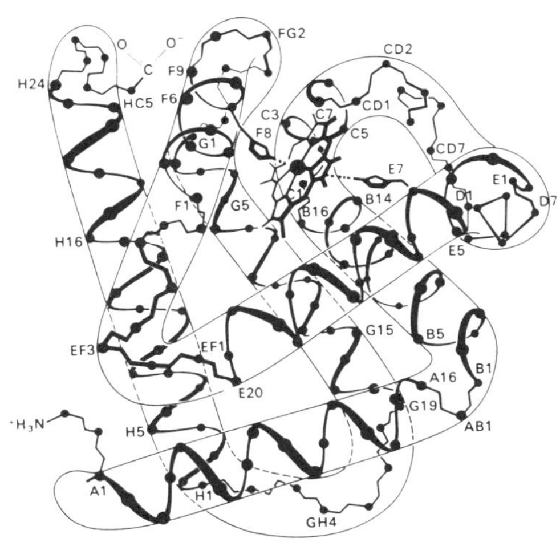

Figure XVI.7. — Model of myoglobin showing the α-carbon positions and the eight α-helical segments. The heme group with its central iron atom is in the centre at the top of the molecule.

The temperature dependence of a number of properties of myoglobin have been studied over a wide range of temperature. These studies have been important in developing more sophisicated ideas about protein dynamics involving the concept of confromational substates. Because of the very many degrees of freedom in a protein molecule there is not one single protein conformation, but a distribution of a very large number of closely related structures or substates separated by energy barriers of 0-10's kJ/mol. At physiological temperatures there can be rapid transitions between different substates whereas at low temperatures (typically below 200K) there is the possibility that an individual molecule becomes 'frozen' into a particular substate.

Evidence for this picture first came from low temperature flash photolysis studies of ligand binding in myoglobin which measure ligand (e.g. oxygen) binding rates to the heme iron (Ansari et al. 1985). It was found that the observed non-exponential rebinding kinetics could only be explained by postulating the existence of a distribution of substates with different rebinding rate constants. Further characterisation of the nature of protein dynamics has come from Mössbauer experiments which revealed a striking dynamic transition at about 200K (Parak, 1988). Above this temperature the mean-square displacement of the heme iron shows a much faster increase with temperature than expected for a harmonic system. This has been interpreted as due to the excitation of new slow modes of motion above the transition temperature and correlates with the ability of small ligands to cross the protein matrix between the heme pocket and the solvent. Recently it has been shown that the dynamic transition in myoglobin near 200K can also be clearly observed in INS experiments and analysis of the data gives some new insights into the nature of picosecond motions in proteins (Doster, Cusack and Petry, 1989, Cusack and Doster, 1990).

XVI.6.2. Inelastic neutron scattering results on myoglobin

A schematic view of the observed incoherent dynamic structure factor $S_{inc}(q,\omega)$ of myoglobin at fixed q and three different temperatures is shown in Figure XVI.8. At very low temperatures (Figure XVI.8, left), there is essentially no motion and the spectrum consists only of the sharp elastic peak. As the temperature is raised (to say 150K, Figure XVI.8, centre), the amplitude of harmonic vibrations increases giving rise to an inelastic vibrational spectrum and a corresponding decrease of the elastic peak. Figure XVI.8 (right) shows the situation above the transition temperature at say 300K. In addition to the increased scattering from the vibrations, new non-vibrational modes of motion have been excited and give rise to a quasielastic spectrum, with a corresponding sharper decrease in the elastic peak.

Experimental measurements have been made on a powder of sperm-whale myoglobin hydrated to 0.33 gD_2O/gprotein. Under these conditions the protein is essentially fully hydrated, whereas the D_2O only contributes about 5-10% of the scattering. Figure XVI.9 shows the temperature variation between 4 and 320K of the elastic intensity $S_{inc}(q,\omega=0)$ at a number of different values of q measured on the backscattering instrument IN13 at the Institut Laue-Langevin in Grenoble, France. This instrument has a resolution of about 10μev, so that motions slower than 10^{-10}s are not resolved. A transition in behaviour, corresponding to a

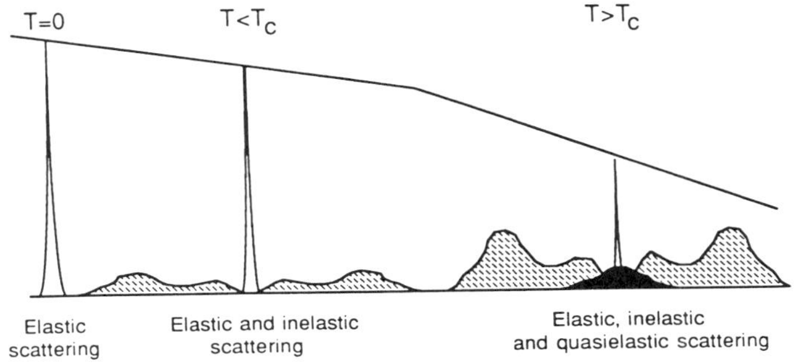

Figure XVI.8. — Schematic diagram of the temperature dependence of the incoherent inelastic neutron scattering spectrum of myoglobin.

sharper decrease in the elastic intensity, is clearly seen at about 200K. These data can be analysed to show that the mean square displacement of the hydrogen atoms increases linearly at low temperatures but then rises steeply above the transition at about 200K, as also observed by Mössbauer experiments and crystallographic measurements of B-factors (Cusack, 1992)

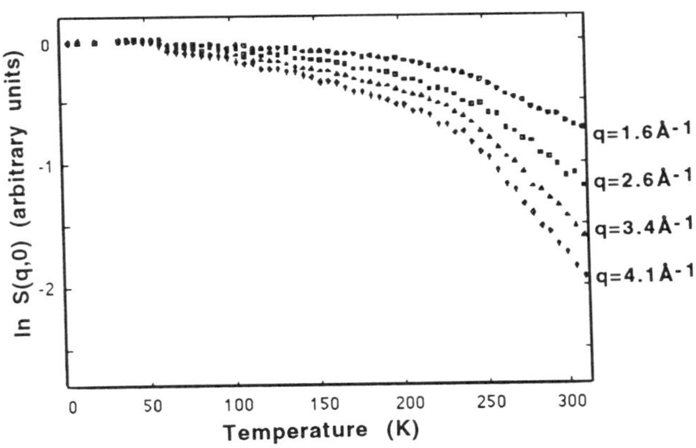

Figure XVI.9. — Normalised elastic intensity S(q,ω=0) for hydrated myoglobin as a function of temperature at a number of different values of q.

Figure XVI.10 shows quasielastic and inelastic spectra obtained on instrument IN6 (resolution 50mev) at temperatures between 100 and 350 K. These show that at low temperatures (180K and below) there is a well-resolved but fairly broad maximum in the structure factor at 25 cm^{-1} and no quasielastic broadening. Above the 180K transition, the maximum becomes less and less resolved due to the rapid increase with temperature of a broad quasielastic band extending to about 30 cm^{-1}. This broad line accounts for the extra decrease observed in the elastic intensity.

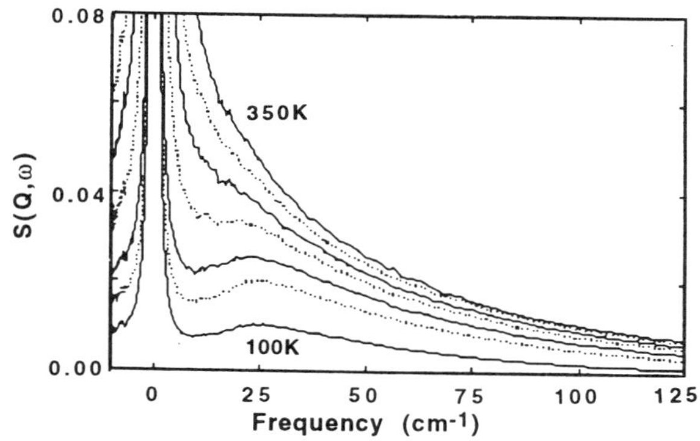

Figure XVI.10. — Inelastic neutron scattering spectra of hydrated myoglobin for a fixed scattering angle at temperatures of 100,180,220,270,300,320,350 K.

Analysis of these INS measurements on myoglobin over a wide temperature range (and similar results on other proteins such as lysozyme) has lead to a general description of the nature of fast internal motions in globular proteins (Doster, Cusack and Petry, 1989, Cusack and Doster, 1990, Cusack 1992). According to this picture, proteins behave as quasiharmonic systems below a temperature of about 180K but above this temperature there is a *dynamic transition* to a state of enhanced mobility due to the excitation of fast non-harmonic motions which are most likely local segmental transitions (Figure XVI.11).

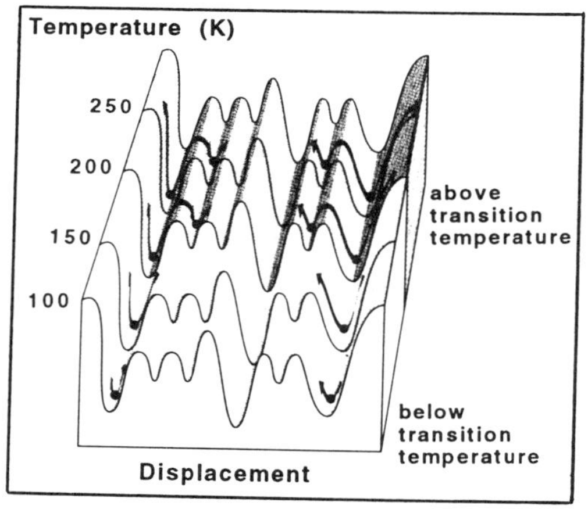

Figure XVI.11. — The 180-220K dynamic transition of proteins illustrated on energy profile diagrams. Below the transition, the protein's motions are confined to harmonic vibrations within a potential well (different molecules can be 'frozen' into different substates, as shown); above the transition, the protein molecules have sufficient thermal energy to 'jump' between potential wells.

XVI.6.3. Comparison of neutron results with molecular dynamics simulations

Support for this interpretation of the neutron results has come from the analysis of molecular dynamics simulations on myoglobin where the detail of the calculations may give further insight into the real nature of the motion observed. Fast (picosecond) transitions between conformational substates were revealed by an early molecular dynamics simulation of myoglobin in vacuum (Elber and Karplus, 1987). More recently, a detailed comparison of the neutron results has been made with simulations of myoglobin as a function of temperature using the CHARMM force-field, one study being made in vacuum with only polar hydrogens explicitly included (Smith et al., 1990) and one with an all hydrogen atomic model and partial hydration (with 350 water molecules) to mimic the neutron experimental conditions (Loncharich and Brooks, 1990). Both studies have shown that the experimental neutron results on myoglobin can be semi-quantitatively reproduced, in particular the change in form of the inelastic spectra as seen in Figure XVI.10, and the corresponding sharp rise in the mean-square hydrogen displacements above 200K. They also both show that the empirical force-fields result in a somewhat softer protein as the main peak in the calculated dynamic structure factor at low temperatures, in the harmonic regime, is below the observed 25 cm^{-1}. This is particular true of the Loncharich and Brooks calculations (where the peak is around 10cm^{-1}) with the result that the observed mean-square vibrational fluctuations are uniformly too large, although the temperature dependence is correct. An interesting result by these authors is that the number of dihedral transitions (defined as jumps in the torsional angles involving heavy atoms) shows a steep rise above the transition temperature in line with the increase in mean-square displacements. The authors have estimated that at room temperature about 40% of the fluctuations are due to local oscillations (vibrations) and 60% due to torsional jumps.

XVI.7. Enzyme activity and the dynamic transition in proteins

A recent publication has addressed the question as to the biological significance of the dynamic transition in proteins (Rasmussen *et al.* 1992, Cusack 1992). The crystal structure of ribonuclease-A was determined at 1.7Å resolution under four conditions (Figure XVI.12). Firstly the crystal was cooled in a cryoprotectant to 212K and a dataset collected. This showed the normal enzyme structure with the sulphate ion that co-purifies with the protein occupying the phosphate-binding subsite in the active site. Secondly, the inhibitor 2'-CMP was flowed over the crystal for six days and another dataset measured. The electron density map was essentially unchanged from the first, indicating that no inhibitor had bound. Thirdly the crystal was warmed to 228K while maintaining the flow of 2'-CMP, and another dataset collected after only a few hours. This showed clear electron density for the 2'-CMP in the enzyme active site. Fourthly, the crystal was cooled back to 212K and washed in inhibitor-free mother liquor for two days. A dataset under these conditions showed that the inhibitor had not been dislodged from the active site. Raising the temperature again to 228K while maintaining the flow of inhibitor free mother liquor it was then shown that the bound inhibitor was rapidly removed. A final control experiment showed that whereas the inhibitor could not be bound to ribonuclease at 212K, it was possible to covalently attach platinum to methionine-29 on the surface of the enzyme at this temperature.

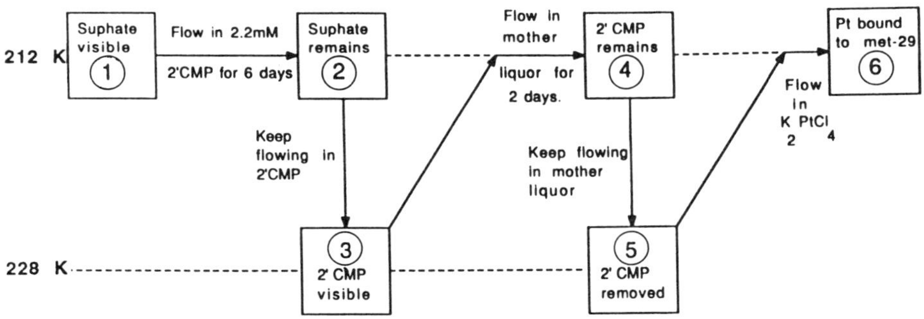

Figure XVI.12. — Crystal conditions for the low temperature X-ray measurements on ribonuclease.

This experiment leads to the conclusion that at sufficiently low temperatures, ribonuclease loses the ability to bind or release a large ligand in its active site; in other words the activation barrier for binding is too high for it to occur in a measurable time. This in itself may not seem so surprising, were it not for the fact that the loss of rapid binding occurs over a narrow 16 degree temperature range which correlates with the dynamical transition temperature of several proteins. These results suggest the following hypothesis which is based on dynamic considerations, since the structure of the enzyme does not change within experimental error. Binding or release of ligands in the active site pocket of ribonuclease A requires small adjustments in the active site residue positions. These motions have higher energy barriers than simple harmonic vibrations and are frozen out below the transition temperature where the protein is locked into a conformation too rigid to permit proper binding.

Clearly this hypothesis, which is consistent with the overall picture of protein dynamics developed above, needs further verification on other enzymatic systems.

General References

McCAMMON, J.A., HARVEY, S.C., 1987 - Dynamics of Proteins and Nucleic Acids - Cambridge University Press.

SCHULZ, G.E., SCHIRMER, R.H., 1979 - Principles of protein structure - Springer (New York).

CANTOR, C.R., Schimmel, P.R., 1979 - Biophysical Chemsitry- Freeman (San Francisco).

CREIGHTON, T.E., 1983 - Proteins - Freeman (New York).

LOVESEY, S. W., 1984 - Theory of neutron scattering from condensed matter- Vol. 1- Clarendon Press (Oxford).

BEE, M., 1988 - Applications of quasielastic neutron scattering to solid-state chemistry, biology and material science- Adam Hilger (London).

REFERENCES

ANDERSON, C.M., ZUCKER, F.H., STEITZ, T., 1979 - Science 204, 375

BRUNGER,A.T., KURIYAN, J., KARPLUS, M., 1987 - Science 235, 458

BROOKS, B.R., BRUCCOLERI, R.E., OLAFSON, B.D., STATES, D.J., SWAMINATHAN, S., KARPLUS, M., 1983 - J. Comp. Chem. 4, 187

COOPER, A., DRYDEN, T.F. 1988 - pp 159 in 'The Enzyme Catalysis Process: energetics, mechanism and dynamics", eds Cooper, A., Houben, J.L., Chien, L.C. Plenum (New York).

CUSACK, S., 1988 - pp 103 in 'The Enzyme Catalysis Process: energetics, mechanism and dynamics", eds Cooper, A., Houben, J.L., Chien, L.C. Plenum (New York).

CUSACK, S., DOSTER, W., 1990 - Biophys. J. 58, 243

CUSACK, S., 1992 - Curr. Bio. 2, 411

CUSACK, S., NASSAR, N., BERTHET-COLOMINAS, C., HARTLEIN, M., LEBERMAN, R. 1990 - Nature 347, 249

CUSACK, S., SMITH, J., FINNEY, J., TIDOR, B., KARPLUS, M. 1988 - J. Mol. Biol. 202, 903

DOSTER, W., CUSACK, S., PETRY, W. 1989 - Nature 337, 754

ELBER, R., KARPLUS, M. 1987 - Science 235, 318

FABER, H.R., MATTHEWS, B.W., 1990 - Nature 348, 263

GERSTEIN, M., SCHULZ, G., CHOTHIA, C., 1993 - J. Mol. Biol. 229, 494

GIBRAT, J-F., GO, N., 1990 - Proteins 8, 258

HUBER, R., BENNETT, W.S., 1983 - Biopolymers 22, 261.

JANIN, J., WODAK, S.J., 1983 - Prog. in Biophys. and Mol. Biol. 42, 21

JOSEPH D., PETSKO, G.A., KARPLUS, M., 1990 - Science 249, 1425

KARPLUS, M., PETSKO, G.A., 1990 - Nature 347, 631

LONCHARICH, R.J., BROOKS, B.R., 1990 - J. Mol. Biol. 215, 439

PARAK,F., 1988 - pp 197 in 'The Enzyme Catalysis Process: energetics, mechanism and dynamics", eds Cooper, A., Houben, J.L., Chien, L.C. Plenum (New York).

RASMUSSEN, B.F., STOCK, A.M., RINGE, D., PETSKO,G.A. 1992 - Nature 357, 423

RINGE, D., PETSKO, G.A., 1985 - Prog. in Biophys. and Mol. Biol. 45, 197.

RUFF, M. et al. 1991- Science 252, 1682

SMITH, J., KUCZERA, K., KARPLUS, M., 1990 - PNAS 87, 1601

SMITH, J. , 1991 - Quart. Rev. Biophys. 24, 227

WADE,R.C., DAVIS,M.E., LUTY,B.A., MADURA,J.D., McCAMMON,J.A., 1993 -

Biophys. J. 64, 9

CHAPTER XVII

APPLICATIONS OF XAS TO BIOLOGICAL AND RELATED CHEMICAL SYSTEMS

C. D. GARNER

XVII.1. Introduction

The fact that many of the metals are present in biological systems only in trace amounts in no way diminishes their significance. Whenever nature has a difficult task to accomplish, a metal ion (or a cluster of ions) is invariably employed. This is especially true for metabolic processes involving small molecules. Thus, the chemical transformations which constitute the oxygen and nitrogen cycles involve catalyses at transition metal centres. Electron flow in natural systems is overwhelmingly from transition metal centre to transition metal centre. Furthermore, Zn^{II}, Mn^{II}, and Fe^{II} cations are central to the control of both metabolic and genetic processes. Therefore, it is essential that we understand the biochemistry of these metals; especially how the chemistry of a metal is modulated by binding to an organic matrix to achieve a specific functionality. This challenge is clearly manifest by asking <u>how</u> symbiosis of transition metal and protein functions produces such elegant specificity.

Structural information alone will not describe protein <u>function</u>. However, even to ask intelligent questions, it is a prerequisite that we know the detailed structural arrangement of a biomolecule and the environment about the transition metal atom.

X-ray crystallography is rightly regarded as the most powerful structural technique to provide the architectural details of a protein molecule. However, sometimes the resolution of crystallographic data is insufficient to draw meaningful conclusions concerning the detailed nature of a metal centre within a protein. Also, some proteins refuse to crystallise or yield crystals suitable for a high resolution structure determination. Furthermore, it is important to establish structural details for proteins, and especially their catalytic centres, when maintained at conditions similar to their working environment - typically, in aqueous media in the presence of substrate, inhibitor, or suitable redox partner.

Since 1975, when the first experiment using a synchrotron radiation source was accomplished (Kincaid et al., 1975), XAS has become established as an important technique for probing the environment of transition metals in proteins. Numerous studies have been accomplished and many significant advances made. The majority of these have been identified and collated in several reviews (Cramer et al., 1979; Lee et al. 1981; Powers, 1982; Cramer, 1983; Scott, 1985; Hasnain, 1987; Hasnain et al., 1987; Cramer, 1988).

The information provided by A. Fontaine in chapter XV of the Hercules volume I gives an excellent basis for understanding the theoretical and experimental basis of XAS spectroscopy. This contribution will concentrate upon the application of XAS to characterise the nature of d-transition metal centres in some biological and related chemical systems.

XVII.2. Information Content

The actual energy of a particular absorption edge of an element depends upon the oxidation state and the nature of the immediate chemical environment of that element. Typically, one unit increase of oxidation state increases the 3d or 4d element edge position by 1–3 eV. Labhardt and Yuen (1979) observed a shift in the position of the iron K-edge and its associated structure of ca. 1.5 eV to higher energy upon oxidation of cytochrome c. The sense and magnitude of this shift are consistent with a redox process which is concentrated at the iron centre.

The excitation of a core electron into the continuum may be convoluted with transitions to the valence levels. These promotions give rise to pre-edge and edge features which can provide information concerning the chemical nature and electronic structure of the primary absorber. Mo=O groups give rise to a low energy edge-inflection due to a transition from the K shell to the Mo=O π^* orbital. This has been a ubiquitous feature of the molybdenum K-edge XAS studies of the oxomolybdoenzymes, as opposed to the nitrogenases (Cramer, 1983; Cramer, 1988; Hedman et al., 1990).

Historically, and because of the different theoretical treatment necessary to interpret the data, it has been customary to classify the oscillations within ca. 50 eV of the edge as the X-ray Absorption Near Edge Structure (XANES) and those which extend beyond this region as the Extended X-ray Absorption Fine Structure (EXAFS). The theoretical basis of the latter is considered mature and interpretations of EXAFS data have been reported confidently for over a decade. In contrast, progress in the understanding and, therefore, the application of XANES has been relatively slow and data in this spectral region are generally not intepreted but used qualitatively to "fingerprint" a metal site.

XAS has many advantages as a probe of transition metal centres in biological materials. Beyond the absence of a requirement for crystalline materials, the major attractions are the specificity, and sensitivity of the technique and the provision of interatomic distances with an accuracy of ±0.02Å within (say) 4Å of the primary absorber. However, it should be noted that:

(i) No angular information is usually obtained.

(ii) Rarely does the structural information extend beyond 4Å.

(iii) The spectrum sums data for all atoms of a particular element and, if the element of interest is present in more than one chemical form, an average environment is obtained.

(iv) The possibility of radiation damage must be anticipated and the integrity of samples should be monitored after, and if possible during, measurement.

(v) XAS is a "sporting method" and the strength of any interpretation will benefit from other information. XAS studies of metal centres in enzymes derive considerable benefit from "calibration" by complementary studies on relevant and structurally characterized chemical analogues. Furthermore, as illustrated by studies on rubredoxin (Shulman et al., 1978) and Cu,Zn-superoxide dismutase (Blackburn et al., 1984), XAS and protein crystallography are especially complementary.

XVII.3. Applications

XVII.3.1. Zinc

Zinc is as common as iron in biology and Williams has argued that the role of zinc is vital, not only for catalysis but also in inhibitory control (Williams, 1984). Zinc is now known to be associated with a large variety of proteins.

Some crystal structures of zinc proteins are available, however, information concerning zinc centres in proteins is rather limited. Because of its filled 3d-shell, zinc is not accessible by spectroscopic techniques such as optical absorption and EPR. Zinc sites in proteins may be studied indirectly by spectroscopy; Cd-substitution and monitoring by ^{113}Cd NMR spectroscopy or Co substitution and monitoring by UV/visible or circular dichroism are popular procedures. Nevertheless, it should be borne in mind that these surrogates may not be faithful reporters of zinc sites.

XAS in general, and EXAFS in particular, offers a unique and direct probe of zinc centres in proteins. So far, only three types of amino acid residues have been identified as ligands of zinc in a protein; an imidazole nitrogen of histidine, the carboxylate oxygen of glutamic or aspartic acid, and the sulphur of cysteine. A tentative classification of zinc enzymes based on distinctions possible by EXAFS may be suggested (Garner et al., 1987):

- Type-A zinc: coordination by sulphur exclusively
- Type-B zinc: coordination by sulphur and nitrogen and/or oxygen
- Type-C zinc: coordination by nitrogen and/or oxygen.

The potential of EXAFS to distinguish between classes, A, B and C derives from the facts that:

(i) The backscattering amplitude of sulphur and nitrogen (or oxygen) are approximately π out of phase when placed at a similar distance from the absorber atom.

(ii) Zn-S bonds are typically, ca. 2.3Å in length whereas Zn-N/O bonds are ca. 0.3Å shorter.

(iii) The amplitude of sulphur backscattering is significantly greater than that from nitrogen (or oxygen).

Type-A and Type-C centres can be readily distinguished. Type-B centres are usually readily identified, but it may be difficult to determine the respective occupancies of the two shells precisely, since the contributions are typically not well resolved in the Fourier transform. Also, it is unfortunate that EXAFS cannot distinguish between carboxylate oxygens and water as ligands of zinc. However, coordination by imidazole is usually clearly evident because of the multiple scattering pathways involving all atoms of the 5-membered ring (Strange et al., 1987).

Type-A zinc sites have been identified in metallothioneins (Abrahams et al., 1986) and aspartate transcarbamylase (Phillips et al., 1982). Both involve four sulphur atoms at ca. 2.33Å from the zinc. Such sites will not be catalytically active as the coordination sphere of the metal is saturated. Therefore, a function for the zinc as a pivot, holding together the regulatory chains of the enzyme in a firm but flexible manner is proposed.

Although Type-B and Type-C zinc sites are known to be catalytically active, in these environments zinc can also play an important structural role. This is especially true for "zinc finger proteins". The transcription factor IIIA of eukaryotic cells is a protein with a distinctive repeat sequence, consisting of two cysteine and two histidine residues. These bind zinc in a tetrahedral site causing a loop or "finger" to be formed. The finger binds in the wide groove of DNA in a sequence specific manner. Thus, the DNA binding capability is controlled by the coordination at the zinc and a similar structural motif has been identified in numerous proteins. Diakun et al. (1986) showed that zinc in transcription factor IIIA from *Xenopus laevis* is coordinated by two sulphur and two nitrogen atoms (at 2.30 and 2.00Å, respectively) consistent with ligation by two cysteine and two histidine residues.

XVII.3.2. Metallothioneins

Metallothioneins are a unique and widely distributed group of proteins. They are characterised by their low molecular weight (ca. 6,000), high cysteinyl content, and the ability to bind substantial numbers of metal ions (Vasak et al., 1988). The proteins bind copper and zinc, thereby providing a mobile pool as part of the normal metabolism of these elements, and offer protection from the invasion of inorganic forms of the toxic elements cadmium, lead and mercury. In addition other metals such as iron and cobalt can be induced to bind. XAS is ideally suited to probe the environment of these different metal atoms (Fig. XVII.1) and the structural interpretations obtained from an analysis of the EXAFS

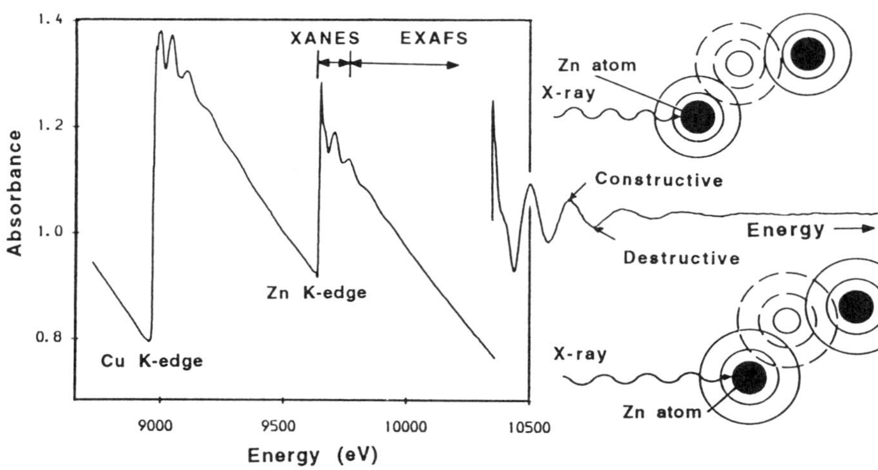

Figure XVII.1. — X-ray absorption profile of Cu_5Zn_3-metallothionein and a representation of the origin of the Extended X-ray Absorption Fine Structure (EXAFS) associated with the zinc edge, from Garner et al. (1993).

data obtained in several such studies are summarised in Table 1. Thus, in each case, the data are consistent with the primary coordination of the metal deriving from the cysteinyl residues.

As indicated in Section II, uncertainties of ca. ±25% in N are inherent in EXAFS from an analysis of the amplitude envelope alone. Therefore, an appeal should be made to other information. For systems as simple as those in Table 1,

TABLE 1

Structural parameters deduced from EXAFS of metallothioneins (Charnok et al., 1989)

Absorber	Scatterer	R / Å
Zn in Zn_7-MT	4S	2.33
Cu in Cu_6Zn_3-MT	3S	2.25
Cu in Cu_5Zn_2-MT	3S	2.25
Pb in Pb_7-MT	2S	2.65
Hg in Hg_7-MT	3S	2.42
Cd in Cd_7-MT	4S	2.51
Cd in Cd_4Co_3-MT	4S	2.51
Co in Cd_4Co_3-MT	4S	2.30
Co in Co_7-MT	4S	2.31
Co in Co_3-MT	4S	2.31
Fe in Fe_7-MT	4S	2.32
Fe in Fe_3-MT	4S	2.32

correlations of bond length with coordination number are helpful. Thus, for copper in pig liver metallothionein, reference may be made to crystallographic information for $Cu(SR)_2$, $Cu(SR)_3$, and $Cu(SR)_4$ centres for which the Cu-S bond lengths are ca. 2.15, 2.24-2.29, and ca. 2.35Å, respectively. Thus, on this basis, copper would appear to be three-coordinate in the Cu,Zn-metallothioneins (Abrahams et al., 1986). Similarly, mercury would also appear to be three-coordinate in Hg-metallothionein.

The first EXAFS study of a cadmium environment in a protein was achieved for Cd_5Zn_2– and Cd_7-Metallothionein from rat liver (Abrahams et al., 1985). The two samples manifest identical EXAFS and the data are consistent with a shell of four sulphur atoms at ca. 2.51Å. These results demonstrate that the inequivalence of cadmium atoms observed by ^{113}Cd NMR studies of metallothioneins (Otvos et al., 1980) does not arise from marked variations in atom type, coordination number, or metal ligand distances within the metals' first coordination sphere.

Clear evidence for the close approach of metal atoms in Cd_7-metallothionein is well established by ^{113}Cd NMR spectroscopy and X-ray crystallography. Indications of the intermetallic distances involved for cadmium, and other metals have been sought from EXAFS. Although some indications of metal back-scattering bound within metallothioneins have been obtained (Abrahams et al., 1986; George et al., 1988), the principal conclusion is that these metal-metal separations are not coherent. Therefore, the backscattered waves, especially with their high frequency at distances > 3Å, engage in destructive interference which effectively renders them 'silent' in the EXAFS.

XVII.3.3. Iron-Storage and Transport Proteins

Iron can be said to be the most chemically versatile of all the elements used by nature. It is essential for dioxygen uptake and transport in the vast majority of living systems, ubiquitous in electron transfer relays and oxygen metabolism, used widely as the catalytic centre in enzymes catalysing chemical changes as diverse as dinitrogen fixation, nitrate reduction, and isopenicillin-N-synthase, and vital for DNA synthesis. However, the problem with storing large quantities of free iron in the body under physiological conditions is that soluble iron(III) ions react to form insoluble iron(III) oxide, the accumulation of which is toxic to cells. To avoid this, evolution has produced two iron storage proteins, ferritin and haemosiderin.

The major iron storage protein, ferritin, has been extensively studied (Ford et al., 1984) and shown to consist of a hollow, spherical proteinaceous shell surrounding an iron(III)-oxide core. The other iron storage protein haemosiderin has received rather less attention but, for "normal" haemosiderin, techniques such as ^{57}Fe Mössbauer spectroscopy and electron diffraction (Bell et al., 1988) indicate the presence of a smaller ferritin-like core.

Iron K-edge XAS studies have been accomplished for ferritin (Islam et al., 1989; Mackle et al., 1991) and haemosiderin (Mackle et al., 1991). These have allowed a direct comparison of structural aspects of the iron-oxide cores of ferritin and haemosiderin, which could only be inferred from other techniques. The EXAFS recorded for horse-spleen ferritin and "normal" haemosiderin are indistinguishable, consistent with them possessing an identical iron-oxide core

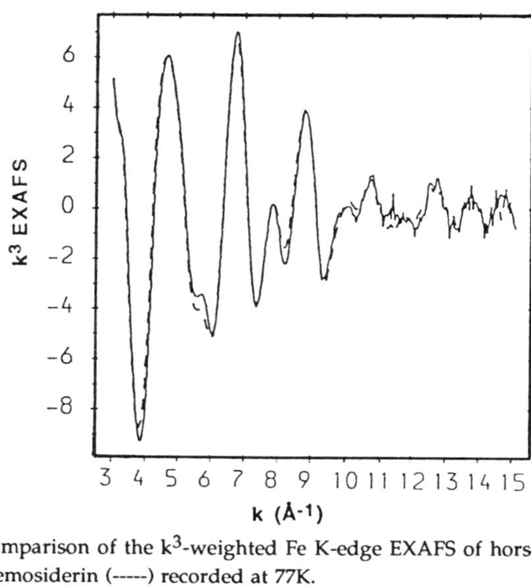

Figure XVII.2. — Comparison of the k^3-weighted Fe K-edge EXAFS of horse spleen ferritin (——) and horse spleen haemosiderin (-----) recorded at 77K.

(fig. XVII.2). The data are indicative of an average environment of the iron(III) consisting of six oxygen atoms at ca. 1.93Å and a split shell of iron atoms at ca. 2.95 and 3.38Å, together with a further oxygen shell at ca. 3.57Å (Mackle et al., 1991). Iron K-edge EXAFS has been used to show that *Azotobacter vinelandii* ferritin and horse spleen ferritin have essentially the same structure and that phosphate influences the local structure about iron. Thus, when the phosphate content is high, the iron atoms appear to possess five to six phosphorus neighbours at ca. 3.17Å. The incorporation of phosphate reduces the number of iron atoms present at ca. 3Å and increases the Fe-----Fe separation to ca. 3.50Å. Therefore, some phosphate would appear to bridge neighbouring iron atoms as well as acting as chain termini (Rohrer et al., 1990).

The clinical condition thalassaemia, i.e. iron over-load, results in serious tissue damage and eventual death. There are two stages in the progression of this disease, primary haemochromatosis and, the stage before death, secondary haemochromatosis. In both of these conditions, the predominant form of stored iorn is haemosiderin, and not ferritin, as in healthy individuals. Iron K-edge XAS studies show that, for primary haemochromatosis, there is a dramatic change in the structure from "normal" haemosiderin. The EXAFS is a damped sinusoidal wave, representative of back-scattering from a single shell of oxygen atoms at ca. 1.94Å. Therefore, the iron-oxide core is amorphous. Secondary haemochromatosis haemosiderin reverts back to an ordered iron-oxide core, with ca. three oxygens at ca. 1.90 and three at ca. 2.03Å and a split iron shell at ca. 2.97 and 3.34Å. Thus, there are also substantial structural differences between this and the iron-oxide core of "normal" haemosiderin (Mackle et al., 1991).

XVII.3.4. Oxomolybdoenzymes

Molybdenum is the most important 4d-transition metal within the biosphere and is essential for the activity of a large group of enzymes (Spiro, 1985). For many of these, clear evidence has been obtained that molybdenum is the site of substrate binding and conversion. These enzymes are found in organisms which range from bacteria to man; in the former, molybdenum is involved in nitrogen fixation and, in the latter, xanthine and sulphite oxidation. There is a clear biochemical distinction between the nitrogenases on the one hand and oxomolybdoenzymes - such as xanthine oxidase and dehydrogenase, sulphite and aldehyde oxidase, and nitrate reductase - on the other. Such a distinction was initially apparent from EPR spectroscopy (Bray, 1988) and subsequently reinforced by the isolation of the iron-molybdenum-cofactor (FeMoco) from the nitrogenases (Shah et al., 1977) and the molybdopterin cofactor (Moco) from the oxomolybdoenzymes (Pienkos et al., 1977; Johnson et al., 1980).

XAS has played a vital role in defining the chemical nature of these molybdenum centres and how they respond to changes in the oxidation level of the protein and/or to the presence of substrates, substrate analogues, or inhibitors of enzymic activity (Cramer, 1983; Cramer, 1988; Hedman et al., 1990). The prefix oxo for this latter group of enzymes is appropriate. Thus, not only does each enzyme catalyse a conversion, the net result of which can be represented as oxygen atom transfer, but also XAS studies have confirmed the presence of at least one terminal oxo ligand (Mo=O) for molybdenum in each of the enzymes.

Sulphite oxidase is responsible for the physiological oxidation of sulphite to sulphate. The molybdenum K-edge EXAFS results achieved (Arber et al. 1989) for this enzyme are the clearest such data and the interpretation achieved represents a prototype for other oxomolybdoenzymes. The molybdenum site has been investigated in each of its three accessible oxidation states((VI), (V), and (IV)) as a function of pH and chloride concentration. The molybdenum(VI) is coordinated to two oxo-groups, at ca. 1.70Å, one oxygen (or nitrogen)- and three sulphur-donor ligands at ca. 2.06 and 2.42Å, respectively. Two of these sulphur atoms would be expected to arise from the molybdopterin (Johnson et al., 1980). The molybdenum(VI) centre is not affected by changing the pH from 6 to 9 or by a variation in the chloride concentration.

The reduced molybdenum centres each possess a single oxo-ligand, at ca. 1.69Å, one oxygen (or nitrogen)- and three sulphur-donor ligands at ca. 2.00 and 2.37Å, respectively. Both of these centres appear to bind a chloride ligand at pH 6 in 0.3M KCl. The results for molybdenum(V) have a special significance, in that they permit a direct comparison of the EXAFS results with EPR data. EPR spectroscopy (Bray, 1988) shows that the centre can exist in two different forms, which are in a pH- and anion-dependent equilibrium. George et al. (1989) concluded that the molybdenum K-edge EXAFS data were consistent with one chloride ligand binding to the low pH form and that the number of oxo-groups remains the same upon transition from the high-pH to the low-pH molybdenum(V) form.

Thus, for sulphite oxidase, a clear picture emerges from the intepretation of the EXAFS data. Reduction of molybdenum(VI) results in the loss of one oxo-group, presumably due to protonation, and the generation of an anion binding site. The behaviour is consistent with the chemistry of molybdenum in its higher oxidation states. A cis-dioxomolybdenum(VI) centre is generally converted to a mono-oxomolybdenum(V) or (IV) centre upon reduction, provided that steric restrictions prevent dimerisation - as would be expected for the molybdenum centres in enzymes. These changes are crucial for the mediation of oxygen atom transfer, which is formally a two-electron process and is difficult to accomplish, except by atom transfer.

XVII.3.5. Nitrogenases

Nitrogen is an essential element for life and is abundant in the Earth's atmosphere in the form of molecular dinitrogen. Dinitrogen is, however, not metabolised by most organisms. Consequently, they must obtain their nitrogen in a chemically combined form, e.g. as ammonia, nitrate, or an organic molecule. Because organic nitrogen is incompletely recycled in living ecosystems and the available ammonia and nitrate are continually metabolised to dinitrogen through nitrification and denitrification, all life ultimately depends upon the biological fixation of dinitrogen.

Nitrogen fixation occurs only in certain prokaryotes and is catalysed by the nitrogenases which have two protein components. These are the Fe-protein, which acts as a specific reductant of the larger MoFe-protein. Within the latter there are at least four metal-containing prosthetic groups, including two iron-molybdenum cofactors (FeMoco) which are the catalytic centres of the enzyme.

One of the earliest successful applications of EXAFS to probe a metalloenzyme was the study of the molybdenum site of nitrogenase. Studies were made on both the *Clostridium pasteurianum* and *Azotobacter vinelandii* MoFe-proteins and on isolated FeMoco (Cramer et al., 1978). These studies showed definitively that molybdenum is present as part of a polynuclear cluster containing sulphur and iron, with Mo-S and Mo-Fe distances of ca. 2.36 and 2.72Å, respectively. This work inspired the successful development of many chemical systems containing Mo-Fe-S clusters and XAS studies of these systems strengthened the basis for the interpretation of corresponding data for the natural system. The most accurate picture of the molybdenum site of FeMoco currently available involves a coordination of ca. three oxygen (or nitrogen), sulphur, and iron atoms at ca. 2.10, 2.37, and 2.70Å, respectively (Conradson et al., 1987).

Genetic suppresion of the "normal", molybdenum-dependent nitrogenase of certain classes of *Azotobacter* allows expression of the vanadium-dependent enzyme. The vanadium and molybdenum nitrogenase systems show many similarities and, in particular, an iron-vanadium cofactor (FeVaco), analogous to FeMoco, has been isolated (Smith et al., 1988). Clear evidence of a strong similarity between active sites in the MoFe- and VFe-proteins has been provided by vanadium K-edge XAS studies. Two VFe-proteins have been investigated, one from *Azotobacter chroococcum* (Arber et al., 1987, Arber et al., 1989) and one from *Azotobacter vinelandii* (George et al., 1988). The edge structure shows a weak, single 1s$\rightarrow$3d transition, the intensity of which precludes the presence of terminal V=O bonds and implies an octahedral coordination around vanadium. The edge and XANES structure is very similar to that recorded for a VFe_3S_4 cubane-like cluster $[NMe_4][VFe_3S_4Cl_3(DMF)_3]$ (Kovacs et al., 1986). The EXAFS results (Arber et al., 1987; Arber et al., 1989) are consistent with vanadium in the VFe-protein of *Azotobacter chroococcum* being ligated by ca. three oxygen (or nitrogen), sulphur, and iron atoms at ca. 2.13, 2.33, and 2.75Å, respectively. The vanadium would thus appear to substitute for molybdenum in FeMoco to form FeVaco. The role of these metals has still to be elucidated. Possibly they facilitate redox reactions within the cluster and, therefore, it is noteworthy that neither the molybdenum nor the vanadium centre experiences any significant structural change as the oxidation level of the cofactor is varied.

The average environment of iron in FeMoco (Arber et al., 1988) and FeVaco (Harvey et al., 1990) has been investigated by XAS (fig. XVII.3). These data provide clear evidence for the two cofactors possessing the same molecular geometry and give further insights into their internal organisation. Specifically, a longer range structural order is identified at ca. 3.68Å corresponding primarily to Fe---Fe backscattering. This situation is in clear contrast to that for the metallothioneins (Section III C) and indicates that these longer Fe---Fe distances are <u>coherent</u>. Furthermore, the iron atoms are ligated essentially by sulphurs at ca. 2.20Å and the cofactors are primarily comprised of $Fe\langle^{S}_{S}\rangle Fe(M)$ (M=Mo,V) rhombs with metal-metal separations of 2.65-2.75Å.

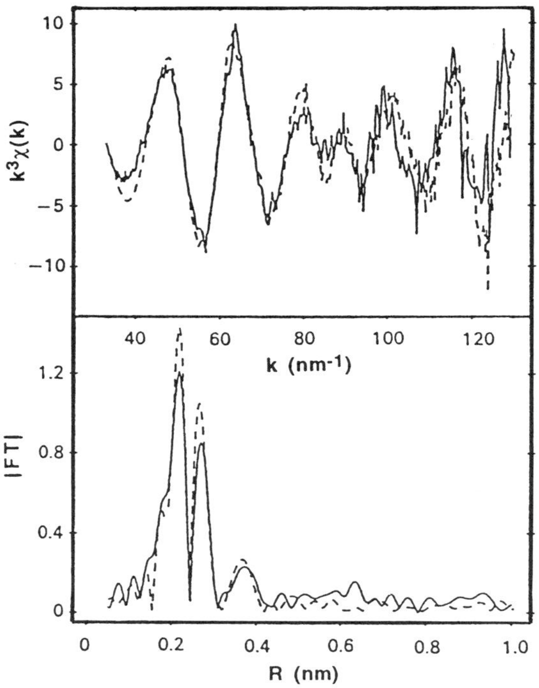

Figure XVII.3. — Comparison of the Fe K-edge EXAFSand Fourier transforms of FeVaco (——) and FeMoco (-----), from Harvey et al. (1990).

REFERENCES

ABRAHAMS I.L., GARNER C.D., BREMNER I., DIAKUN G.P., HASNAIN S.S. 1985 - J. Amer. Chem. Soc., 107, 4596-4597.

ABRAHAMS I.L., GARNER C.D., BREMNER I., DIAKUN G.P., HASNAIN S.S. 1986 - Biochem. J., 236, 585-589.

ARBER J.M., DOBSON B.R., EADY R.R., HASNAIN S.S., GARNER C.D., MATSUSHITA J., NOMURA M., SMITH B.E. 1989 - Biochem. J., 258, 733-737.

ARBER J.M., DOBSON B.R., EADY R.R., STEVENS P., HASNAIN S.S., GARNER C.D., SMITH B.E. 1987 - Nature, 325, 372-374.

ARBER J.M., FLOOD A.C., GARNER C.D., GORMAL C.A., HASNAIN S.S., SMITH B.E. 1988 - Biochem. J., 252, 421-425.

BELL S.H., WEIR M.P., DICKSON D.P.E., GIBSON J.F., SHARP G.A., PETERS T.J. 1984 - Biochim. Biophys. Acta, 787, 227-236; DICKSON D.P.E., REID N.M.K., MANN S., WADE V.J., WARD R.J., PETERS T.J. 1988 - Biochim. Biophys. Acta, 957, 81-90.

BLACKBURN N.J., HASNAIN S.S., BINSTED N., DIAKUN G.P., GARNER C.D., KNOWLES P.F. 1984 - Biochem. J. . 219, 985-990.

BRAY R.C. 1988 - Quart. Rev. Biophys., 21, 299-329.

CHARNOCK J.M., GARNER C.D., ABRAHAMS I.L., ARBER J.M., HASNAIN S.S., HENEHAN C., VASAK M. 1989 - Physica B, 158, 93-94.

CONRADSON S.D., BURGESS B.K., NEWTON W.E., MORTENSON L.E., HODGSON K.O., 1987 - J. Amer. Chem. Soc., 109, 7507-7515.

CRAMER S.P. 1983 - Advances in Inorganic and Bioinorganic Mechanisms (Ed. SYKES A.G.) 259-316, Academic Press, (London).

CRAMER S.P. 1988 - X-Ray Absorption Spectroscopy (Eds. Koningsberger D.C., PRINS R.) Ch. 7, Wiley Interscience (New York).

CRAMER S.P., HODGSON K.O. 1979 - Prog. Inorg. Chem., 25, 1-39.

CRAMER S.P., HODGSON K.O., GILLUM W.O., MORTENSON L.E. 1978 - J. Amer. Chem. Soc., 100, 3398-3407; CRAMER S.P., GILLUM W.O., HODGSON K.O., MORTENSON L.E., STIEFEL E.I., CHISNELL J.R., BRILL W.J., SHAH V.K. 1978 - J. Amer. Chem. Soc., 100, 3814-3819.

DIAKUN G.P., FAIRALL L., KLUG A. 1986 - Nature, 324, 698-699.

FORD G.C., HARRISON P.M., RICE D.W., SMITH J.M.A., TREFFRY A., WHITE J.L., YARIV J. 1984 - Phil. Trans. Roy. Soc. (London), 304B, 551-565.

GARNER C.D., CHARNOCK J.M. 1993 - Biomolecular Spectroscopy Part B, Advances in Spectroscopy Vol. 21, p. 319, Editors R.J.H. Clark and R.E. Hester, John Wiley.

GARNER C.D., FEITERS M. 1987 - Biophysics and Synchrotron Radiation (Eds. BIANCONI A., CONGIU-CASTELLANO A.) 136-146, Springer-Verlag (Berlin).

GEORGE G.N., BYRD J., WINGE D.R. 1988 - J. Biol. Chem., 263, 8199-8203.

GEORGE G.N., COYLE C.L., HALES B.J., CRAMER S.P. 1988 - J. Amer. Chem. Soc., 110, 4057-4059.

GEORGE G.N., KIPKE C.A., PRINCE R.C., SUNDE R.A., ENEMARK J.H., CRAMER S.P. 1989 - Biochemistry, 28, 5075-5080.

HARVEY I., ARBER J.M., EADY R.R., SMITH B.E., GARNER C.D., HASNAIN S.S. 1990 - Biochem. J., 266, 929-931.

HASNAIN S.S. 1987 - Life Chem. Rep., 4, 273-331.

HASNAIN S.S., GARNER C.D. 1987 - Prog. Biophys. Mol. Biol., 50, 47-65.

HEDMAN B., HODGSON K.O., GARNER C.D. 1990 - Synchrotron Radiation and Biophysics (Ed. HASNAIN S.S.) 43-62, Ellis Horwood (Chichester).

ISLAM Q.T., SAYERS D.E., GORUM S.M., THEIL E.C. 1989 - J. Inorg. Biochem., 36, 51-62.

JOHNSON J.L., HAINLINE B.E., RAJAGOPALAN K.V. 1980 - J. Biol. Chem., 255, 1783-1786.

KINCAID B.M., EISENBERGER P. 1975 - Phys. Rev. Lett., 34, 1361-1364.

KOVACS J.A., HOLM R.H. 1986 - J. Amer. Chem. Soc., 108, 340-341.

LABHART A., YUEN C. 1979 - Nature, 277, 150-151.

LEE P.A., CITRIN P.H., EISENBERGER P., KINCAID B.M. 1981 - Rev. Mod. Phys., 53, 769-806.

MACKLE P., GARNER C.D., WARD R.J., PETERS T.J. 1991 - Biochim. Biophys. Acta, 1115, 145-150.

OTVOS J.D., ARMITAGE I. 1980 - Proc. Natl. Acad. Sci. USA, 77, 7094-7098.

PHILLIPS J.C., BORDAS J., FOOTE A.M., KOCH M.J.H., MOODY M.F. 1982 - Biochemistry, 21, 830-854.

PIENKOS P.T., SHAH V.K., BRILL W.J. 1977 - Proc. Natl. Acad. Sci. USA, 74, 5468-5471.

POWERS L. 1982 - Biochem. Biophys. Acta, 683, 1–38.

ROHRER J.S., ISLAM Q.T., WATT G.D., SAYERS D.E., THEIL E.C. 1990 - Biochemistry, 29, 259-264.

SCOTT R.A. 1985 - Meth. Enzymology, 117, 414-459.

SHAH V.K., BRILL W.J. 1977 - Proc. Natl. Acad. Sci. USA, 74, 3249-3253.

SHULMAN R.G., EISENBERGER P., TEO B.K., KINCAID B.M., BROWN G.S. 1978 - J. Mol. Biol., 124, 305-321.

SMITH B.E., EADY R.R., LOWE D.J., GORMAL C. 1988 - Biochem. J., 250, 299-302.

SPIRO T.G. 1985 - Molybdenum Enzymes - Wiley Interscience (New York).

STRANGE R.W., BLACKBURN N.J., KNOWLES P.F., HASNAIN S.S. 1987 - J. Amer. Chem. Soc., 109, 7157-7162.

VASAK M., KAGI J.H.R. 1988 - Met. Ions Biol. Sys., 15, 213-273.

WILLIAMS R.J.P. 1984 - Endeavour, 8, 65-70.

Index

Impression : EUROPE MEDIA DUPLICATION S.A.
F 53110 Lassay-les-Châteaux
N° 3121- Dépôt légal : février 1994